Onions and Allied Crops

Volume II

Agronomy, Biotic Interactions, Pathology, and Crop Protection

Editors

Haim D. Rabinowitch, Ph.D.
Chairman
Department of Field and Vegetable Crops
The Hebrew University of Jerusalem
Rehovot, Israel

James L. Brewster, Dr.Phil.
Principal Scientific Officer
National Vegetable Research Station
AFRC Institute for Horticultural Research
Wellesbourne, Warwick, U. K.

CRC Press, Inc.
Boca Raton, Florida

Library of Congress Cataloging in Publication Data

Onions and allied crops / editors, Haim D. Rabinowitch, James L. Brewster
p. cm.
Includes bibliographies and index.
ISBN 0-8493-6300-0 (v.1)
ISBN 0-8493-6301-2 (v.2)
ISBN 0-8493-6302-4 (v.3)
1. Onions. 2. Allium. I. Rabinowitch, Haim D. II. Brewster, James L.
SB341.O62 1990
635'.25 — dc20 89-25330

This book represents information obtained from authentic and highly regarded sources. Reprinted material is quoted with permission, and sources are indicated. A wide variety of references are listed. Every reasonable effort has been made to give reliable data and information, but the author and the publisher cannot assume responsibility for the validity of all materials or for the consequences of their use.

Direct all inquiries to CRC Press, Inc., 2000 Corporate Blvd., N.W., Boca Raton, Florida, 33431.

International Standard Book Number 0-8493-6300-0 (v.1)
International Standard Book Number 0-8493-6301-2 (v.2)
International Standard Book Number 0-8493-6302-4 (v.3)

Library of Congress Card Number 89-25330
Printed in the United States

INTRODUCTION

The edible alliums are of major economic and dietary importance in all parts of the world. The common onion *(A. cepa)* and garlic (*A. sativum)* are grown, traded and consumed in most countries (Tables 1 to 3). Because of their economic importance, great efforts have been made in the selection and breeding of locally adapted cultivars, and in the development of cultural techniques. As a result these crops can be cultivated from the tropics to subarctic regions. Other edible alliums are more localized in economic significance although they are still of great importance.

It is estimated that the value of the world production of bulb onions alone approaches $5 billion annually, more than 90% of which is consumed within the countries of production. Bulb onions are an easily transportable commodity and between $400 and $500 millions worth (at 1984 prices) are traded internationally each year. The crop is a major export earner for some economies. In addition, the common onion is also an important salad crop when eaten green (scallions). Statistics on this crop are harder to come by, but by way of example, in the U.K., the monetary value of this crop is approximately 65% of that of the dry bulb crop. Leeks (*A. ampeloprasum)* are a valuable crop in Europe (Table 4). Although total European production is well below that for bulb onions, the price per unit weight of leeks is higher than for bulb onions, and in monetary terms the value of the leek crop approaches that of the dry bulb onion crop. Several other alliums are of major economic importance in far eastern countries (Table 5). The annual production of the Japanese bunching onion (*A. fistulosum)* in Japan is approximately 50% of that for dry bulb common onions. However the mean wholesale prices (yen per kilogram obtained during the period January 1983 to December 1985 in Tokyo for common onion, Japanese bunching onion, Chinese chive (*A. tuberosum)*, and rakkyo (*A. chinense)*, respectively, were 93, 216, 344 and 271.[6] Thus the monetary value of the Japanese bunching onion crop is probably similar to that of the common bulb onion in Japan. Japanese bunching onion, Chinese chive, and rakkyo are also important crops in China.[5] Hence, the alliaceous crops are clearly a major agricultural asset and have a great impact on many economies.

It is a quarter of a century since the last comprehensive account of the edible *Allium* crops was written by Jones and Mann.[7] Since then, continued and increased efforts have been made to better understand the physiology, chemistry, genetics, and pathology of these important crops. Some of these investigations have involved topics which were unexplored as regards alliums in 1963, for example, tissue culture, the effects of plant growth regulators, the biochemistry of carbohydrates and flavor compounds, and mycorrhizal associations, to name but a few. Scientific information concerning virtually all aspects of these crops has been accumulating at an accelerating rate in recent years. The extent of this knowledge means that it is no longer feasible for any individual to competently review the alliaceous crops. Consequently, in this book, we have brought together contributions from experts in many disciplines. It has been the policy of the editors to allow authors to freely express their views on matters on which there is still some scientific uncertainty or debate. We, therefore, make no apologies for any difference of emphasis or views which may be apparent between chapters. We are confident however, that these volumes consitute an up to date and soundly based summary of the current knowledge on the edible *Allium* crops.

TABLE 1
Total Production and Areas Harvested of Dry Bulb Onions in 1988[1] and Exports in 1987[2]

	Total production 1000 MT	Area harvested 1000 ha	Exports Amount 1000 MT	Exports Value 10^6\$
World	25496	1787	2009	467
Africa	2088	159	67	32
N. and Central America	2286	67	256	57
S. America	1990	136	29	7
Asia	12120	1003	520	93
Europe	4817	233	1063	260
Oceania	178	5	73	19

TABLE 2
The Leading Ten Producer, Exporter, and Importer Countries for Dry Bulb Onions in 1987[2,3]

Producer country	Total production 1000 MT	Area harvested 1000 ha
China	3600	237
India	2790	280
USSR	2000	184
U.S.	1993	49
Turkey	1500	84
Japan	1260	29
Spain	1104	32
Brazil	856	75
Iran	740	42
Poland	615	31
Exporter country	**Quantity 1000 MT**	**Value 10^6\$**
Netherlands	474	106
Spain	287	59
India	210	38
Mexico	141	23
Turkey	135	13
U.S.	94	26
Poland	110	20
Italy	60	29
Australia	44	11
Pakistan	49	3
Importer country	**Quantity 1000 MT**	**Value 10^6\$**
FR Germany	377	101
UK	242	62
U.S.	168	64
France	142	38
Malaysia	105	28
Canada	98	37
United Arab Emirates	85	15
Belgium	81	18
Singapore	53	18
Kuwait	44	12

TABLE 3
Garlic Production in 1987[4]

	Production 1000 MT	Area harvested 1000 ha
World	2662	421
Africa	237	8
N. and Central America	137	15
S. America	140	32
Asia	1612	264
Europe	488	90
Oceania	4	—
Leading Ten Countries		
China	615	71
Korean Republic	280	40
Spain	230	40
India	217	60
Egypt	215	5
Turkey	98	14
Thailand	98	29
Brazil	76	18
USA	71	5
Pakistan	57	7

TABLE 4
Production of Leeks Within Countries of the European Economic Community[5]

	Production[a] 1000 MT	Area harvested 1000 ha
EEC Total	601	26.8
Major Producer Countries		
France	225	9.8
Belgium	107	3.3
Netherlands	75	2.8
U.K.	61	2.7
Spain	55	2.9
FRG	43	1.7
Greece	37	1.8
Italy	30	1.3

[a] Data are for 1987.

TABLE 5
Production of Japanese Bunching Onion (*A. fistulosum*), in Japan and South Korea and Production of Rakkyo (*A. chinense*) and Chinese Chives (*A. tuberosum*) in Japan[6]

Crop	Production[a] 1000 MT	Area 1000 ha
Japan		
Japanese bunching onion	563	24.1
Rakkyo	31.5	2.3
Chinese chives	66.1	2.6
South Korea		
Japanese bunching onion	432	18.9

[a] Data are for the year 1984.

REFERENCES

1. **FAO,** *Quarterly Bulletin of Statistics,* Food and Agriculture Organization, Rome, 1988.
2. **FAO,** *Trade Yearbook,* Food and Agriculture Organization, Rome, 41, 149, 1988.
3. **FAO,** *Production Yearbook,* Food and Agriculture Organization, Rome, 41, 190, 1988.
4. **FAO,** *Production Yearbook,* Food and Agriculture Organization, Rome, 41, 192, 1988.
5. **Eurostat,** *Crop Production Quarterly Statistics 4-1987,* Statistical Office of the European Communities, Luxembourg, 1987, 35.
6. **Komochi, S.,** personal communication, 1987.
7. **Jones, H. A. and Mann, L. K.,** Onions and Their Allies, Leonand Hill Books, London, and Interscience Publishers, New York, 1963.

THE EDITORS

Haim D. Rabinowitch, Ph.D., is an Associate Professor at the Faculty of Agriculture, The Hebrew University of Jersalem, Rehovot, Israel and currently leads the Group of Vegetable Physiology and Breeding.

Prof. Rabinowitch graduated from the Hebrew University of Jersalem, in 1965, obtained his M.Sc. (*suma cum laude)* in 1967, and received his Ph.D. in 1973 from the same university. He was appointed Lecturer in Vegetable Production at his alma mater in 1976, promoted to Senior Lecturer in 1981, and to Associate Professor in 1986, at Rehovot.

He did post-doctoral work at the Department of Applied Genetics, the John Innes Institute, Norwich England, from 1974 to 1976. In 1981—1982 and in 1986—1987 he spent sabbatical leaves at the Department of Biochemistry, Duke University Medical Center Durham, North Carolina

Prof. Rabinowitch is a member of the American Society for Horticultural Science, The International Society for Horticultural Science, The Scandinavian Society for Plant Physiology, and EUCARPIA. He is a member of the National Vegetable Commodity Committee and of the Professional and Research Committees for Onion Breeding and Production, and for Seed Production, of the Israel Ministry of Agriculture; Member of the Working Group on *Allium,* the European Cooperative Group for the Conservation and Exchange of Crop Genetic Resources, the International Board for Plant Genetic Resources (IBPGR); served as a consultant to the IBPGR on Germplasm Conservation of *Allium* species; in charge of the IBPGR Global Field Bank for vegetatively propagated short-day *Allium* species; and recipient of the American Society for Horticultural Science, National Food Processing Association Award in raw products research for 1983.

Prof. Rabinowitch published more than 45 scientific papers, developed 2 tomato cultivars, and 3 short-day F_1 onion cultivars for fresh market and for the dehydration industry. His current research interests are in physiology and breeding of alliaceous and some *Solanaceae* crops, and in the protective mechanisms against oxygen toxicity in plants.

James L. Brewster, D.Phil is a Principal Scientific Officer in the Crop Production Division, Institute of Horticultural Research, Wellesbourne, Warwick, U.K.

Dr. Brewster graduated in 1966 from Wye College, University of London, with a B.Sc (Honors) degree in Agricultural Science and obtained his D.Phil degree from the University of Oxford in 1971. He remained at Oxford working on the quantitative modeling of nutrient uptake by plant roots until transferring to Wellesbourne in 1974.

Dr. Brewster is a member of the Society of Experimental Biology, Association of Applied Biologists, and Institute of Horticulture. He is an Associate Editor of the *Journal of Horticultural Science.* He is a member of the Onion panel for the National Institute of Agricultural Botany, U.K. He has frequently presented lectures on aspects of his work at scientific and technical meetings in the U.K. and Europe.

Throughout his research career he has been concerned with aspects of the physiology of alliums and has authored 33 research papers covering aspects of the nutrition, flowering, growth rate, control of bulbing, and agronomy of these crops. The main theme of his research is to obtain quantitative relationships for the environmental control of physiological processes and to apply these to understand and predict growth in the field.

ACKNOWLEDGMENTS

As editors and also as contributors, we would like to thank firstly the heads of our respective institutes, namely Professor T. J. Swinburne and Professor J. K. A. Bleasdale, of the Institute of Horticultural Research, Wellesbourne, U.K. for J. L. Brewster, and Professor Ilan Chet, the Dean of the Faculty of Agriculture, Hebrew University of Jerusalem, Rehovot, Israel, for H. D. Rabinowitch for their support for the production of this book. Prof. Rabinowitch also wishes to thank Professor Irwin Fridovich, Duke University Medical Center, Durham, North Carolina, with whom he spent a sabbatical. Without the help and facilities available at the above institutes, the task would not have been possible. We would also like to acknowledge the support of the contributing authors to these volumes, who have so generously given of their time and expertise. We thank them and the institutions in which they are based. We would also like to thank Dr. Suren S. Apte and other staff at the International Agricultural Centre, Wageningen, Netherlands for arranging, and the Government of the Netherlands for providing scholarships for 1 month of intensive joint editing work at Wageningen. Again without this generous support, it is difficult to imagine this project being completed. We would like to thank many individuals who contributed their assistance and expertise. In particular Miss Heather A. Butler of IHR Wellesbourne, for meticulously redrawing some figures, for checking reference abbreviations, and her careful and rapid work in many other respects. Also, we would like to thank Mrs. Margaret Jonas of IHR Wellesbourne for much secretarial assistance with the production of papers and with correspondence. We would like to acknowledge the help of our computer departments, particularly Mrs. Julie Rose at Wellesbourne and Mrs. Zehavit Regev at Rehovot. Without the BITNET and EARN communications between U.K. and Israel and U.K. and U.S., it is difficult to envisage how we as editors could have completed this task. We acknowledge the help of a number of colleagues in the reviewing of chapters on subjects on which our own expertise is limited. In particular we would like to thank Dr. Lesley Currah, Mr. B. D. Dowker, Dr. R. J. Whenham, Dr. D. J. Ockendon, and Professor Y. L. Bar-Sinai.

DEDICATION

Prof. Haim D. Rabinowitch would like to dedicate these volumes to his wife, Shoshie and to his mother, Sara. Prof. James L. Brewster would like to dedicate these volumes to his parents.

To my daughter Shlomit

— HDR

CONTRIBUTORS

James L. Brewster, D.Phil.
Principal Scientific Officer
National Vegetable Research Station
AFRC Institute for Horticultural Research
Wellesbourne, Warwick, U. K.

Joe N. Corgan, Ph.D.
Professor
Department of Agronomy and Horticulture
New Mexico State University
Las Cruces, New Mexico

Peter R. Ellis, Ph.D.
Department of Entomology
National Vegetable Research Station
Wellesbourne, Warwick, U. K.

A. R. Entwistle, Ph.D.
Department of Plant Pathology
National Vegetable Research Station
AFRC Institute for Horticultural Research
Wellesbourne, Warwick, U. K.

C. D. Green, Ph.D.
Director
Plant Pathology Branch
Department of Agriculture
Biological and Chemical Research Institute
New South Wales, Australia

Kedar Nachum, Ph.D.
Professor
Department of Field and Vegetable Crops
The Hebrew University of Jerusalem
Rehovot, Israel

R. B. Maude, D.Sc.
Department of Plant Pathology
National Vegetable Research Station
AFRC Institute for Horticultural Research
Wellesbourne, Warwick, U. K.

Haim D. Rabinowitch, Ph.D.
Chairman
Department of Field and Vegetable Crops
The Hebrew University of Jerusalem
Rehovot, Israel

Baruch Rubin, Ph.D.
Senior Lecturer
Department of Field and Vegetable Crops
The Hebrew University of Jerusalem
Rehovot, Israel

S. K. Soni, B.Sc.
Department of Entomology
National Vegetable Research Station
AFRC Institute for Horticultural Research
Wellesbourne, Warwick, U. K.

D. P. Stribley, Ph.D.
Department of Soil Microbiology
Rothansted Experimental Station
Harpenden, Herts., U. K.

David G. A. Walkey, Ph.D.
Department of Plant Pathology
National Vegetable Research Station
AFRC Institute for Horticultural Research
Wellesbourne, Warwick, U. K.

J. O. Uzo, Ph.D.
Professor
Department of Crop Science
University of Nigeria
Nsukka, Nigeria

TABLE OF CONTENTS

VOLUME II

Chapter 1

CULTURAL SYSTEMS AND AGRONOMIC PRACTICES IN TEMPERATE CLIMATES

J. L. Brewster

TABLE OF CONTENTS

I. INTRODUCTION

This review emphasizes recent developments in onion culture in temperate regions. As much as possible, reference is made to scientifically conducted trials and comparisons, upon which cultural techniques are increasingly based. For discussion of those aspects of onion culture where the practices are long established or have developed more informally by farmer innovation and experience, reference is made to advisory literature, particularly that of Britain,[1,2] Holland,[3,4] and Ontario, Canada.[5] Crop protection, namely weed, pest and disease control, which forms an important part of agronomic practice is not considered since this is fully discussed in other chapters.

II. GENERAL CONSIDERATIONS FOR ONION CULTURE

A. TYPES OF ONION GROWN AND THE MARKETS AVAILABLE

Temperate regions, being at fairly high latitudes, have long photoperiods during the summer months. Consequently cultivars requiring a long photoperiod for bulbing, frequently termed ''long daylength cultivars'' predominate. A much smaller area is devoted to growing overwintered crops of ''intermediate daylength cultivars''. Within ''long daylength cultivars'' there remains considerable variation in response to photoperiod. For example, cultivars adapted to the north-eastern USA have a shorter photoperiod requirement than the ''Rijnsburger'' types that predominate in Britain and Holland.

Within temperate regions onions are grown both to harvest as green shoots for salad use and to harvest as bulbs. The desired marketable product largely determines the cultural techniques that are appropriate. Bulb crops can be divided into three broad uses, those that are sold as whole bulbs to the consumer, bulbs sold for use in food manufacturing or processing and small bulbs (sets) to be used as propagules for a subsequent crop. Whole bulbs sold for direct consumption can be subdivided by a number of properties. Aspects that a grower might consider include bulb size, shape, color, flavor, desired storage life, earliness of bulbing, and dry-matter percentage. Many of these properties are genotypically determined and they therefore determine the choice of cultivar, but some, notably size are modified by cultural techniques. Bulb onions are normally sold in graded diameter classes, for example in Britain diameters of <15 mm are regarded as unmarketable, 15 to 40 mm are sold for pickling, the preferred range being 28 to 35 mm, 40 to 60 mm are sold for fresh consumption, and >60 mm are sold for fresh consumption or for use in processed food. Within these size groups European Community regulations lay down fairly narrow ranges in diameter that may occur within a single pack.[6] Bulb size can be modified by varying plant density. Bulb shape is varietally determined but is also modified by cultural conditions particularly plant density. Onion skin color may vary from white through yellow to brown, and red or purplish cultivars exist. Market preferences for skin color vary from country to country. Skin color is primarily a varietal characteristic although the extent of brown coloration can vary with soil conditions and can be modified by curing treatments just after harvest (see Section VI.A). Flavors differ primarily in the degree of pungency and sweetness. Cultivars differ greatly in their natural storage life. Early maturing cultivars are normally grown for rapid consumption and frequently have innately short dormancy. Earliness can be encouraged by using transplants or sets so this is not wholly a matter of cultivar. In developed economies increasing market emphasis is being put on the 'quality' aspects of bulbs. Price premiums or even legal requirements are made for uniformity of size, shape, and skin color. Bulbs must be free from sprouting, loss of dry outer skins, cracking of skins, bruising, soiling, staining, rooting, and rotting, and must be free of noxious taints or toxins.[6]

Onions used in food manufacture are frequently the quality "grade-outs" from the bulb crops grown primarily for sale as fresh bulbs. Thus oversized or blemished bulbs from such a crop may be diverted for processing. Frequently bulbs graded out as too small from a fresh bulb crop will be sold for pickling. On the other hand some crops are deliberately sown at very high densities to produce small bulbs for pickling, or for use as very small pickled "silverskin" and "cocktail" onions. Onions for dehydration should ideally have a high dry-matter content (15 to 20%) and in some instances crops of appropriate cultivars are specifically grown for this market.

In addition to the above major markets there remains considerable production of vegetatively multiplied, small-bulbed strains of *Allium cepa,* notably shallots *A. cepa var. ascalonium,* and multiplier or potato onions *A. cepa var. aggregatum.* A higher proportion of these types are probably traded in very local markets or grown for home consumption than is the case with the larger bulb types.

B. CULTURAL SYSTEMS

Climate, notably the length of the growing season and the winter temperatures, has a considerable influence on the cultural systems used. Direct sowing of dry seed is the most common and lowest cost system for bulb and salad onion production.[1,3,5] Recent research on seedbed preparation, soil conditioners, seed drilling (sowing) machinery, and on the use of pregerminated seed is likely to result in continuing improvements in direct sowing techniques. However a shorter growing season, more reliable establishment and greater yields can frequently be achieved with transplants. Recently, the transplanting of small "modules" of compost, each one containing several onion seedlings has been developed into a highly practical system of crop establishment.[7,8] An alternative for easy crop establishment and a short growing season is the use of sets (small bulbs less than 21 mm diameter) as the propagule.[1,4,5] Techniques of set raising, storage, and heat treatment to prevent flowering (bolting) have been thoroughly researched. A disadvantage of sets is the risk of spreading soil-borne diseases and pests via contaminated sets. In regions with a long growing season, direct sowing of dry seeds is the common system for bulb production. The use of multiseeded module transplants originated in Norway[7] and these, or the use of sets or multiplier onions, make it possible to raise bulb crops in high latitude regions with a short growing season, e.g., Finland.[9] Transplants and sets are used in other regions for early bulb production. An alternative for early season bulb production is the use of dry seed in autumn-sown overwintered crops.[10,11] These are successful in regions of fairly mild winter temperatures although some cultivars are extremely hardy and can overwinter satisfactorily in Central Europe.[12]

For salad onions (sometimes termed "bunching onions" or "spring onions") direct sowing is the common practice. In climates with a mild winter such crops can overwinter to provide salad onions throughout the year.[2] In climates with more severe winters spring salad onions may be produced from sets or, for use in winter and early spring, produced by harvesting the green shoots from bulbs planted in a greenhouse.[5]

C. SOILS FOR ONION PRODUCTION

Bulb onions can be grown on a range of soils from loams with some clay content to quite light sands.[1] Peat (muck) soils are very suitable. Bulb maturity is usually later on peat soils than on light loams.[5] In England deep silts and peats with a high available water capacity can frequently keep the crop supplied with moisture throughout the growing season and can produce excellent crops in the absence of irrigation.[1] The need for a fine, clod-free seed bed coupled with problems of clods at harvest can make clay soils difficult for onion production. Stones in the soil also cause difficulties in seed drilling and mechanical harvesting.[1] Good drainage is essential, waterlogging can result in failure of overwintered crops.[1] In a study of the effect of water table depth on onions grown in an organic soil, yields were depressed with water tables of 10, 20, or 40 cm compared to 70 cm. Onions were much more sensitive than carrots to the effects of a high water table.[13]

Loamy sands to fine sandy loams are recommended for salad onion production in Britain.[2] These soils have a fairly high available water capacity but are quickly draining and can be cultivated satisfactorily for much of the year.

Recommendations for pH are: in Britain a minimum of 6.5 on mineral soils and 5.8 on peat soils,[1] in Holland a minimum of 5.5 on mineral soils,[3] and in Ontario 5.8 to 7.0, but pHs as low as 4.0 on peats have been satisfactory.[5] Experiments on acid organic soils showed that yields were not increased by liming above pH 4.0. At this pH the concentration of Ca and Mg in onion shoots was about 86 mEq per 100 g tissue, typical of the value found in productive crops on other soils.[14]

D. PLANT DENSITY AND ARRANGEMENT

The optimum population density for a bulb onion crop depends on the size of bulb required and on the potential yield to be expected. The relationship between mean bulb weight and population density (plants per unit area) can be described by the equation:[15]

$$(\text{bulb weight})^{-\theta} = \alpha + \beta \times (\text{density}) \tag{1}$$

Where θ, α, and β are fitted parameters. The parameter θ has values $\theta \leq 1$. A value of $\theta = 1$ gives an asymptotic relation between yield per unit area and plant density, whereas $\theta < 1$ gives an overturning curve with maximum yields per unit area at a particular density and lower yields at higher densities. At extremely low densities the yield per plant tends to $1/\alpha$, so this can be regarded as the maximum size attainable by a plant in the crop. With $\theta = 1$ the yield per unit area at very high densities approaches $1/\beta$, and this can be considered to measure the yield potential of the environment in which the crop is grown.[16]

The majority of studies with bulb onions have found that θ does not differ significantly from unity, indicating an asymptotic relationship between bulb yield and plant population,[17-19] but others found $\theta = 0.8$ gives the best fit.[15,20] The majority of studies have involved densities of less than 500 plants m^{-2}, but a recent study using densities appropriate for pickling onions, considered densities up to 1600 plants m^{-2} in conditions of high fertility with irrigation.[21] Total bulb yield was maximum at 400 plants m^{-2} and decreased progressively to levels significantly below the maximum as density increased beyond this.[21]

Values of β vary significantly between years at the same site, indicating year to year differences in potential bulb yield.[20] Values of α do not vary significantly between seasons but do vary between varieties.[17] Wide differences in β have been found between experiments from different localities[18] and between different levels of soil phosphate at one site.[22] Values of $1/\beta$ of 9 to 11 kg m^{-2} were found for irrigated onion trials in New Zealand and Australia and lower values for nonirrigated trials in Europe.[18] In several trials with irrigated well-fertilized crops actual yields of in excess of 10 kg m^{-2} (100 ha^{-1}) have been attained at high plant densities.[21,23,24]

If the parameters α and β are known, then it is possible to estimate the yield of bulbs in a particular size grade.[19] Bulbs are normally graded by diameter rather than weight but assuming a specific gravity of 1 g cm^{-3} and an equation,

$$\log_{10}(\text{diameter (mm)}) = 1.02 + 0.364 \log_{10} (\text{bulb fresh weight (g)}) \qquad (2)$$

enabled weights to be converted to diameters.[19] The tendency of bulb shape to change with bulb weight probably results in a small inaccuracy in this approach. Bulb length/bulb diameter decreases with bulb weight[20] and increases as plant density increases.[25]

A major problem with the above curve fitting approach to predicting optimum densities lies in predicting the expected or potential yield which may vary from year to year with weather and soil conditions. In practice, advisory literature tends to specify crop densities appropriate for a particular locality and the required bulb sizes, based on accumulated experimental and field experience.[1,3,5] In the U.K., for a maximum yield of bulbs greater than 40 mm diameter, a density of 90 plants m^{-2} is recommended.[1] However, plant densities of 70 to 80 plants m^{-2} are generally advised to avoid an excess of small bulbs in low yielding crops.[1,26] In Holland a density of 100 to 120 plants m^{-2} is recommended,[3] in Norway one of 60 plants m^{-2}.[7] If larger-sized bulbs are required, then these can be obtained by reducing plant densities further, but below 100 plants m^{-2} total yields will decrease markedly with density.[15,23]

With the advent of good herbicidal weed control for vegetable crops there is no longer the need for wide inter-row spacing to facilitate mechanical weed control.[27] The possible benefits of narrow row spacings have therefore been investigated. The most uniform distribution of plants over an area of ground will be achieved when the distance from plant to plant within a row equals that between rows. For a low growing, unbranched plant like an onion it might be expected that the exploitation of the available light, water, and nutrients will be maximized by such an arrangement, and that the plants would have insufficient plasticity in growth to compensate fully for a less uniform planting arrangement. Several studies have examined the influence of rectangularity (the between-row distance/mean within-row distance) on bulb onion yield and have shown considerable yield reductions at high as compared to low rectangularities. At a density of 75 plants m^{-2} reducing between row spacing from 45 to 22 cm increased yields by 10 to 30% depending on experiment.[15] In Norway reducing row spacing from 70 to 35 cm increased yields by 12.5%.[7] In Tasmania, a yield reduction of 10% was recorded when a rectangularity of 8 was compared to that of 1 at a plant density of 65 plants m^{-2}.[18] In English overwintered crops yield reductions of 13 to 17% occurred at rectangularities of 8 compared to 1 at densities of 43, 86, and 129 plants m^{-2}.[23] On English peat soils an average yield reduction of 15% was observed when a group of treatments with rectangularity 6 to 14 was compared with treatments of rectangularity less than 6 and it was recommended that inter-row widths should not exceed 25 cm when growing bulbs at a typical commercial density of 80 plants m^{-2}.[26]

A uniform between-plant distance within a row as opposed to a random within-row spacing does not significantly increase total yields or yields of bulbs greater than 40 mm diameter in crops grown at densities of 86 to 108 plants m^{-2}.[28] Neither is uniformity of bulb size increased by uniform within row spacings. The same conclusions were drawn from a similar experiment using sets rather than seeds.[28]

For salad onion crops a plant density of 430 m^{-2} is recommended for spring-drilled crops in Britain and 260 m^{-2} for overwintered crops.[2] These crops are frequently grown in rows 30 cm apart but closer rows, down to 10 cm apart, are recommended to produce more uniform onions.[2]

Both bulb and salad crops are increasingly grown on the "bed" system, where the plants are sown in closely spaced rows within "beds" the width of the distance between tractor wheels. For example in Holland, five rows are sown 27 cm apart separated by a path 42 cm wide. All tractor operations involve driving down the wide rows which are 1.5 m apart[3] leaving the soil between unconsolidated.

III. SOIL PREPARATION

A. CULTIVATION

On mineral soils autumn or early winter ploughing is recommended, followed by shallow cultivation (approximately 3 cm depth) in the spring.[1,3,5] On peat soils late winter ploughing followed by rolling to break clods and form a level seedbed works well.[1,5] Satisfactory establishment and yields of autumn and spring-sown onions have followed direct seed drilling into burnt cereal stubble with no initial ploughing or cultivation.[29] Direct seed drilling into unburnt cereal stubbles is less satisfactory.[29]

B. FERTILIZER INCORPORATION

Phosphorus and potassium are normally incorporated into the soil prior to sowing. In most instances the majority of the nitrogen is also incorporated at this stage, but additional nitrogen is also applied during crop growth. Comparative studies with other vegetable crops have shown that onions are among the most responsive to applied phosphorus and potassium.[30,31] In these experiments, on a nutrient depleted sandy loam, optimum levels of fertilizer were 206, 105, and 119 kg ha^{-1} of N, P, and K respectively for spring-sown bulb crops, and 209 and 138 kg ha^{-1} of P and K respectively for autumn-sown bulb crops. Recent experiments in Poland indicated optimum applications of 133, 57, and 166 kg ha^{-1} N, P, and K[32] and in Germany an optimum of 83 kg ha^{-1} K was found.[33] Studies of nutrient requirements on sphagnum peat soils in Canada indicated optimum requirements of 275, 40, and 70 kg ha^{-1} N, P, and K for soils that had not recently been cropped and fertilized.[34] In Denmark bulb yields reached a maximum with 120 kg ha^{-1} N.[35] Onion emergence can be reduced by an excessive concentration of fertilizer solutes in the seedbed, for example, increasing the presowing N application from 75 to 150 kg ha^{-1} as calcium nitrate, reduced seedling emergence, and lowered plant stand by 10 to 20%.[36]

Advisory recommendations to growers of bulb onions emphasize the importance of soil nutrient status in determining the appropriate fertilizer requirements. In Britain applications of P ranging from 26 kg ha^{-1} on high P status soils to 129 kg ha^{-1} on low P status soils are recommended.[1] The corresponding range for potassium is 50 to 250 kg ha^{-1}.[1] Recommendations for salad onions are similar.[2] In Holland the recommended range for P application is from 0 to 106 kg ha^{-1} depending on soil analysis and soil type and from 0 to 365 kg ha^{-1} for K.[3] In Ontario (Canada) the range of P recommended is from 0 to 79 kg ha^{-1} on mineral soil and from 0 to 44 kg ha^{-1} on peat, depending on the soil analysis. Fertilizer K can range from 0 to 191 kg ha^{-1} on either type of soil depending on analysis.[37] In Britain the amount of nitrogen recommended depends on previous cropping.[1] After a cereal crop 100 kg ha^{-1} N is recommended on mineral soils and 60 kg ha^{-1} on peat soils.

Recommended N levels are lower following peas, beans, or potato crops or after the application of manure. In Holland it is recommended that the available nitrogen in the top 0 to 60 cm layer of soil in the spring plus the added fertilizer should equal 180 kg ha^{-1}.[3] In Ontario up to 110 kg ha^{-1} N is recommended for both mineral and peat soils with two thirds of this applied before sowing and one third when the plants are 15 cm tall.[37] In Holland also, it is recommended that one third of the N fertilizer is applied when plants are 10 cm tall.[3]

Recent research in Denmark has shown 5 to 10% higher yields and 5 day earlier maturity in spring-sown bulb onion crops when 40 kg ha^{-1} P was placed as a band 5 cm below the seeds and 3 to 5 cm to the side of the seed drill rather than simply broadcast and mixed into the seedbed.[38] In Britain increases of up to 60% in plant weight per meter of row in late spring have resulted from similar banding of applied P rather than broadcasting and up to 19% increases in final bulb yields also resulted from the placement treatment.[39] Fertilizer banding 5 to 8 cm below the seed is recommended as the most efficient mode of application for Ontario.[5]

IV. PLANT ESTABLISHMENT

As with many small-seeded vegetable crops, the successful establishment of the desired population of plants is probably the most difficult and unpredictable stage of culture. As noted above, yield and bulb size are largely determined by plant population density. Hence, populations of the appropriate density must be established to produce bulbs in the desired size range. As well as causing differences in bulb size, wide variations in population density within a field will result in uneven bulb development, with the areas of low population being later to mature than areas of high population. This is important because the timing of the application of sprout suppressant chemicals and modern direct harvesting techniques are closely tied to the stage of crop maturity (see Sections V.B and VI.A), and uniform high quality bulbs can be expected only from an evenly maturing crop. Herbicide applications (see chapter on "Weed Competition and Herbicides") are determined by the stage of seedling growth since seedlings below a certain size are liable to damage from herbicides. Hence, for successful herbicide use it is important that seedlings are all at a similar stage of development. Therefore, with direct-sown crops, seedling emergence should occur over a short duration. Consequently, the establishment of an evenly distributed population of seedlings at the appropriate plant density and at a uniform stage of development is the prerequisite for the success of all subsequent cultural operations. The importance of successful crop establishment has prompted considerable research and development work on sowing and planting techniques.

A. SOWING SEED

1. Seed Quality and Grading and Sowing Rates

Onion seed rapidly loses viability after harvest. The loss of viability is more rapid as temperature and seed moisture content increase and the rate of loss of viability is more rapid than for most other crop species[40] (see chapter on "Physiology of Crop Growth and Bulbing"). It is, therefore, generally recommended that only fresh seed is used.[5] Normally only seed of high germination percentage is sold.[3] Laboratory germination percentages correlate well with percentage emergence in the field. However, under dry field conditions a smaller proportion of live seeds emerge from samples of low than of high laboratory germination.[41] Seed size, laboratory germination percentage, the rate of radicle emergence at 15°C and a number of biochemical attributes of seeds do not always correlate among themselves or correlate consistently with emergence in the field.[42]

The seed trade sells both ungraded and graded seed of stated minimum percentage laboratory germination. Various size diameter grades are available.[1,3] Seed graded to narrow size limits, e.g., 2.25 to 2.5 mm or 2.50 to 2.75 mm, is necessary for use in the belt and cell mechanisms of precision-spacing seed drills. Seed is also available pelleted to increase its diameter to 3.25 to 4.00 mm[3] or 3 to 3.5 mm,[3] for certainty of single seed sowing by precision-spacing seed drills. Seed is now normally sold with an insecticide coating for protection against onion fly and with fungicide coating for protection against neck rot (*Botrytis allii*) and damping-off diseases.

The weight of seed to sow is calculated according to the desired final population, the number of seeds per gram weight, the laboratory germination and a ''field factor'' which attempts to predict the proportion of viable seeds sown that emerge in the field.

$$\begin{array}{c}\text{The seed requirement}\\ (\text{kg ha}^{-1})\end{array} = \frac{1000 \times \text{number of plants m}^{-2}}{\begin{pmatrix}\text{No. of seeds}\\ \text{per gram}\end{pmatrix} \times \begin{pmatrix}\text{\% laboratory}\\ \text{germination}\end{pmatrix} \times \begin{pmatrix}\text{field}\\ \text{factor}\end{pmatrix}}$$

The ''field factor'' depends on soil conditions. Recommended values are for cold soil and poor tilth, 0.5 or 0.6; for average conditions, 0.7; for good conditions, 0.8, and for ideal conditions, 0.9.[1]

2. Seedbed Preparation, Sowing Depth, Soil Conditioners and Mulches

Advisory literature emphasizes the need for a fine, moist, firm, level, and clod-free seedbed for onions.[1,3,5] Good contact between the small onion seed and soil is thought to be essential for satisfactory germination. Dutch advice recommends the seedbed be prepared in such a way that seed is sown on a moist, firm underlying layer.[3] Recent research at Wellesbourne, U.K. has confirmed the importance of these firm seedbed conditions. Emergence was compared from sowings into (1) a spring-prepared seedbed cultivated to a depth of 100 mm with a power harrow, (2) an autumn-harrowed bed harrowed to a depth of 30 mm in spring then lightly rolled, and (3) an autumn harrowed bed with no spring cultivations. In a dry spring, treatments 2 and 3 gave five- or sixfold greater emergence than treatment 1.[43] In both a dry spring and a wet spring the best emergence was obtained when cultivation prior to seed drilling was no deeper than the depth of drilling.[44] Soils cultivated in this way show a distinctly higher penetration resistance immediately below seed depth, indicating a consolidated zone which is thought to enhance emergence because its improved capillarity will supply water more easily to the seed and germinating seedling than will loose soil.[44]

Depths of sowing recommended are 12 to 15 mm,[1] 2 cm[3], and 0.5 to 2.5 cm[3]. Experiments indicate that differences in drilling depth can lead to differences of several days in the mean time to seedling emergence, hence an even depth is important for uniform emergence.[43] Deeper sowing is recommended if soil conditions are dry[1] or on peat soils in which wind or water erosion is a problem.[5] Shallower drilling is advisable on light mineral soils that slake and form a hard surface crust or ''cap''.[3,5] An experimental study in Germany examined drilling depths from 1 to 6 cm and found that 2 to 3 cm gave optimal emergence.[45] Depths of 4 cm or more reduced emergence and resulted in a higher proportion of elongated bulbs.[45] In another study percentage emergence from pots decreased with depth from 50% at 1 cm to 7.5% at 7 cm, but 2 cm was recommended for outdoor sowing to avoid the seed drying out.[46]

On many soils heavy rain can cause the soil to slump and form a hard surface crust or ''cap'' on drying, which prevents seedlings from emerging. Crops with a long emergence time, like onions and leeks, are statistically more likely to experience heavy rainfall between sowing and emergence than those with a short emergence time. Recently soil conditioners have been developed which, when sprayed on the surface of the soil after sowing, can prevent the formation of a ''cap''.[47] The soil conditioner cellulose xanthate was shown to increase the percentage emergence of onions and leeks by up to 40% and to have no detrimental effects.[47,48] In another trial an alternative soil conditioner [Soiltex (Allied Colloids Ltd.)] increased the emergence of onions by 66 to 80% and final yields by 29 to 35%.[49] Such benefits will only occur where an impenetrable soil cap forms between sowing and emergence. The likelihood of this depends on the soil, the climate and the duration of the interval between sowing and emergence. However, for onions on many soils, particularly with early spring sowings, such soil conditioners are an economically viable precaution against poor emergence.

Experiments in Norway have shown that the covering of April-sown onions with clear polyethylene tunnels can increase yields by about 10% and advance maturity by about 10 d.[7] It was necessary to remove the plastic tunnels soon after seedling emergence, otherwise high temperatures damaged the crop and reduced yields. Larger increases in yield from tunnels were obtained on a heavy soil than on a light soil. By using plastic mulches or tunnels, onion production in Norway can be extended to heavier soils than the sandy and peat soils normally considered suitable.[7] Similar benefits were also obtained from a clear plastic mulch laid directly on the ground until germination.

3. Factors Affecting the Percentage and Rate of Emergence

Studies of emergence in near-ideal conditions namely, moist irrigated seedbeds treated with anticapping agent, gave a mean emergence of 82% of the seeds sown in experiments in sterilized soil[50] and 77% in other experiments on nonsterilized soil.[51] Eight percent of these seeds produced abnormal seedlings when germinated on paper and were therefore unlikely to emerge in soil, a 15% loss therefore occurred for other reasons, possibly soil slumping.[51] In the absence of irrigation a much lower percentage final emergence can occur. Percentage seedling emergence from seed sown into identically prepared seedbeds at a sequence of dates ranged from 5.5 to 87% depending mainly on soil moisture conditions.[50] When onion seed was sown into dry soil and not irrigated, the seedlings emerged only after a significant rainfall.[52] The percentage emergence from viable seeds tends to decrease as soil temperature decreases.[53] Once onion seeds have germinated, viability can be reduced by periods of 2 d or more at -2°C.[54] The damage is more severe the lower the temperature and the longer the duration of low temperature.[54] The probability of such temperatures occurring in central England for sufficient duration to be damaging to germinating onions, was calculated as falling from approximately 13 occasions per 100 for sowings made in February to approximately 3 occasions per 100 for sowings made in April.[54] In Holland field experience has shown late March or early April to be the best time for spring sowing, percentage emergence from earlier sowings is frequently too low.[3] In England a March drilling is recommended, but only if a good seedbed can be prepared.[1] The occurrence of prolonged periods of temperatures below -2°C at seed level may be why earlier sowings sometimes give poor emergence.[54]

In constant temperatures with a plentiful supply of moisture, onions need 219 day-degrees above a base temperature of 1.4°C after sowing to reach the stage at which 50% of seedlings have a straight cotyledon.[55] Other temperate vegetables have a similar base temperature but onion (and leeks) need more day-degrees than most for emergence, for example approximately double the number required for cabbage emergence.[55] Field sowings into moist soil indicate that:[51]

$$1/\text{time to 50\% emergence} = -0.01 + 0.00713 \times (\text{mean air temp °C})$$

This corresponds to a requirement for 140 day-degrees above a base temperature of 1.4°C for 50% emergence. In this instance emergence was recorded as soon as the seedling was visible. In the absence of irrigation the correlation between emergence rate and temperature is much weaker than in moist soil with irrigation.[50]

4. The Use of Pregerminated Primed Seed

The slow emergence and establishment of onion seedlings has prompted considerable research into the sowing of pregerminated and primed seeds, in the hope that these will give more rapid establishment. Increases in bulb yields and earlier maturity have been recorded using pregerminated seeds.[56] However, in U.K. conditions the results have been variable and the emergence percentage is frequently lower than that for dry seed.[57,58] Sowing pregerminated seeds in a fluid gel carrier (fluid drilling) consistently gives an earlier date of 50% emergence than sowing dry seed. However, the final percentage emergence from fluid drilling is lower than from dry seed under dry soil conditions, although it can be higher under moist soil conditions.[50,51,57] Pregerminated seeds are sown when the longest radicles are about 4 mm long, but at this stage radicles have emerged on only about 20% of seeds.[57] In experiments where all seeds were selected to have radicles about 2 mm long and then sown into moist soil and given irrigation after sowing, selected pregerminated seeds had more rapid emergence, less spread of emergence, and a higher percentage emergence than dry seed.[50] Using such selected seed, rapid, high-percentage emergence can be ensured, provided the seeds are drilled into moist soil followed by a single seedbed irrigation if conditions are dry and a soil conditioner is used to prevent capping.

Accelerated germination and emergence can be attained using osmotically "primed" onion seed.[59,60] Seeds were treated for 14 d at 15°C in solutions of polyethylene glycol (mol. wt. 6000) with nominal osmotic potentials of -10 and -15 Bars. After these treatments seeds were either washed with water or washed and dried in air for 24 h. Primed undried seeds germinated 3 to 5 d sooner and emerged from a field sowing 3 to 9 d sooner than untreated seed.[59,60] Drying the primed seed increased germination time on paper by about 1 d compared to undried primed seed but it did not delay field emergence. Compared with dry, untreated seed the spread of emergence time was reduced by seed priming but the percentage emergence was not affected. Twelve weeks after sowing, primed seedlings were on average 39% heavier than plants from unprimed seeds.[60] Similar acceleration of emergence and increases in seedling weight were attained using primed leek seeds in comparison with dry seeds.[61] With leeks it was shown that priming did not increase seedling relative growth rates and the higher weights of seedlings from primed seeds were solely due to their earlier emergence. Primed and dried onion seeds can be stored for up to 18 months at 10°C with 9% moisture content without loss of viability and without an increase in their germination time.[62] The pretreatment of seeds in osmotica like polyethylene glycol can reduce the spread in germination time on their return to solutions of low osmotic pressure.[58,63] Such techniques combined with techniques for separating ungerminated from germinated seeds could be applied to produce uniformly germinated seeds for sowing.[51]

B. TRANSPLANTING

There are several reasons for using transplants. First, bulb onion crops need a fairly long season of growth and at high latitudes the length of growing season may be insufficient for reliable production of field drilled crops. By raising transplants in glass or plastic houses, seeds can be sown several months earlier than in the field, and the available growing season is lengthened. For example, transplants are used in Ontario (Canada) to ensure a sufficiently long growing season for the production of sweet-Spanish type onions.[5] New techniques of transplanting have outperformed sets as a mode of production in the short growing season of Norway.[64] Second, transplanted crops normally mature earlier than field drilled crops of the same cultivar. This can be important in supplying the demand for onion bulbs at a season of the year when supplies of stored onions from the previous year are at an end. Finally, the yields from transplanted crops are usually higher than from directly sown crops. This is only partly because of the difficulty of ensuring good seedling emergence from seed sown in the field in early spring. For example, in a trial in central England, February sown transplants of the long-day cultivar Rijnsburger Bola transplanted in late April produced a yield of 68 t ha^{-1} with 66 t ha^{-1} greater than 45 mm diameter. A directly drilled crop sown in late February and thinned to the same plant density as the transplants (86 plants m^{-2}) yielded 45 t ha^{-1} of which only 27 t ha^{-1} were larger than 45 mm.[65] Experiments in eastern England produced average yields of 45 t ha^{-1} from transplants as against 30 t ha^{-1} for directly sown crops.[1]

1. Single Plant Transplants

The technique recommended in Ontario[5] is to sow under glass at a rate of about 4500 seeds m^{-2}, either broadcast or in rows 7 to 10 cm apart at a depth of 3 to 12 mm. Day temperatures of 18 to 15°C and night temperatures of 10°C produce robust seedlings. To minimize the risk of damping-off diseases, watering is minimal on cloudy days and water is applied in the morning on sunny days so that leaves and the soil surface are dry by nightfall. Ten to twelve weeks after sowing seedlings have pseudostems approximately 3.5 mm thick and are suitable for transplanting. Before transplanting, seedlings are hardened by exposing them to temperatures of 4 to 7°C for 7 to 10 d coupled with less frequent watering. To facilitate handling at transplanting the shoots are frequently trimmed to a length of 10 to 15 cm. Yields from trimmed or pruned seedlings are slightly lower than unpruned seedlings, and the more severe the pruning the greater the reduction in yield.[66,67]

In Ontario[5] seedlings are transplanted as early in spring as soil and weather conditions allow. The base of the stem is planted 2 to 2.5 cm below the soil surface. Plants are spaced 10 to 15 cm apart in rows 41 to 43 cm apart (15 to 24 plants m^{-2}), the closer spacing giving higher yields but smaller bulbs.

Deep transplanting of Sweet Spanish cultivars, i.e., 10 or 5 cm deep, produces more elongated bulbs with a significantly lower diameter/length ratio than in bulbs planted 2.5 cm deep.[67,68] The more elongated deeper planted bulbs have fewer internal branches and hence produce more unsplit, single-centered onion "rings" when the bulbs are sliced transversely.[68] This effect can also be attained by planting at the normal 2.5 cm depth followed by banking up soil round the plants 60 d after transplanting.[68]

Transplanting by hand involves making a hole and firming the soil back about the roots, or placing seedlings along the side of a shallow furrow and then covering and pressing the soil against them.[5] Various semi-automatic vegetable transplanters may be used. To facilitate handling in these machines trimmed transplants are necessary. Such machines frequently plant seedlings nonvertically. In an experiment, horizontally placed transplants gave only 56% of the bulb yield of vertically placed transplants.[66]

Experiments in Norway, with its high latitude and consequent long late-spring and summer photoperiods, showed that artificial shortening of the photoperiod during transplant raising, prolongs growth after transplanting and increases bulb yields.[69] This is particularly important for later transplanted seedlings.[69] If conditions during plant raising cause seedlings to bulb strongly they may fail to re-root and resume growth upon transplanting to the field.[70]

2. Multiseeded Module Transplants

The labor of transplanting has been considerably reduced by the development and adoption in Norway and England of techniques for transplanting groups of seedlings. Between three and eight seedlings depending on the required final bulb size are raised in small cells or "modules" of compost.[7,8,64,71-75]In recent years several systems of producing "modular" transplants have developed. Each uses plastic trays moulded into small cells which hold 9 to 15 cm^3 of compost.[8] As an alternative, small brickettes of compressed peat "blocking compost" of various volumes can be used.[76] These systems have proved convenient for the mass raising of many types of vegetable for transplanting. Experiments in Norway first showed that high yields of early maturing bulbs could be produced by raising between three and seven seedlings in small pots and transplanting the whole group to the field.[7] Thereby the number of transplantings to achieve a given plant population was reduced. During development the bulbs push each other apart as they swell, and they are not deformed by pressure from neighboring bulbs.

Small cells filled with 9 to 15 cc compost are now the most popular form of raising such transplants in England, and various aspects of plant production in this system have been researched.[72] Tests on various compost media have shown that well structured but finely milled sphagnum peat, free of bark and fiber, mixed with about 10% by volume of sand is the most suitable medium.[72] In preparing the compost the following amounts of liming and nutrient substances are incorporated per cubic meter: 2.25 kg calcium magnesium carbonate (magnesium lime), 2.25 kg calcium carbonate (ground chalk), 2.25 kg calcium phosphate (single superphosphate), 1.0 kg potassium nitrate, and 0.4 kg WN 255 fritted trace elements.[8] With these incorporations the compost can supply all the necessary phosphate and trace elements needed for a 6- to 7-week raising period (up to the 1 to 2 true leaf stage) and sufficient nitrogen up to the first true leaf stage.[72] From the first true leaf stage onward a liquid feed containing 100 mg/l nitrogen and 200 mg/l potassium should be applied two or three times. Various alternative modes of supplying nutrients have been tested and can be as successful as the above system.[72]

Soil and weather conditions do not always allow seedlings to be planted as soon as they reach the appropriate size. Modules containing plants 5 or 6 weeks from sowing can be stored at 0.5°C, 95% relative humidity, in the dark for up to 2 weeks without adverse effects on establishment or final bulb yield.[72] Storage for 3 weeks in these conditions resulted in 66% plant losses upon planting out.[72]

Peat blocks have been used as an alternative module for raising groups of onion transplants.[76] Block sizes 27, 32, 38, and 43 mm cubed sown with 7 to 8 seeds to produce approximately 6 plants per block at harvest time have been used. In some experiments final yields from 43 mm blocks were significantly higher than those from 32 mm blocks,[74] but in other trials differences between 27 and 43 mm blocks were not significant. The 27 mm cuboid blocks are made up by compressing 50 cc of compost and they therefore contain considerably more compost and hence a higher reserve of nutrients than the 9 to 15 cc typically used in loose filled cells. For this reason, nutrient reserves in peat blocks are sufficient for the production of a group of 1 to 2 true leaved seedlings with only one liquid feed during propagation.[75] A liquid feed just prior to planting out is recommended for both blocks and loose filled tray modules.[8] With the larger-sized blocks seedlings can be grown on for longer than with small blocks which is useful if weather or soil conditions are unsuitable for planting.

Various temperature regimes for raising multiple seeded modules have been used. Germination at 15°C followed by 10 d at 12°C then night frost protection with day ventilation at 10°C, produced transplants in 6 to 7 weeks.[72] By germinating at 18°C followed by gradual hardening to lower temperatures, plants at the one to two true leaf stage, suitable for transplanting, were obtained after 200 day-degrees above 6°C had accumulated since sowing.[75] Higher temperatures are favored by some raisers, i.e., germination at 21°C followed by 3 weeks at 15°C then 10°C for 2 weeks followed by hardening off to ambient springtime temperatures,[74] a similar temperature regime has been used in Norway to produce seedlings ready for planting out after 6 weeks.[7,64] In England sowing in late January to mid-February is satisfactory for producing transplants by late March or early April when conditions are suitable for planting.[73,77,78] In Norway such seedlings are planted between mid-April and mid-May.[64] If transplanting is delayed until May, then reducing the photoperiod during plant raising to 11.5 or 12 h prevents premature bulbing after transplanting and increases yields.[64] Short photoperiod treatment is most important during the last half of the propagation period and has more effect on early maturing than on late maturing cultivars.[7,79] Seedlings transplanted in April in Norway do not benefit from short photoperiods during plant raising.[64]

Several studies have examined the question of the optimum number of seedlings per module, and how this interacts with the spacing of modules after planting out, in its effect on yield and final bulb size.[7,74,77] For early production in Norway, groups of three seedlings per module were 11 d earlier than groups of 5 plants, in reaching 100 g weight.[64] For maincrop bulb production in Norway 5 plants per module planted to give 60 plants m^{-2} (12 modules m^{-2}) was optimal and produced almost 50 tha^{-1} of bulbs.[7] In English experiments 7 plants per group gave a high proportion of small onions (less than 40 mm diameter) and 5 to 6 plants per group was optimal.[74,77] With 5 to 6 plants per block the yield of bulbs of 40 mm diameter or greater reached a maximum of 45 to 55 t ha^{-1}, depending on experiment, at a planting density of 10 blocks m^{-2}.[74,77] At this density 60 to 70% of the bulbs were greater than 60 mm in diameter. Densities of 14.5 blocks m^{-2} gave a high proportion of bulbs in the 40 to 60 mm size grades.[77] Yields were significantly reduced at planting densities below 7.5 blocks m^{-2}.[77]

Fertilizer N, P and K should be incorporated into the soil before planting at rates similar to those for a directly sown crop. Supplementary nitrogen at rates of 60 kg ha^{-1} 1 or 2 weeks after planting out has been beneficial.[73] Irrigation is recommended to aid establishment after planting,[7,64,73] although in some localities satisfactory establishment without irrigation is possible.[72] If peat blocks become very dry, they can harden and become difficult to rewet and they can entrap the seedlings in the hard, dry block. To avoid this, the top of the blocks should be buried and irrigation applied until the seedlings have rooted out of the blocks. Herbicide treatments are similar to those for direct sown onions, but after planting out herbicide treatments should be avoided until the seedlings have hardened and acquired a normal waxy coat on the leaves.[1]

Studies using groups of four leek seedlings raised and transplanted in modules of different shapes, but containing the same volume of compost, showed that seedling growth and final yields were somewhat greater from pyramid shaped modules than from cubes, cylinders, or inverted pyramids.[80] It was also shown that, when planting into unirrigated soil, water saturated blocks increased root outgrowth, plant survival, establishment, and subsequent yield of leeks compared to dry blocks.[81] If irrigation was applied after planting, dry and wet blocks were equally satisfactory. Seedlings and root systems of leeks and onions are similar and similar responses to those for leeks might be expected of onions.

Raising and planting transplants is expensive compared with the cost of establishing a field of onions by direct sowing. Excluding costs common to both systems, e.g., land preparation, the cost in England of establishing a hectare of multiseeded peat blocks was about 3.5 times that of establishing a directly drilled crop.[74] The extra cost must therefore be justified by the higher yields, earlier maturity or improved quality that can follow from transplanting as opposed to direct sowing.

C. GROWING FROM SETS

1. Producing Bulbs from Sets

Planting sets (small bulbs <25 mm in diameter) is probably the easiest way of establishing an onion crop, since a set is a large propagule which is less sensitive to soil conditions for establishment than a seed or transplant. Because of ease of crop establishment sets are widely used by amateur growers. The set has a large reserve of stored assimilates compared to a seed and it more rapidly develops into a large plant and needs a shorter growing season to produce a mature bulb. For this reason sets are frequently used for producing early bulb crops.[64,77,82] A disadvantage with sets is the possibility that they can be contaminated with soil borne pests and diseases, e.g., stem and bulb eelworm (*Ditylenchus dipsaci*) (see chapter on "Nematode Pests of Allium Species") or onion white rot (*Sclerotium cepivorum*) (see chapter on "Root Diseases"), or virus diseases (see chapter on "Virus Diseases"). The spread of white-rot in recent years has discouraged the use of sets in favor of transplants.

Marketable bulb yields have been found to increase with plant density up to about 86 sets m^{-2} in England,[83] but a higher proportion of large-sized bulbs are produced at lower planting densities[77] and planting densities of 30 sets m^{-2} are recommended in Britain,[1] 36 to 60 m^{-2} in Norway,[64] 60 to 80 m^{-2} in Holland[4] and 25 to 60 m^{-2} in Canada.[5] The weight of sets needed to plant at a given density depends on the set size, and the number of sets per kilogram must be known. Sets of 1 to 2 cm diameter are recommended for planting,[1,5] those less than 16 mm diameter rarely flower.[1] Flowering (bolting) is increased by exposure to long periods at cool temperatures (7 to 12°C) before planting or after planting and before bulbs start to form. Bolting is therefore increased by early planting[64] and cool seasons.[9] In the absence of bolting larger set sizes will give higher yields.[84,85] Because of higher planting costs though, large sets are not economically jusitifed except for the rapid production of early bulbs at high latitudes.[64,85] A disadvantage of larger sets is the tendency for the production of double or split bulbs to increase as set size increases, although this characteristic also depends on cultivar.[86,87] In Holland it is recommended that sets should be graded into 8 to 14 mm and 14 to 21 mm groups and that these should be mechanically planted separately.[4] Otherwise, with mixed set sizes, planting is uneven within the rows and this results in uneven growth and a variable bulb size at harvest.[4] Soil preparation prior to planting is similar to that for a seed sown crop. Ideally sets should be planted the right way up and just covered with soil. Planting by hand into shallow furrows is common.[1] Set planting machines have been developed[4] but yields are reduced by mechanized planting compared with hand planting because planting sets other than vertically reduces yields. Experiments in Norway,[88] showed that deeper planting and increasing set angle slows emergence and bulb yields decrease as the planting depth and set angle (the angle between the vertical and the central axis of the onion from the root plate to the top) increase. These effects are larger with set angles exceeding 90° than with angles smaller than 90°. Yield losses of between 10 and 30% occur with machine planting compared with hand planting.[83,88]

2. Producing Salad (Bunching) Onions from Sets

Sets can be used to produce green salad onions.[5,89] For this purpose sets should be planted deeper than for a bulb crop, 2 to 5 cm apart in rows 30 cm apart.[5] Experiments in Poland showed that September or October planted sets 17 to 25 mm in diameter survived until spring better than sets 11 to 17 or 5 to 11 mm diameter.[89] Autumn planted sets produced salad onion crops 14 to 18 d earlier than spring-planted sets. However, the greatest yields of salad onions was obtained by spring planting followed by covering the plots with perforated plastic for 16 to 30 d after planting.[89]

V. GROWING ON

After establishment of the seedlings or sets, steady growth and the attainment of a high leaf area index (LAI) is needed for high yields, to encourage early bulb maturity and to minimize the number of onions failing to mature normally and consequently producing "thick-necks".[90-92] Early and even maturity facilitates harvesting and the production of a high-quality, long-storing crop. Onion seedlings have an innately low relative growth rate compared to many other crop and weed seedlings[93] (see chapter on "Physiology of Crop Growth and Bulbing"). The crop therefore takes a long time to develop a high LAI and is vulnerable to weed competition over a long period.[94,95] Efficient weed control is therefore essential (see chapter on "Weed Competition and Herbicides"). With modern herbicides, onion crops can normally be produced satisfactorily without, or with little, mechanical and hand weeding.

A. IRRIGATION AND SUPPLEMENTARY NITROGEN APPLICATIONS

The majority of experimental studies show yield increases from irrigation, but increases may vary from more than double in a dry season to a nil effect in a wet year.[96] The increased yields from irrigation (other then seedbed irrigation) result from larger onions rather than from higher plant populations, hence yield increases of the more commercially valuable large bulbs are even greater than the increases for total yield.[96] Bulb yield increases in response to irrigation have been reported in a number of countries including Hungary,[97] Germany,[98-101] England,[24,92,96] Poland,[102] Norway,[103] north eastern U.S.,[104] Canada,[105] and USSR.[106] In Canada frequent irrigation with small amounts of water is recommended.[5] In England 25 mm of irrigation whenever the soil moisture deficit falls to 35 mm is recommended.[1]

Irrigation after sowing is often essential to ensure rapid and even emergence of late summer-sown crops[11] and the emergence of spring-sown crops is frequently improved by irrigation.[50,51] Irrigation is also important to aid seedling establishment after transplanting.[81]

The effect of controlled drought during different phases of growth of a bulb crop raised from sets has been reported from Norway.[103] The highest yields were obtained from crops irrigated to field capacity whenever the moisture tension at 15 cm depth exceeded 0.04 MPa. Plots subjected to drought for 3 weeks after planting showed 15 to 20% lower yields and slow root development. Three weeks of drought during the mid growing season reduced yields 13 to 21%, and with 3 weeks drought late in the natural growing season yields were 2 to 12% lower. Plots receiving only rainfall throughout the season yielded 41 to 44% less than fully irrigated controls.[103]

High levels of rainfall and irrigation can lead to leaching of available nitrate out of the root zone.[107] Therefore, nitrogen as well as water availability must be considered in relation to irrigation. In wet seasons irrigation without the application of additional nitrogen during growth can lower growth rate and final bulb yield.[99,105] Several studies have shown that extra nitrogen is needed to get the maximum yield gain from irrigation.[99,100,92] Similarly, nitrogen top dressings in the absence of irrigation may give no increase in yield.[96,105,108] It is important to have no limitation in the supply of water and nitrogen during the main phase of bulb growth,[101] however applications of nitrogen late in bulbing do not increase yield.[105] In both England[1] and Germany[98] a basal application of 80 kg ha^{-1} N with a top dressing of 50 kg ha^{-1} N in June is the general recommendation for irrigated spring-sown onions.

Irrigation frequently delays bulb maturity[91,101,104,105] although in some conditions no such delay occurs.[109] Delays in maturity of up to 19 d between nonirrigated and heavily irrigated treatments have been recorded.[104] To encourage ripening and avoid excessive late growth with consequent bulb splitting it is advised that irrigation ceases three weeks before harvest.[1,5] In Ontario it is recommended that the water table on peat soils is maintained at least 60 cm below the soil surface but that it should be lowered further, 2 to 3 weeks before harvest, to aid bulb ripening.[5] Nitrogen applications late in bulb development may cause poor ripening and thick-necked bulbs[1] but nitrogen applied early in crop growth accelerates bulb maturity on irrigated crops.[105] In conditions of high rainfall, in the absence of heavy nitrogen fertilizer applications, nitrogen deficiency can delay maturity.[108,110] In other experiments no significant effect on maturity date of nitrogen, including some very high levels, has been observed.[92,108]

Irregular growth, caused by periodic water stress followed by rainfall or irrigation late in bulb development, can impair bulb quality by causing splitting of the outer skins and swollen sheaths.[1] However, most experiments show no difference in the harvest quality of bulbs between nonirrigated and irrigated treatments,[97,101,110] although in one study, irrigation in a wet season increased the number of doubled bulbs and other defects at harvest.[104] In the majority of studies irrigation has been found to have no detrimental effect on the storage life of bulbs.[97,99,101,110] Some studies have found increased storage life as a result of irrigation, particularly in dry seasons.[104,111] Increases in nitrogen application rate both as base and top dressings can accelerate sprouting in storage particularly if irrigation is applied as well.[105] However, in other studies no detrimental effects, and in some years a beneficial reduction in storage losses occurred with increases in nitrogen.[111]

The majority of the experimental evidence then, indicates that irrigation increases bulb yield and has no detrimental effect on quality or storage. Plant physiological studies show that the growth and photosynthesis of onion plants is more sensitive to water stress than other crop species and this, coupled with the shallow rooting depth of onions (see chapter on "Physiology of Crop Growth and Bulbing"), shows why they are responsive to irrigation. The difficulty in specifying irrigation regimes precisely, lies in the unpredictability of natural rainfall and of evaporative demand in most temperate regions. Irrigation requirements and scheduling also depend on the available water capacity of the soil.[1,112] More frequent irrigation in smaller amounts is needed on soils of low available water capacity (e.g., sands) than on those of high available water capacity (e.g., silts or peats). It is essential that irrigation is applied evenly, otherwise growth and nitrogen leaching will be variable and variable growth and uneven maturity will result.

For salad (green) onion crops irrigation is important to (1) wet the soil prior to sowing and after sowing if soil conditions are dry, to ensure quick and uniform emergence, (2) to ensure steady growth in the spring, and (3) to soften the soil as an aid to harvesting, if the surface soil is dry and hard.[1]

B. TREATMENT WITH SPROUT SUPPRESSANTS

Onion crops intended for long term storage are frequently treated with maleic hydrazide (MH) 1 to 2 weeks before harvest. Experiments have shown that MH can be absorbed only if green leaf tissue is present.[113] A minimum level of 20 ppm in the center shoot of the bulb is needed to suppress sprout growth.[114] However, the rate of application needed to achieve this varies from season to season.[114] Findings vary slightly on what is the optimum stage and rate to apply MH. Application at approximately 50% foliage fall-down has been advocated[113,115] whereas others advocate application at about 10% fall down.[1,116] Application when the foliage is still totally erect leads to a high proportion of killed, brown shoots in the treated bulbs.[116] In Britain a treatment at the rate of 8.4 l ha^{-1} of 36% ai MH is optimal.[1,116] The Dutch recommendation is 12.5 l ha^{-1} applied in at least 500 l of water when the foliage is just beginning to fall.[3] A dry period of at least 10 h after application is needed for MH absorption after spraying.[3]

VI. HARVESTING

A. BULB HARVESTING

With the increasing emphasis on the quality of bulbs in developed economies,[6] harvesting techniques have been improved in recent years to ensure that high quality bulbs are stored and sold. Traditional techniques which relied to some extent on the natural drying and curing of windrowed bulbs in the field have been replaced by direct harvesting techniques in which onions are lifted from the field and dried and cured using warm air and forced ventilation.[1,3] The effective control of neck rot with fungicidal seed treatments (see chapter on "Storage Diseases of Onions") has made foliage removal and direct harvesting a viable system for producing high quality bulbs.[117] In Britain, only the direct harvesting technique is now recommended[1] and in Holland nearly all bulb crops are harvested in this way.[3]

The traditional technique of harvesting involves pulling the onions by hand or undercutting and severing the roots by machine and then laying the onions out in rows (windrows) in the field. This is best done when the foliage has collapsed on about 60 to 70% of the plants.[3,5] Onions are usually left in the windrow 1 to 2 weeks before being gathered up and taken in for further drying at about 18°C[117] or in warm climates the foliage may be removed at this stage and the onions packed into nets or boxes.[5] If bulbs remain wet in the windrow for even a few days, microbial activity leads to unattractive dark staining of the outer skins and consequent loss of quality. For this reason windrowing and field curing is very risky in wet climates.[1,3]

The direct harvesting system has less risk of skin staining, hence its increasing popularity in areas where there is a high risk of rain at harvest time. In this technique the leaves are removed by cutting 8 to 10 cm above the bulb using a flail, and are blown or augered to the side of the crop. In fine weather a delay of 2 h allows sap released by leaf removal to dry away, then the crop is undercut and either elevated directly into a trailer or pushed into a windrow from which an elevator gathers it into a trailer. The onions are then taken to a store and piled to a depth not more than 3.5 to 4 m, otherwise the lower ones are damaged. Before loading the store it is most important that loose soil is removed from the bulbs. During all picking up and store loading operations care must be taken to minimize damage by padding projections and points of hard contact and movement on machinery, and by avoiding long drops from elevators into trailers or stores.

As soon as possible after harvest, the surface moisture must be removed from the onions by ventilating through the stack with warmed air. In Holland a temperature of 25°C is recommended for surface drying,[3] in England 30°C is recommended at an air flow rate of 425 m^3 h^{-1} t^{-1}.[1] Surface moisture must be rapidly removed at this stage to prevent skin staining and air of a low relative humidity (25 to 35%) may be used. When the onions are dry outside to the extent that loose skins "rustle", then, rather than continuing to blow heated outside air into the store, air is recirculated within the store and outside air is only introduced when the relative humidity exceeds 75%. A continuing slow drying is needed until the neck (pseudostem) tissue is fully dry. This normally takes 10 to 15 d at a temperature of 25 to 30°C and relative humidities of 70 to 75%.[1] In Holland this continued slow drying of the necks is accomplished at lower temperatures.[3] Drying at 30°C under a high airflow reduces the incidence of neck rot (*Botrytis allii*) compared to drying at ambient temperatures.[118] Exposure to temperatures above 21°C causes the outer skins to darken, and the higher the temperature above 21°C the shorter is the period needed to reach a given depth of color.[119] Drying at 25 to 30°C for 10 to 15 days is favored in England because it produces deep brown colored bulbs as well as removing moisture from the onion necks.[1] After complete drying of the necks, temperatures are reduced for long-term storage. Relative humidities of around 75 to 80% are maintained in the store to prevent cracking of the outer onion skins.[1,3] Water condensation on the bulbs causes skin staining and must be avoided.[1]

Studies on the date of harvest have shown that yields can increase by as much as 30 to 40% between the stage when foliage begins to collapse until it is fully senesced and non-green.[102,120-127] As harvesting is delayed the neck thickness decreases but more bulbs show cracked outer dry skins which easily peel off in store.[123,124,126] The optimum time of harvest for maximum storage life before the bulbs sprout, occurs while the onion foliage is still partially erect long before maximum yields are attained.[121,122,124-126] Dowker and Fennel[127] showed that storage life is longest if onions are harvested when about 50% of the foliage has collapsed. Rod[125] found longest storage when they were harvested at 60 to 70% foliar collapse. The optimum time for harvest is therefore a compromise between increasing yields and the decreasing bulb quality and storability that result from delay in harvest.

In modern harvesting systems onion foliage is normally removed at harvest time to minimize the bulk of waste material carried into store and to reduce the drying requirements after harvest. However, bulbs can be stored with the foliage intact. Before neck-rot disease (*Botrytis allii*) was effectively controlled (see chapter on "Storage Diseases of Onions") cutting off the foliage increased neck-rot in store.[118,125,128] Cutting off foliage also decreases bulb yield compared to leaving it intact.[125,128] As an alternative, various chemical desiccants have been applied to onion foliage prior to harvest. These kill the foliage and cause it to dry out rapidly, thereby reducing the bulk of waste material entering the store.[129-131] Both mechanical foliage removal and chemical desiccants shorten the storage life of onions[128,132-134]

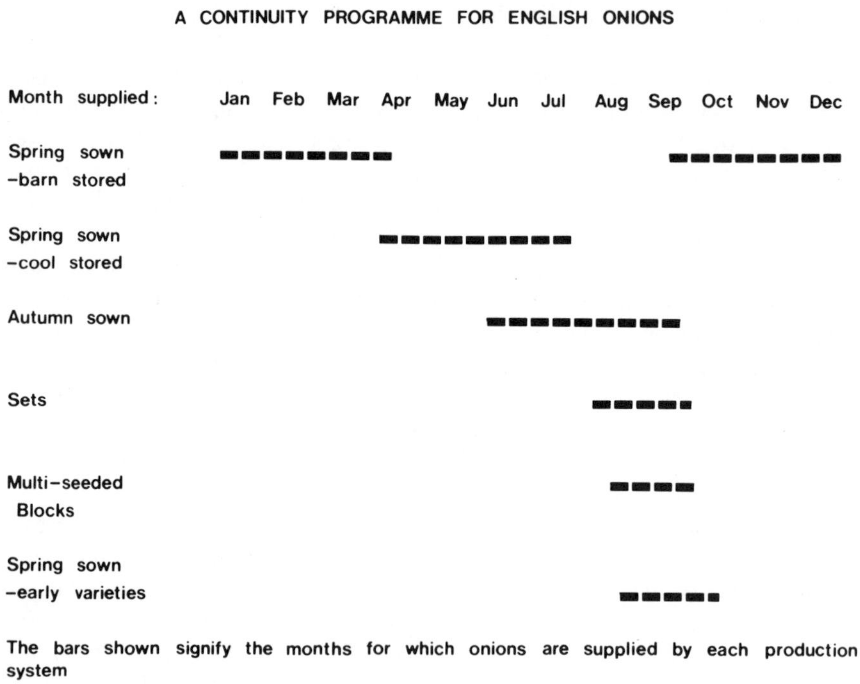

FIGURE 1. A continuity program for English onions. The bars shown signify the months for which onions are supplied by each production system.

and small increases in bulb rotting have followed the use of chemical desiccants,[134] but in other instances rotting decreased slightly.[133] Despite the disadvantages of slight loss of yield and some reduction in storage life as a consequence of the removal of foliage before harvest, the improved efficiency of store utilization, drying, and curing possible as a result, have outweighed the disadvantages.

B. SALAD ONION HARVESTING

Salad or green bunching onions are pulled and tied into bunches in the field. The bunches are washed and roots, decaying leaves, and excessive leaf length are removed before they are packed loosely into containers.[1,5] Salad onions deteriorate rapidly after harvest, and storage for only 2 to 3 d at 0°C and 95 to 98% relative humidity is recommended. The time for harvesting, cleaning, and packing before cooling and placing in store should be minimal.[1,5]

VII. SPECIAL ASPECTS OF ONION CULTURE

A. MAINTAINING CONTINUITY OF SUPPLY OF BULB ONIONS

Although bulb onions are a "world commodity" that are transported easily from producing to consuming regions it is possible in many temperate regions, using the range of cultural and storage techniques now available, to supply local markets throughout the year with good quality onions. For example, a sequence of production systems suitable for England is shown in Figure 1.[117]

Maincrop stored onions can be kept satisfactorily from harvest time in September until March or April in stores cooled by ambient air, provided a suitable long storing variety treated with maleic hydrazide sprout suppressant is used. Similar crops can be stored satisfactorily until late June in stores refrigerated at 0 to −2°C[1] (see chapter on 'Bulb Dormancy and Storage Physiology' for details of Storage). From early June onward bulbs from autumn sowings or other techniques of early production become available.

Various production systems for early crops have been compared in Britain[82] and in Norway.[64] In Britain high yields and early maturity (mid to late June) are obtained when intermediate daylength cultivars (IDCs) are directly sown in late summer or sown then transplanted in autumn. Transplanting late summer sown IDCs in spring greatly reduces yield and slightly delays maturity. Spring sowing of IDCs in the field results in very low yields and maturity delayed 2 to 3 weeks compared with autumn sowing. Spring planted sets of IDCs are high yielding but 1 month later maturing than autumn-sown crops of IDCs. Autumn-planted sets of IDCs show a higher percentage of bolting than autumn-sown crops from seed. Sets and transplants of long day cultivars (LDCs) mature late July to early August, 6 to 7 weeks later than autumn sown IDCs but up to 4 weeks earlier than spring-sown LDCs. For economical and reliable early production in Britain the autumn-sowing of intermediate daylength cultivars is the most appropriate system.[82] Studies in Norway compared sets and transplants with and without plastic tunnels for the production of early bulb crops.[64] Covering with plastic tunnels for 1 month after planting advanced the time of harvest by 13 days. In the absence of plastic tunnels, early transplanting of groups of three seedlings gave marketable bulbs sooner than did sets.[64]

B. THE CULTURE OF OVERWINTERED CROPS

In order to be suitable for overwintering, cultivars must be winter-hardy and bolting resistant when sown in late summer. Trials have shown that a number of Japanese and some European cultivars have these attributes.[12,135,136] Studies in Czechoslovakia showed that more than 80% of plants of these cultivars survived when the minimum temperature fell below 0°C for more than 100 d and when an absolute minimum of −20°C occurred.[12] Experience in Britain has shown that these cultivars are most susceptible to winter killing in waterlogged soil. Freezing in these conditions can lead to "frost heaving" where seedlings are entrapped by a collar of ice at the same time as ice formation raises the soil surface by 5 to 15 mm.[137] This results in breakage of roots. Large seedlings are more resistant to root breakage. Small seedlings can be raised out of the soil by successive occurrences of this process, and can die from desiccation on the soil surface.[137] Frost heaving is particularly severe on peat soils, and for overwintered production free draining light mineral soils are preferable.

Cultivars with a shorter photoperiod requirement (IDCs) than typical spring-sown cultivars are needed for early bulb production from overwintered crops. Spring-sown cultivars sown in autumn show a high percentage of bolting and may mature no earlier than from a spring sowing.[11,82]

The sowing date in the late summer or autumn is critical. If sown too early, large seedlings develop in the warm autumn growing conditions. These can reach the critical size necessary for inflorescence initiation[138] by late autumn and will be subject to flower inductive temperatures throughout the winter months. As a result, a high proportion of such early sown plants will flower (bolt). On the other hand if sown too late, seedlings are small by the time severe winter weather starts and a high proportion of winter losses occur.[10,11] If sowings are made at the right time, a high percentage of seedlings survive the winter and very few flower.

In English conditions it appears that seedlings should be about 0.45 g shoot dry weight by March 1 in order to attain maximum yields.[139] The logarithm of shoot dry-weight is proportional to the thermal time above 6°C accumulated since seedling emergence.[139] A shoot dry weight of 0.45 g requires approximately 474 accumulated day-degrees above 6°C between emergence and March 1. The day-degrees accumulated depends both on the emergence date and on autumn and winter temperatures. If conditions are dry at sowing time, irrigation is necessary to ensure rapid emergence.[11] Warm autumns and mild winters result in larger seedlings during the winter months than do cool autumns and winters and consequently there is more risk of bolting in the former case. Records of past seasons' temperatures were used to predict the sowing date which gave the highest probability of 474 day-degrees accumulated from emergence to March 1, and hence the best chance of a high bulb yield.[139] In central England mid- to late August is the optimum sowing date whereas further north, in Scotland, an early August sowing is optimal.[10,11,139] Similar principles govern autumn-sowing dates in warmer regions also (see chapter on "Cultural Systems and Agronomic Practices in Subtropical Climates").

In winter, rainfall frequently exceeds evapotranspiration and leaching of nitrate from the soil occurs. Consequently, it is important to apply supplementary nitrogen in spring so that nitrogen is available to allow rapid leaf growth as temperatures increase. In trials in eastern England one application of 100 kg ha^{-1} N in early spring gave maximum yields.[140] With the European cultivar Presto, a lack of nitrogen during the winter and early spring causes a high percentage of bolting,[141] and with this variety an adequate nitrogen supply throughout the winter and spring is needed. However, Japanese cultivars are less sensitive to low nitrogen in this respect, and these types are now the most widely grown.

The bulb yield depends critically on the leaf area index (LAI) established during bulb development.[24] It is, therefore, important that rapid leaf growth occurs and a high LAI develops before photoperiods are sufficient for bulb initiation and the consequent slowing and cessation of leaf blade growth. The larger a seedling in the early spring, the more rapid its leaf growth rate and, in the absence of bolting, high yields develop from the large seedlings that result from early sowings. This is particularly true with early-maturing varieties which have only a short time available for leaf growth in the spring. Fortunately the best early-maturing cultivars (e.g., cv. Express Yellow) are more resistant to bolting than later maturing cultivars (e.g., cv. Senshyu) and they can be sown slightly earlier in the summer.[139]

In some conditions it is more convenient to transplant in the autumn. This allows more time for the harvest of previous crops and land preparation for the onions. Recent experiments in northern England have shown that high-yielding, high-quality crops are obtained from group transplants of multiseeded modules sown in mid-August, the normal date for direct sowing in the field, and transplanted to the field in late September.[142,143]

Overwintered crops have a long growing season during most of which they are weakly competitive with weeds. Hence, it is important that good weed control is established at emergence and maintained throughout the autumn, winter, and spring. Herbicide spray programs have been devised that meet this objective.[142,144]

Overwintered cultivars have rather thin outer skins.[145] Bulb growth is rapid as they approach maturity in early summer and as a consequence the outer dry skins are easily cracked causing their peeling and loss after harvest.[126] This can be avoided by direct harvesting (see Section VI.A) followed by artificial drying and curing. In this way a high quality skin finish can be produced with these cultivars.[142] To avoid skin splitting, bulbs must be harvested at an early stage of maturity just before the foliage starts to collapse.[142]

C. CULTURAL PROBLEMS

1. Trace Element Deficiencies

Onions are grown on drained peat soils in many temperate regions. When brought into cultivation these soils may be deficient in certain trace elements. The deficiencies may vary with soil pH.[146] The commonest deficiency in this situation is copper. Copper deficient plants show white chlorotic leaf tips that twist or bend at right angles to the rest of the plant.[147-149] Bulbs formed on copper deficient peat soils show thin lightly colored skins that break off easily in handling.[5,150,151] Copper is normally incorporated into such soils when bringing them into cultivation. On Irish peatland an application of 44 kg ha^{-1} of copper is sufficient for most vegetable crops for up to 10 years. Deficiency symptoms show up when the level of copper in foliage decreases to 3 μg g^{-1}.[146] In Canada up to 336 kg ha^{-1} of powdered copper sulfate are added to such soil.[5] Copper deficiencies identified during growth can be corrected by foliar sprays. In New Zealand two sprays of 0.25% copper sulfate in water were successful.[149] In the U.K. a foliar spray of 2.2 kg ha^{-1} of copper oxychloride plus wetter applied once or, in severe cases, twice separated by three week intervals, is recommended.[1] Concentrations of copper in the plant dry matter of 20 ppm or more are phytotoxic.[152]

On peat soils at high pH (6.5 and above), manganese deficiency can be severe on onions.[1,153,146] In Holland this also occurs on chalky clay mineral soils and on soils containing much phosphate.[3] Symptoms are severely reduced growth and a striped chlorosis of the outer leaves.[147] Growing crops can be treated by foliar sprays of manganese sulfate solution. In Britain 9 kg ha^{-1} is recommended[1] and in Holland 15 kg ha^{-1} in 1000 l of water.[3] Experiments in Minnesota, U.S. showed that foliar sprays of 34 kg ha^{-1} were as effective as soil applications of 168 kg ha^{-1} manganese sulfate in increasing yields on manganese deficient peat.[154] Severe deficiencies may require two foliar sprays of manganese sulfate.[1,5]

Zinc deficiency causes leaves to be yellow striped and twisted and the plants to be stunted.[147] Zinc deficiency occurs on peat soils in Ireland at a pH of 6 or more, particularly in cool wet weather. In these conditions routine foliar sprays of zinc sulfate solution are recommended.[146]

On peat soils in Ireland of pH 5.4 or less, molybdenum deficiency can result in the poor emergence and death of onion seedlings. Coating the seed with molybdenum improves the percentage emergence in these conditions.[146]

2. Bolting

The influence of sowing date on the bolting (flowering) of overwintered crops has been discussed above. Bolting is occasionally a problem in spring-sown or planted onions, particularly in onions grown from sets or transplants. The cultural conditions likely to induce bolting can be inferred from a knowledge of the physiology of inflorescence initiation. Onion seedlings or sets below a certain weight cannot initiate inflorescences. The critical weight varies with variety but is about 0.06 g shoot dry weight for seedlings of Rijnsburger, a spring-sown Dutch cultivar[138] and probably about 4 g fresh weight for sets of spring-sown cultivars.[155] In seedlings a minimum time of about 40 days at 9 to 12°C at 16 h photoperiods is required to produce an inflorescence initial and at least another 40 days of cool temperatures and long photoperiods are required for the inflorescence to elongate and become visible.[156] Thus, in the field, bolting will be favored if large seedlings or sets are exposed to long periods at temperatures around 9°C before they start to develop bulbs. Such conditions can occur if large sets are planted early. Bolting is also likely if transplants are planted early and a warm early spring allows rapid initial growth, after which there occurs a long period of cool temperatures in the late spring and early summer.

3. Thick Necks

Onions which fail to ripen normally and which continue leaf blade production rather than forming a bulb with a narrow pseudostem are termed "thick-necks" or "bull-necks" in England and "scallions" in North America. Various degrees of this problem can often be seen in the same field. Some plants will form normal bulbs, others will form bulbs but with unusually wide fleshy pseudostems and others may produce little sign of bulbing and remain essentially like large seedlings. The occurrence of even a small proportion of such plants in a crop makes the mechanized harvesting and drying of the bulbs difficult, since the thick pseudostems cannot be satisfactorily dried. Such bulbs also sprout rapidly and may rot in store. Analysis of data from variety trials conducted at several sites over many seasons showed that the occurrence of a high percentage of "thick-necked" plants was associated with sites, seasons or cultivars of late maturity date.[92] These trials, and general experience in England, show that this problem is usually worse on peat soils than on mineral soils. A high proportion of thick-necked plants occur in spring-sown crops in cool, wet, late seasons in England. Thick-necked plants fail to make the transition from leaf blade production to bulb scale production during bulb development (see chapter on "Physiology of Crop Growth and Bulbing"). This transition is delayed in crops of low leaf area index (LAI) compared with crops of high LAI.[90,91] Hence cultural conditions which tend to reduce leaf area index, like late sowing and low plant density, delay bulb scale initiation and bulb maturity, and may result in thick-necked plants.[91] Other factors which might reduce leaf area index are herbicide or hail damage to young seedlings or nutritional or water stress in the crop at an early stage. It is important that a uniform high percentage of seedling emergence occurs, otherwise patches of low density, late-maturing, thick-necked plants in the crop will be a severe problem at harvesting time, especially if mechanical harvesting and bulk storage are practiced. Thus, to minimize the risk of thick necks, a high percentage of seeds should emerge, emergence should occur over a short interval of time, and the crop should be grown without checks or damage during the development of its leaf canopy. Thick-necked plants may also be increased in number by prolonging irrigation towards harvest time,[5] or by applying nitrogen late in the growing season.[1]

4. Effects of Leaf Damage

Loss of functional leaf area during crop growth may occur because of foliar disease (see chapter on "Leaf Diseases of Onions"), because of ozone damage causing leaf blast[157] or because of hail damage. The effects of foliage loss at various stages of crop development have been studied by several authors in order to assess the likely influence on yield of hail damage. The yield reduction from a given percentage of foliage removal is maximal when defoliation occurs during the early stages of bulb development.[158-160]

D. MINOR MODES OF PRODUCTION

1. Producing Onion Sets

Onion sets are produced similarly to bulbs for eating except that they are grown at very high plant density.[4] Restrictions of water late in the growing season have also been used to limit size.[161] Descriptions of systems for set production are given in several publications.[4,161–136] Soils for set production must be well structured and free from lumps as these are troublesome during harvest and grading.[4] The soil must be free of pest nematodes (*Ditylenchus dipsaci*, see chapter on "Nematode Pests of *Allium* Species") and sclerotia of the white rot fungus (*Sclerotium cepivorum*, see chapter on "Root Diseases"). In Holland, the main producer country of sets for export, soil is inspected for absence of these organisms before a certifiable crop of sets can be grown.[4] Nematode free seed is sown in late March or early April at a rate of 9 to 11 g seed m^{-2}.[4] The seed is sown in beds in rows 15 to 20 cm apart using seed drills that spread the seed into a broad band 7 to 10 cm wide. Under this system about 1300 plants m^{-2} are established and a production of about 20 t ha^{-1} of sets in the range 8 to 21 mm in diameter, and 2.5 t ha^{-1} above 21 mm is expected.[4] Experiments in Poland indicated an optimum seed rate of 18 g m^{-2} in rows 15 cm apart.[164] In Holland, to encourage hard growth, no basal nitrogen fertilizer is applied and up to 40 kg ha^{-1} N is applied during growth, depending on the appearance of the plants. Weed control is essentially the same as for a bulb crop.[4]

At harvest the leaf blades are mown off in the field, the crop is undercut and lifted and may be windrowed and later picked up, or it may be lifted and mechanically picked up in one operation.[4] Harvested sets are stored in bulk up to a depth of 3 m. Immediately upon placing in store they must be rapidly dried using warmed, fan-driven air at a temperature of 20 to 25°C.[4] Thereafter they are ventilated with unheated air for 2 to 3 weeks to complete drying. Sets are stored either in large boxes or in bulk stores similarly to bulbs for eating.

There has been extensive research on the effects of temperature during set storage on flowering and growth in the following season.[9,86,162,165-167] For most onion cultivars storage at temperatures around 27 to 28°C from November to March prevents or greatly reduces the tendency to flower (bolt) in the following season. In addition, nearly all workers have noted that this treatment increases the vigor of growth after planting out, delays bulb maturity, and increases bulb yield quite apart from its effect on bolting. In Holland, Rijnsburger type cultivars are stored at 25.5 to 28°C and 60 to 70% RH from mid-October until early March to prevent bolting.[4] Large sets (19 to 24 mm diameter) must be stored under these conditions until early April.[4] The bolting resistant cultivars Stuttgarter and Sturon are stored at 3 to 4°C in stores ventilated by outside air. High humidities (60 to 70% RH) are necessary to prevent excessive moisture loss during 28°C storage, even so, about 10 to 25% weight loss and a 12% loss in numbers occurs during such storage.[161]

2. Multiplier Onions *Allium cepa* L. *var agregatum* and Shallots *Allium cepa* L. *var ascalonicum*

Onion cultivars grown in Finland and the northern Soviet Union are normally vegetatively propagated.[9] These cultivars divide into a few large bulbs or many small bulbs, depending on cultivar, and they may flower poorly. Those from the more northerly regions of Finland produce many small bulbs,[9] those from south Finland fewer, larger daughter bulbs. Such cultivars are termed multiplier or potato onions (*A. cepa var. aggregatum*).[9,168] These cultivars behave similarly to onion sets in their flowering and growth responses to storage temperatures.[9] Storage of bulbs at 9 to 14°C or 3 to 5°C during the winter results in a high percentage of bolting in the following season. Storage of bulbs at −3 to 0°C prevents flowering but delays sprouting upon planting out. The highest yields, with little flowering, occur after storing the bulbs at 20 to 22°C through the winter.[9] Storage at 28 to 30°C gives yields almost as high as 20 to 22°C. Temperatures of 28 to 30°C delay sprouting after planting out and both this treatment and storage at 20 to 22°C results in several weeks delay in bulb ripening compared to 3 to 5 or 9 to 14°C storage. As with onion sets, leaf growth is more vigorous and the growing season is longer after storage at 28 to 30°C.[9]

Studies on shallots (*A. cepa* L. *var ascalonicum*) in Denmark, Holland, England and France have also shown important influences of bulb storage temperature on subsequent growth and yield. Warne[169] showed that storage of bulbs at 30°C during the winter delayed sprouting, increased vigor and delayed maturity compared to bulbs stored at room temperature. In one season, yields were doubled by high temperature storage and in another they were 71% higher. The increased yield was the result of increases in bulb size rather than in the number of daughter bulbs. A temperature of 0 to 4°C for 88 d followed by 71 d at room temperature increased yield to an even greater extent (106%) compared to continuous storage at room temperature.[169] In Holland, Bruinsma,[170] found that storage of shallots at 25.5 to 28°C during the winter delayed ripening in the following season compared to storage in an unheated barn at 1 to 10°C. The longer the proportion of the total storage period that was at 25.5 to 28°C as opposed to 1 to 10°C, the more ripening was delayed and the larger the leaf blades grew in the following season.[170] A 3-month period at 25.5 to 28°C was optimal, this increased yields by 40 to 50% or 70 to 80% depending on season, again by increasing the size of the individual daughter bulbs.[170] In Denmark shallot storage at −2°C from mid-November to April 1 increased daughter bulb weight and number compared to storage at 5 to 10°C or 30°C.[171] The best treatment was −2°C storage with 2 weeks at 30°C in mid-December. This increased yields 64 to 86% compared to 5 to 10°C storage, prolonged the growing season and prevented bolting.

Studies on the Jersey shallot in France[170] in which bulbs were stored at −2, 2, 9, 20, and 30°C showed that sprouting was most rapid after planting bulbs stored at 30°C and slowest from bulbs stored at 9°C. This is opposite to the effect noted by several workers for onion sets and other cultivars of shallot, where 30°C storage delays sprouting.

In a study of the effect of planting density and mother bulb weight on the yield of daughter bulbs, it was shown that yields increased from 25.5 t ha^{-1} when bulbs of 10 g weight were planted at a density of 14 m^{-2} to 32.9 t ha^{-1} when bulbs of 20 g weight were planted at 29 m^{-2}.[173] In the following season the yields of these two treatments were 28.1 and 41.1 t ha^{-1}, respectively. Small yield increases occurred as planting density increased through 14, 16, 22, and 29 plants m^{-2} and also with increases in the weight of the planted bulb, using 10, 15, and 20 g bulbs.[173] The larger the number of bulbs harvested per unit area, the smaller the mean weight of the harvested bulbs. For a given mother bulb weight, the number of bulbs harvested per unit area increased approximately in proportion to the number of bulbs planted per unit area. The number of daughter bulbs produced increased as mother bulb weight increased. Bulbs of 10, 15, and 20 g multiplied an average of 7.1, 9.4, and 11.8 times respectively, and this rate of multiplication was little affected by the planting density. Hence, the maximum number of bulbs harvested per unit area occurred with 20 g bulbs planted at the highest density (29 plants m^{-2}) and this treatment gave the highest proportion of daughter bulbs in the smaller size grades.[173] The size and spacing at which mother bulbs were grown had some influence on the multiplication rate of daughter bulbs in the following season. Daughter bulbs from mother bulbs grown at high density in the previous season multiplied less than daughter bulbs of the same weight from mother bulbs grown at low density in the previous season.[173]

3. Producing Salad Onions by the Greenhouse Forcing of Bulbs

In cold climates green salad onions are produced for winter and early spring production by planting mature storage bulbs side-by-side in a glasshouse or cold frame.[5] About a 5 cm soil depth is required and smaller bulbs are preferable since they produce more green shoots per unit area. Water should be applied as needed and temperatures maintained at 15 to 18°C. When green shoots are of marketable size, the bulbs are cross sectioned with a knife slightly below the center and the upper half of the bulb is removed and discarded in cleaning the green shoots for market. The lower halves of the bulbs can remain in the soil to produce a second crop.

4. Growing Giant Onion Bulbs

In Britain a minor industry and much amateur effort is devoted to the production of giant prize winning onion bulbs.[174] The primary purpose of the growers is to produce the maximum bulb weights, but the bulbs must satisfy certain "quality" criteria, for example, they must be single unsplit bulbs. There are a number of cultivars specially bred as large exhibition onions. The large bulb size of these cultivars is achieved at the expense of a low bulb dry matter content.[175] Competition growers raise their prize plants from early sowings made under glass or in growth chambers. Seedlings are transplanted to a wide spacing in heavily enriched soils. The plants may be grown under a well aerated polythene structure to minimize wind damage. Individual leaves are sometimes supported to prevent folding and damage. Regular irrigation and liquid fertilizer are applied during bulb development. There is considerable secrecy and rivalry among competitors concerning precise cultural practices. The world record weight for an onion bulb with foliage trimmed is 3.997 kg.[176]

ACKNOWLEDGMENTS

I would like to thank Miss H. A. Butler for much help in the preparation of this review, particularly in collecting and ordering the references.

REFERENCES

1. **ADAS/MAFF,** *Bulb Onions.* Reference book 348, Grower Books, London, 1982.
2. **ADAS/MAFF,** *Salad Onions.* Reference book 349, Grower Books, London, 1982.
3. **Blikman, L. E., de Geus, C., Hoeh, D., and Hooghiemstra, D.,** *Growing Bulb Onions from Seed,* Teelthandleiding no 2, Proefstation voor de Akkerbouw en de Groenteteelt in de Vollegrond, Lelystadt, Netherlands, March, 1985.
4. **Koert, J. L., de Geus, C., Hoeh, D., and Hooghiemstra, D.,** *The Culture of Onion Sets,* Teelthandleiding no 9, Proefstation voor de Akkerbouw en de Groenteteelt in de Vollegrond, Lelystadt, Netherlands, 1979.
5. **Riekels, J. W., Tiessen, H., and Nonnecke, I. L.,** *Onions,* Ont. Dep. Agric. Food Publ., 1976, 486.
6. **MAFF,** *EEC Standards for Onions,* G. B. Ministry Agriculture Fisheries and Food Technical Bulletin.
7. **Vik, J.,** Experiments on the development of a new cultural technique for the production of transplanted onions, *Meld. Nor. Landbrukshoegsk.,* 51, 1, 1972.
8. **Hiron, R. W. P. and Symonds, W.,** *Vegetable Propagation in Cellular Trays,* Leaflet 909, Ministry of Agriculture Fisheries and Food, Alnwick, U. K., 1985.
9. **Aura, K.,** Studies on the vegetatively propagated onions cultivated in Finland with special reference to flowering and storage, *Ann. Agric. Fenn.,* 2, 1, 1963.
10. **Salter, P. J. and James, J. M.,** The performance of Japanese and European cultivars of onions from autumn sowing for early production, *J. Natl. Inst. Agric. Bot. GB.,* 13, 367, 1975.
11. **Salter, P. J.,** Early production of bulb onions from overwintered crops in the United Kingdom, *ADAS Q. Rev.,* 19, 101, 1975.
12. **Tronickova, E.,** The Winter-hardiness of onion varieties sown in August, *Sbornik UVTIZ Zahradnictvi,* 1, 13, 1971.
13. **Millette, J. A.,** Effect of water table depths on the growth of carrots and onions on an organic soil in vitro, *Can. J. Plant Sci.,* 63, 739, 1983.
14. **Van Lierop, W., Martel, Y. A., and Cescas, M. P.,** Optimal soil pH and sufficiency concentrations of N, P and K for maximum alfalfa and onion yields on acid organic soils, *Can. J. Soil Sci.,* 60, 107, 1980.
15. **Bleasdale, J. K. A.,** The effects of plant spacing on the yield of bulb onions (*Allium cepa* L.) grown from seed, *J. Hortic. Sci.,* 41, 145, 1966.
16. **Willey, R. W. and Heath, S. B.,** The quantitative relationships between plant population and crop yield, *Adv. Agron,* 21, 281, 1969.
17. **Dowker, B. D. and Mead, R.,** Yield comparisons in onion variety trials, *J. Hortic. Sci.,* 44, 155, 1969.
18. **Frappell, B. D.,** Plant spacing of onions, *J. Hortic. Sci.,* 48, 19, 1973.
19. **Rogers, I. S.,** The influence of plant spacing on the frequency distribution of bulb weight and marketable yield of onions, *J. Hortic. Sci.,* 53, 153, 1978.
20. **Dowker, B. D. and Fennell, J. F. M.,** Heritability of bulb shape in some European onion varieties, *Ann. Appl. Biol.,* 77, 61, 1974.
21. **McGeary, D. J.,** The effects of plant density on shape, size, uniformity, soluble solids content and yield of onions suitable for pickling, *J. Hortic. Sci.,* 60, 83, 1985.
22. **Nicholls, M. A.,** A note on a plant density and fertilizer experiment with onions in New Zealand, *Hortic. Res.,* 7, 144, 1967.
23. **Brewster, J. L. and Salter, P. J.,** The effect of plant spacing on the yield and bolting of two cultivars of overwintered bulb onions, *J. Hortic. Sci.,* 55, 97, 1980a.
24. **Brewster, J. L., Mondal, M. F., and Morris, G. E. L.,** Bulb development in onion (*Allium cepa* L.). IV. Influence on yield of radiation interception, its efficiency of conversion, the duration of growth and dry matter partitioning, *Ann. Bot. (London),* 58, 221, 1986.
25. **Mondal, M. F.,** Studies on the Control of Bulbing in Onions *(Allium cepa* L.), Ph.D. thesis, University of Birmingham, U.K., 1985.
26. **Rickard, P. C. and Wickens, R.,** Effect of row arrangement and plant population on the yield of ware sized bulb onions, *Exp. Hortic.,* 31, 1, 1979.
27. **Bleasdale, J. K. A.,** Some problems and prospects in plant spacing, *J. R. Agric. Soc. Engl.,* 134, 89, 1973.
28. **Brewster, J. L. and Salter, P. J.,** A comparison of the effects of regular versus random within-row spacing on the yield and uniformity of size of spring sown bulb onions, *J. Hortic. Sci.,* 55, 235, 1980.
29. **Whitlock, A. J.,** Tuning up the herbicide programme for onions, *Grower,* 91, 16, 1979.
30. **Greenwood, D. J., Cleaver, T. J., Turner, M. K., Hunt, J., Niendorf, K. B., and Loquens, S. M. H.,** Comparison of the effects of phosphate fertilizer on the yield, phosphate content and quality of 22 different vegetable and agricultural crops, *J. Agric. Sci.,* 95, 457, 1980.
31. **Greenwood, D. J., Cleaver, T. J., Turner, M. K., Hunt, J., Niendorf, K. B., and Loquens, S. M. H.,** Comparison of the effects of potassium fertilizer on the yield, potassium content and quality of 22 different vegetable and agricultural crops, *J. Agric. Sci.,* 95, 441, 1980.

32. **Malachowski, A.,** Indices of onion nutrition with nitrogen, phosphorus and potasium. II. Field experiments. *Rocz. Nauk Roln. Ser. A.,* 100, 25, 1974.
33. **Bottcher, H. and Kolbe, G.,** The effect of mineral fertilizers on the yield, quality and storage properties of onions. II. The effects of potassium and phosphorus on yield and quality, *Arch. Gartenbau.,* 23, 161, 1975.
34. **Bishop, R. F., Chipman, E. W., and MacEachern, C. R.,** Effect of nitrogen, phosphorus and potassium on yields and nutrient levels in onions grown on a sphagnum peat soil, *Commun. Soil Sci. Plant Anal.,* 3, 97, 1972.
35. **Henrikson, K.,** Nitrogen fertilizing in seed onions *(Allium cepa)* at high soil moisture content, *Tidsskr. Planteavl,* 88, 621, 1984.
36. **Henrikson, K.,** Sowing depth and N-fertilizing in relation to yield and uniformity in onion crops *(Allium cepa* L.), *Acta Hortic.,* 72, 69, 1978.
37. **Ontario Ministry of Agriculture and Food,** *Vegetable Production Recommendations,* Ontario Department of Agriculture Food Publication, 363, 1983.
38. **Henrikson, K.,** Startgodning til kepalog, *Statens Planteavlsforsog Meddelese,* 1652, 1982.
39. **Costigan, P., Rowse, H. R., Heaviside, M., and Gray, H.,** Starter fertiliser for maincrop onions, *Rep. Natl. Veg. Res. Station 1986(1987),* 1987.
40. **Ellis, R. H. and Roberts, E. H.,** A revised seed viability nomograph for onion, *Seed Res.,* 5, 93, 1977.
41. **Bedford, L. V. and Mackay, D. B.,** The value of laboratory germination and weight measurements in forecasting emergence of onion and carrot seed in the field, *J. Natl. Inst. Agric. Bot. GB.,* 13, 50, 1973.
42. **Kretschmer, M. and Perl, M.,** Biochemical activities in onion *(Allium cepa)* seeds and their relation to seed quality, *Gartenbauwissenschaft,* 48, 265, 1983.
43. **Rowse, H. R., Stone, D. A., and Goodman, D.,** Seedbed cultivations, *Rep. Natl. Veg. Res. Station 1984(1985),* 128, 1985.
44. **Rowse, H. R., Stone, D. A., and Goodman, D.,** Cultivation and seed drill design, *Rep. Natl. Veg. Res. Station 1985(1986),* 98, 1986.
45. **Fiedler, B.,** Studies on the effect of drilling on the crop yield of onions, *Arch. Gartenbau.,* 23, 295, 1975.
46. **Kathan, J. G.,** Suppression of onion seed germination. Dependence on depth of sowing, *Dtsch. Gartenbau.,* 38, 572, 1984.
47. **Page, R. E.,** Cellulose xanthate as a soil conditioner: field trials, *J. Sci. Food Agric.,* 31, 718, 1980.
48. **Hiron, R. W. P.,** Crop establishment studies, *Rev. Kirton, Exp. Hort. Station 1979(1980),* 19, 1980.
49. **Rowse, H. R., Stone, D. A., and Goodman, D.,** Soil conditioner, *Rep. Natl. Veg. Res. Station 1984(1985),* 131, 1985.
50. **Finch-Savage, W. E.,** Effects of soil moisture and temperature on seedling emergence from natural and pre-germinated onion seeds, *J. Agric. Sci.,* 107, 249, 1986.
51. **Finch-Savage, W. E.,** A comparison of seedling emergence and early seedling growth from dry-sown natural and fluid-drilled pregerminated onion *(Allium cepa* L.) seeds in the field, *J. Hortic. Sci.,* 62, 39, 1987.
52. **Roberts, H. A.,** Crop and weed emergence patterns in relation to time of cultivation and rainfall, *Ann. Appl. Biol.,* 105, 263, 1984.
53. **Whitwell, J. D. and Davies, A. C. W.,** Some effects of temperature on the emergence of bulb onion seed stocks, *Exp. Hortic.,* 28, 41, 1976.
54. **Gray, D. and Steckel, J. R. A.,** Viability of onion *(Allium cepa)* seed as influenced by temperature during seed growth, *Ann. Appl. Biol.,* 104, 375, 1984.
55. **Bierhuizen, J. F. and Wagenvoort, W. A.,** Some aspects of seed germination in vegetables. I. The determination and application of heat sums and minimum temperature for germination, *Sci. Hort.,* 2, 213, 1974.
56. **Lipe, W. N. and Skinner, J. A.,** Effect of sowing pre-germinated onion seeds in cold soil on time of emergence, maturity, and yield, *HortScience,* 14, 238, 1979.
57. **Finch-Savage, W. E.,** Effects of fluid drilling germinating onion seeds on seedling emergence and subsequent plant growth, *J. Agric. Sci.,* 102, 461, 1984.
58. **Gray, D.,** Fluid drilling of vegetable seeds, in *Horticultural Reviews* Janick J., Ed., AVI Publishing, Westport, CT, 1981.
59. **Brocklehurst, P. A. and Dearman, J.,** Interactions between seed priming treatments and nine seed lots of carrot, celery and onion. I. Laboratory germination, *Ann. Appl. Biol.,* 102, 577, 1983.
60. **Brocklehurst, P. A. and Dearman, J.,** Interactions between seed priming treatments and nine seed lots of carrot, celery and onion. II. Seedling emergence and plant growth, *Ann. Appl. Biol.,* 102, 585, 1983.
61. **Brocklehurst, P. A., Dearman, J., and Drew, R. L. K.,** Effects of osmotic priming on seed germination and seedling growth in leek, *Sci. Hort.,* 24, 201, 1984.
62. **Dearman, J., Brocklehurst, P. A., and Drew, R. L. K.,** Effects of osmotic priming and ageing on onion seed germination, *Ann. Appl. Biol.,* 108, 639, 1986.

63. **Heydecker, W. and Coolbear, P.,** Seed treatments for improved performance-survey and attempted prognosis, *Seed Sci. Technol.*, 5, 353, 1977.
64. **Vik, J.,** Experiments with onion group transplants, onion sets and other factors influencing an early onion crop *(Allium cepa* L.), *Meld. Nor. Landbrukshoegsk.*, 53, 1, 1974.
65. **Salter, P. J.,** Is transplanting the answer to higher yields of sown onions?, *Hortic. Ind.*, July, 508, 1977.
66. **Lipe, W. N. and Thomas, D.,** Effects of onion transplant pruning and orientation on plant survival, yield and size distribution, *HortScience,* 15, 28, 1980.
67. **Sabota, C. M. and Downes, J. D.,** Onion growth and yield in relation to transplant pruning, size, spacing, and depth of planting, *HortScience,* 16, 533, 1981.
68. **Chipman, E. W. and Thorpe, E.,** Effects of hilling and depth of plant setting on the incidence of multiple hearts and shape of sweet Spanish onions bulbs, *Can. J. Plant Sci.*, 57, 1219, 1977.
69. **Bremer, A. H.,** Korleis skal ein i praksis nytta ut dei reaksjonar vanleg lauk *(Allium cepa)* har for ymse daglengder, *Meld. Nor. Landbrukshoegsk.*, 30, 185, 1950.
70. **Ramtohul, M. and Splittstoesser, W. E.,** The effect of bulb and neck size upon establishment of transplanted onions, *HortScience,* 14, 738, 1979.
71. **Hiron, R. W. P.,** Dry bulb onion: Spring sown-multi-seeded peat block transplants 1978, *Rev. Kirton Exp. Hort. Station 1978(1979),* 55, 1979.
72. **Hiron, R. W. P.,** Multi-seeded onions in hassy trays, *Rev. Kirton, Exp. Hort. Station 1982(1983),* 22, 1983.
73. **Senior, D.,** Bulb onion production in the north of England, *Rev. Stockbridge House Exp. Hort. Station 1979(1980),* 32, 1980.
74. **Senior, D.,** Bulb onion production in the north of England, *Rev.Stockbridge House Exp. Hort. Station 1981(1982),* 36, 1982.
75. **Wood, M. B.,** Onions-Transplanting does more than fill a gap, *Rev. Arthur Rickwood Exp. Husbandry Farm 1983(1984),* 20, 1984.
76. **Anon.,** *Blocks for Transplants,* Grower Guide No. 10, Grower Books, London, 1980.
77. **Hiron, R. W. P.,** Supplying the market with quality onions in August and early September, *Rev. Kirton. Exp. Hort. Station 1980(1981),* 19, 1981.
78. **Wood, M. B.,** Blocking the onion gap, *Rev. Arthur Rickwood Exp. Husbandry Farm 1981(1982),* 18, 1982.
79. **Nes, A.,** Effects of sowing date and daylength before transplanting in onion *(Allium cepa), Forsk. Fors. Landbruket,* 36, 37, 1985.
80. **Cox, E. F.,** The effect of shape of compost blocks on the propagation, transplant establishment and yield of four vegetable species, *J. Hortic. Sci.*, 59, 205, 1984.
81. **Cox, E. F.,** The effects of irrigation on the establishment and yield of lettuce and leek transplants raised in peat blocks, *J. Hortic. Sci.*, 59, 431, 1984.
82. **Salter, P. J.,** Comparative studies of different production systems for early crops of bulb onions, *J. Hortic. Sci.*, 51, 329, 1976.
83. **Rickard, P. C.,** Opportunities for onions from sets, *Rev. Arthur Rickwood Exp. Husbandry Farm 1971(1972),* 48, 1972.
84. **Kossowski, M. and Tendaj, M.,** Cultivar trials of onions grown in the Lublin region. 1. Morphological characteristics and evaluation of cultivars grown from sets, *Ann. Univ. Mariae Curie-Sklodowska. Sect. C.*, 29, 351, 1974.
85. **Salokangas, K.,** Growing onions from sets, *Maatalous Koetoiminta,* 21, 10, 1967.
86. **Daugerod, H. and Apeland, J.,** Results of experiments on the storage and use of onion sets, *Gartneryrket,* 55, 304, 1965.
87. **Rabinowitch, H. D.,** Doubling of onion bulbs as affected by size and planting date of sets, *Ann. Appl. Biol.*, 93, 63, 1979.
88. **Holmoy, R.,** Planting of onions, *Meld. Nor. Landbrukshoegsk.*, 53, 19, 1974.
89. **Kossowski, M. and Tendaj, M.,** Cropping of bunching onions in relation to date of set planting and the use of perforated plastic film, *Biul. Warzywniczy,* 27, 97, 1984.
90. **Mondal, M. F., Brewster, J. L., Morris, G. E. L., and Butler, H. A.,** Bulb development in onion (*Allium cepa* L.). I. Effects of plant density and sowing date in field conditions, *Ann. Bot. (London),* 58, 187, 1986.
91. **Mondal, M. F., Brewster, J. L., Morris, G. E. L., and Butler, H. A.,** Bulb development in onion (*Allium cepa* L.). III. Effects of the size of adjacent plants, shading by neutral and leaf filters, irrigation and nitrogen regime and the relationship between the red:far-red spectral ratio in the canopy and leaf area index, *Ann. Bot. (London),* 58, 207, 1986.
92. **Brewster, J. L., Lawes, W., and Whitlock, A. J.,** The phenology of onion bulb development at different sites and its relevance to incomplete bulbing ('thick-necking'), *J. Hortic. Sci.*, 62, 371, 1987.
93. **Brewster, J. L.,** The response of growth rate to temperature in seedlings of several *Allium* crop species, *Ann. Appl. Biol.*, 93, 351, 1979.

94. **Hewson, R. T. and Roberts, H. A.,** The effect of weed removal at different times on the yield of bulb onions, *J. Hortic. Sci.,* 46, 471, 1971.
95. **Hewson, R. T. and Roberts, H. A.,** Some effects of weed competition on the growth of onions, *J. Hortic. Sci.,* 48, 51, 1973.
96. **Wood, M. B.,** Irrigation of spring sown onions, *Rev. Arthur Rickwood Exp. Husbandry Farm 1978(1979),* 35, 1979.
97. **Barnoczki, A.,** Effect of irrigation on the variability of onion cultivars, *Zoldsegtermesztesi Kutato Intezet Bulletinje,* 15, 15, 1982.
98. **Bottcher, H., Frohlich, H., and Hubner, C.,** Results regarding the complex influence of sprinkler irrigation, plant density and fertilization on yield, quality and storage quality of onions. I. Influence on yield and quality, *Arch. Gartenbau.,* 27, 283, 1979.
99. **Dreibrodt, L. and Karch, G.,** Onion overhead irrigation trials, *Dtsch. Gartenbau.,* 17, 68, 1970.
100. **Zimmerman, H., Engel, E., and Banholzer, G.,** Problems of the value of sprinkler irrigation in one year continuous onion cultivation, *Dtsch. Gartenbau.,* 17, 192, 1970.
101. **Selinger, F. and Hubner, C.,** Maximal use of irrigation for onions in the LPG unit for plant production "Egelner Mulde" at Gross Bornecke, *Gartenbau,* 31, 165, 1984.
102. **Sypien-Perlowska, M., Adamicki, F., and Kepkowa, A.,** The effect of harvesting time on yield, quality, respiration rate and storageability of onions, *Biul. Warzywniczy,* 27, 487, 1984.
103. **Dragland, S.,** Nitrogen and water requirements in onions, *Forsk. Fors. Landbruket.,* 26, 93, 1974.
104. **Drinkwater, W. O. and Janes, B. E.,** Effects of irrigation and soil moisture on maturity, yield and storage of two onion hybrids, *Proc. Am. Soc. Hortic. Sci.,* 66, 267, 1955.
105. **Riekels, J. W.,** Nitrogen-water relationships of onions grown on organic soil, *J. Am. Soc. Hortic. Sci.,* 102, 139, 1977.
106. **Vaseckij, V. F.,** The effect of irrigation on the yield, quality and storage properties of bulb onions, *Vestn. S. kh. Nauki, (Moscow),* 16, 56, 1971.
107. **Burns, I. G.,** A simple model for predicting the effects of leaching of fertilizer nitrate during the growing season on the nitrogen fertilizer need of crops, *J. Soil Sci.,* 31, 175, 1980.
108. **Riekels, J. W.,** The influence of nitrogen on the growth and maturity of onions grown on organic soil, *J. Am. Soc. Hortic. Sci.,* 97, 37, 1972.
109. **Eeden, F. J. Van and Myburgh, J.,** Irrigation trials with onions, *Agroplantae,* 3, 57, 1971.
110. **Sozzi, A., Testoni, A., and Meazza, C.,** Irrigation and the storage quality of winter onions, *Annali Istituto Sperimentale Par la Valorizzazione Technol. Proc. Agric.,* 14, 143, 1983.
110. **Wilson, A. L.,** Relation of nitrate nitrogen to the carbohydrate and nitrogen content of onions, *Memoirs of the Cornell Agricultural Experimental Station,* 156, 1934.
111. **Bottcher, H., Frohlich, H., and Hubner, C.,** Results regarding the complex influence of sprinkler irrigation, plant density and fertilizer application on yield, quality and storability of onions (*Allium cepa* L.). II. Storability, *Arch. Gartenbau.,* 27, 427, 1979b.
112. **Sypien, M., Fajkowska, H., and Szwonek, E.,** Influence of soil type and irrigation on height and structure of onion yields and on the keeping quality of this crop, *Biul. Warzyniczy,* 23, 331, 1979.
113. **Isenberg, F. M.,** The use of maleic hydrazide on onions, *Proc. Am. Soc. Hortic. Sci.,* 68, 343, 1956.
114. **Whitwell, J. D., Frith, L., and Williams, J. H.,** Experiments on the use of maleic hydrazide as a sprout suppressant on spring sown bulb onions, *Exp. Hortic.,* 25, 87, 1973.
115. **Kepkowa, A.,** The influence of the time and rate of application on the effectiveness of maleic hydrazide in the storage of onions under different conditions, *Rocz. Nauk. Roln Ser. A.,* 91, 361, 1966.
116. **Shipway, M. R.,** Use of maleic hydrazide as a sprout suppressant during long term storage, *Rev. Kirton, Exp. Hort. Stn. for 1978(1979),* 67, 1979.
117. **Shipway, M. R.,** Developments in onion drying and storage, *Rev. Kirton. Exp. Hort. Station 1979(1980),* 33, 1980.
118. **Maude, R. B., Shipway, M. R., Presly, A. H., and O'Connor, D.,** The effects of direct harvesting and drying systems on the incidence and control of neck rot (Botrytis allii) in onions, *Plant Pathol.,* 33, 263, 1984.
119. **Bleasdale, J. K. A., Tucker, W. G., and Hough, K.,** Bulk storage of onions, onion skin colour and keeping quality, *Rep. Natl. Veg. Res. Station 1969(1970),* 78, 1970.
120. **Romanowski, R. R.,** Effect of time of harvest on yield, colour development, scale retention, storage losses and firmness of the onion (*Allium cepa* L.), *Diss. Abstr.,* 22, 3798, 1962.
121. **Cervato, A. and Barilli, A.,** The time of sowing and harvesting onions in relation to the storage quality of the product, *Ann. Fac. Agrar. Univ. Studi, Milan,* 13, 182, 1973.
122. **Kepka, A., Sypien, M., and Smolinska-Stepien, M.,** The effect of harvest and topping dates on the yield and storability of the onion cultivar Wolska, *Biul. Warzywniczy,* 12, 377, 1971.
123. **Kepka, A., Sypien, M., and Smolinska-Stepien, M.,** The effect of the harvest date and time of topping on the morphological characteristics and marketability of the onion cultivar Wolska, *Biul. Warzywniczy,* 14, 259, 1973.

124. **Rickard, P. C. and Wickens, R.,** The effect of time of harvesting of spring sown dry bulb onions on their yield, keeping ability and skin quality, *Exp. Hortic.*, 29, 45, 1977.
125. **Rod, J.,** The influence of time and method of the harvest of onion (*Allium cepa*) on its yield and storability, *Bulletin Vyzkumny a slechtitelsky Ustav Zelinarsky Olomouc,* 23/24, 39, 1979/80.
126. **Tucker, W. G. and Drew, R. L. K.,** Post harvest studies on autumn-drilled bulb onions. The effect of harvest date, conditioning treatments and field drying on skin quality and on storage performance, *J. Hortic. Sci.*, 57, 339, 1982.
127. **Dowker, B. D. and Fennell, J. F. M.,** Some responses to agronomic treatments of different genotypes of onion bulbs, *Allium cepa* L., *J. Hortic. Sci.*, 49, 1, 1974.
128. **Rickard, P. C. and Wickens, R.,** Effect of peri-harvest treatments on the yield, storage characteristics and keeping quality of dry bulb onions, *Exp. Hortic.*, 29, 52, 1977.
129. **Bottcher, H., Schulzke, D., and Hoffman, G.,** Killing onion tops with chemicals. I. Killing efficiency, crop yield, quality and residues, *Arch. Gartenbau.*, 25, 159, 1977.
130. **Isenberg, F. M. and Abdel-Rahman, M.,** Effectiveness of a new onion top killer, *HortScience,* 7, 471, 1972.
131. **Pospilova, J. and Janyska, A.,** Effect of chemical desiccants on onion quality, *Sb. Cesk. Akad. Zemed. Ved. Rostl. Vyroba,* 24, 869, 1978.
132. **Stow, J. R.,** The effect of defoliation on storage potential of bulbs of the onion *(Allium cepa), Ann. Appl. Biol.*, 84, 71, 1976.
133. **Bottcher, H., Schulzke, D., and Hoffman, G.,** Killing onion tops with chemicals. II. Keeping quality, *Arch. Gartenbau.*, 25, 217, 1977.
134. **Bubl, C. E., Richardson, D. G., and Mansour, N. S.,** Preharvest foliar desiccation and onion storage quality, *J. Am. Soc. Hortic. Sci.*, 104, 73, 1979.
135. **Tronickova, E.,** The suitability of onion varieties grown throughout the world for autumn sowing. I. Sowing in August, *Sb. Cesk. Akad. Zemed. Ved. Rada C.*, 12, 283, 1966.
136. **Tucker, W. G. and Hough, K.,** Autumn-sown onions for ware bulb production in the UK., *J. Natl. Inst. Agric. Bot. GB.*, 13, 41, 1973.
137. **Gill, P. A. and Waister, P. D.,** Winter hardiness in autumn-sown onions (*Allium cepa* L.), *Crop Res.*, 23, 17, 1983.
138. **Brewster, J. L.,** The influence of seedling size and carbohydrate status and of photon flux density during vernalization on inflorescence initiation in onion (*Allium cepa* L.), *Ann. Bot. (London),* 55, 403, 1985.
139. **Brewster, J. L., Salter, P. J., and Darby, R. J.,** Analysis of growth and yield of overwintered onions, *J. Hortic. Sci.*, 52, 335, 1977.
140. **Shipway, M. R.,** Nitrogen nutrition trial, *Rev. Kirton. Exp. Hort. Stn. for 1978(1979),* 57, 1979.
141. **Cleaver, T. J. and Turner, M. K.,** Fertilizer requirements of overwintered bulb onion crops, *Rep. Natl. Veg. Res. Station 1974(1975),* 58, 1975.
142. **Whitwell, J. D.,** Experiments on overwintered bulb onions in the north of England, *Rev. Stockbridge House Exp. Hort. Station 1984(1985),* 29, 1985.
143. **Whitwell, J. D.,** Overwintered bulb onions grown in multi-seeded blocks, *Stockbridge House Exp. Hort. Station Grower Notes,* Feb., 1986.
144. **Whitwell, J. D.,** Weed control programmes in Autumn sown onions, *Rev. Stockbridge House Exp. Hort. Station 1980(1981),* 24, 1981.
145. **NIAB,** *Varieties of Bulb Onions,* Vegetable growers list No. 6, National Institute of Agricultural Botany, Cambridge, U.K., 145.
146. **MacNaeidhe, F. S.,** Some observations on trace element deficiencies in vegetables, in *Diagnosis of Mineral Disorders in Plants,* Vol. 2, *Vegetables,* HMSO, London, 1983, 88.
147. **Scaife, A. and Turner, M.,** *Diagnosis of Mineral Disorders in Plants,* Vol. 2, *Vegetables,* HMSO, London, 56, 1983.
148. **Adams, A. F. R.,** Copper deficiency of onions grown on peat. I. Preliminary report, *N. Z. J. Sci. Technol. Sect. A.*, 30, 105, 1948.
149. **Adams, A. F. R.,** Copper deficiency of onions grown on peat. II, *N. Z. J. Sci. Technol. Sect. A.*, 31, 34, 1949.
150. **Knott, J. E.,** Some factors affecting the color and thickness of onion scales, *Proc. Am. Soc. Hortic. Sci.*, 28, 318, 1931.
151. **Purvis, E. F. and Carolus, R. L.,** Nutrient deficiencies in vegetable crops, in *Hunger Signs in Crops,* Spraque, H. B., Ed., David McKay, New York, 1964, 286.
152. **Mathur, S. P., Belanger, A., Valk, M., Preston, C. M., Knibbe, E., and Sanderson, R. B.,** A study of onions grown in microplots on three organic soils each containing four levels of copper, *Can. J. Soil Sci.*, 63 221, 1983.
153. **Knott, J. E.,** The response of onions to manganese on unproductive peat soils, *Proc. Am. Soc. Hortic. Sci.*, 37, 803, 1939.

154. **Nyland, R. E.,** The response of onions to soil and foliar applications of manganese and to soil applications of other trace elements, *Proc. Am. Soc. Hortic. Sci.*, 60, 283, 1952.
155. **Brewster, J. L.,** Vernalization in onion-a quantitative approach, in *Manipulation of Flowering,* Atherton, J. G., Ed., Butterworths, London, 1987, 171.
156. **Brewster, J. L.,** Effects of photoperiod, nitrogen nutrition and temperature on inflorescence initiation and development in onion (*Allium cepa* L.), *Ann. Bot. (London),* 51, 429, 1983.
157. **Eagle, R. L., Gabelman, W. H., and Romanowski, R. R. Jr.,** Tipburn, an ozone incited response in onion (*Allium cepa* L.), *Proc. Am. Soc. Hortic. Sci.*, 86, 468, 1965.
158. **Hawthorn, L. R.,** Simulated hail injury on Yellow Bermuda onions, *Proc. Am. Soc. Hortic. Sci.*, 43, 265, 1943.
159. **Baker, R. S. and Wilcox, G. E.,** Effect of foliage damage and stand reduction on onion yield, *Proc. Am. Soc. Hortic. Sci.*, 78, 400, 1961.
160. **Van der Vliet, M., and Scheele, J.,** The economic importance of leaf damage in onions, *Verslagen van Landbouwkundige Onderzoekingen 669,* p. 27, 1966.
161. **Smith, F. G. and Jones, A. G.,** Production of onion sets, *Agriculture (London),* 66, 298, 1959.
162. **Smith, F. G.,** Cultural treatments affecting the production of onions from sets, *Exp. Hortic.*, 4, 31, 1960.
163. **Tinker, M. A. H., Brown, F. C., Heath, O. V. S., and Holdsworth, M.,** The production of onion sets, *J. R. Hortic. Soc.*, 70, 135, 1945.
164. **Kossowski, M. and Tendaj, M.,** The effect of different seed rates on yield of onions grown for sets, *Biul. Warzywniczy,* 27, 127, 1984.
165. **Heath, O. V. S. and Mathur, P. B.,** Studies in the physiology of the onion plant. II. Inflorescence initiation and development, and other changes in the internal morphology of onion sets, as influenced by temperature and daylength, *Ann. Appl. Biol.*, 31, 173, 1944.
166. **Lachman, W. H. and Michelson, L. F.,** Effects of warm storage on the bolting of onions grown from sets, *Proc. Am. Soc. Hortic. Sci.*, 75, 495, 1960.
167. **Bacvarov, S.,** Pre-planting heat treatment of onion sets of the variety Makovski, *Gradinar. Lozar. Nauka.*, 4, 63, 1967.
168. **Jones, H. A. and Mann, L. K.,** *Onions and Their Allies,* Leonard Hill Ltd., London; Interscience Publishers, New York, 1963.
169. **Warne, L. G. G.,** The effect of some pre-planting storage treatments on the growth of shallots, *J. R. Hortic. Soc.*, 73, 230, 1948.
170. **Bruinsma, F.,** *Effects of Heated Storage on Shallot Sets,* Mededeling No. 5, Proefstation voor de Groenteteelt in de Volleground in Nederland, Alkmaar, 1957.
171. **Knoblauch, F.,** Combined cool and warm treatment of shallots for use as sets, *Tidsskr. Planteavl.*, 70, 22, 1966.
172. **Cohat, J.,** The effect of storage conditions on the sprouting of shallot bulbs (*Allium cepa* var. aggregatum), *Agronomie,* 2, 905, 1982.
173. **Cohat, J.,** Effect of weight and planting of shallot bulbs on yield characteristics and rate of multiplication of daughter bulbs, *Agronomie,* 6, 85, 1986.
174. **Calderbank, D. A.,** *Growing Onions and Shallots,* Ross Anderson Publications, Bolton, U.K. 1986.
175. **Brewster, J. L., Hardwick, R. C., and Hardaker, J.,** Onions, *Rep. Natl. Veg. Res. Station 1974(1975),* 68, 1975.
176. Royal Horticultural Society, Newsletter 16, RHS, London, December, 1987, 4.

Chapter 2

ONION CULTIVATION IN SUBTROPICAL CLIMATES

Joe N. Corgan and Nachum Kedar

TABLE OF CONTENTS

FIGURE 1. Areas of concentrated short and intermediate day onion production in the U.S.

I. INTRODUCTION

In the context of this chapter, subtropical onion culture is defined as culture of short or intermediate day onion cultivars that are planted in the fall, winter, or early spring for harvest in spring or summer. Both direct seeded and transplanted crops are included. In addition a review of research on short day cultivars planted in the summer for fall or early winter harvest is included. Emphasis is on culture of these crops under intensive management as practiced in Israel and the warmer portions of the U.S. An attempt is made to update practices in cultural management, with emphasis on specific production related problems of the short and intermediate day onion industries as they occur in Israel and the southern parts of the U.S. In Israel there are about 3240 hectares (ha) of onions, including seed production fields. About 300 tons of seeds and 80,000 tons of onions are produced of which 15,000 to 25,000 tons are exported annually. Fall seeded or transplanted short and intermediate day onions in the U.S. encompasses more than 15,000 ha with a total bulb production of over 450,000 t.[1] The most important production area in the U.S. is in the southernmost part of Texas, with approximately 6000 ha (Figure 1). These cultivars are economically important in most countries situated within latitudes of about 30° north to 30° south of the equator. Some tendencies toward standardization of cultural practices have occurred, but even defined in the relatively narrow context of this chapter, the industry is diverse.

II. TYPES AND CULTIVARS

In the U.S. the short day cultivars that are sold as dry bulbs (not processed) are relatively mild, nonstorage types. In the past, the flat, mild (nonpungent), Bermuda types predominated. These have been replaced, for the most part, by cultivars that have greater bulb depth. This change began with the introduction of the Valencia Grano from Spain in the 1920s. A selection from Valencia Grano was named Early Grano in 1931 by the New Mexico Agricultural Experiment Station,[2] and this cultivar is still grown extensively in the Southwest U.S. as New Mexico Yellow Grano. An early maturing selection of Early Grano was released in 1944 by the Texas Agricultural Experiment Station as Texas Grano[3] and further selection resulted in the naming of Texas Grano 502 in 1947. Strains of Texas Grano 502 are still planted extensively, both in the U.S. and other countries.

White selections from Early Grano were developed by J. V. Enzie of the New Mexico

Agricultural Experiment Station and some of these were informally released and grown as White Grano. A selection of White Grano was formally released in 1979 as Temprana.[4] Hybrid cultivars are grown extensively. Hybrids between grano and bermuda types, called Granex, are grown extensively in many parts of the world. These "Granex" hybrids are very mild, intermediate in depth between bermuda and grano type cultivars, usually very uniform, and highly productive. White granex type cultivars are also available and are produced to some extent in the U.S.

Recent public releases which are grown extensively include Texas Grano 1015Y, a yellow grano type with a high level of pink root resistance released by Texas A&M University,[5] and NuMex BRl a yellow grano type with bolting and pink root resistance released by New Mexico State University.[6] In addition, several grano and granex type cultivars were developed and released by U.S. seed companies during the past 20 years. The list of short and intermediate daylength cultivars is extensive.[7-9]

Onion cultivars in Israel are mostly similar to those described above. Until about 1970, local onion breeding was mostly done by farmers, selecting for bulbs with desired characters. Several good varieties like the early Bet Alpha, the later maturing Egyptian and others were released. Modern onion breeding is done by the agricultural research institutes and by the seed companies which also breed and produce hybrids. The most famous local cultivars are the very early Bet Alpha which can also be produced in the fall for December harvest, the very early cv. Ori, different selections of Early Grano, and the medium late Haemek (which is a good storage onion with brown skin). The latest maturing or intermediate cultivars are Riverside types as "Ben Shemen", selected for February and March seeding and harvest in July for prolonged storage. Ben Shemen is also grown in some of the U.S. short day areas as a spring planted cultivar.

Early onions for production under subtropical conditions usually cannot be stored for prolonged periods. As storage ability is a most important character in tropical and subtropical climates where the late maturing good storage cultivars of Europe and the U.S. cannot be produced, serious efforts have been made in Israel to combine earliness with good storage ability. Several very promising rather early hybrids of this type should be released in 1987.

III. CROP SCHEDULES

A. DIRECT SEEDED ONIONS

Much of the U.S. and about 80% of the Israel short and intermediate day onion crop is direct seeded. Seeding dates vary by location and cultivar (Tables 1 to 3). Within any area, the earlier plantings tend to have higher yield potential (Figure 2). Earliest possible planting date for a given cultivar is determined by its bolting characteristics, and the area or region in which it is grown. Bolting, or flowering, occurs in response to low temperature, but a juvenility phase in small plants makes them insensitive to cold for flowering. Therefore, onions may pass the winter as small plants and not bolt. If overwintering plants become too large they can perceive the cold stimulus for flower induction, and then bolt. The quantitative nature of the cold stimulus is thought to be a number of hours over several days or weeks below some maximum threshold temperature. Flower induction may occur during winter months if plants are large enough, or it may occur in response to inductive temperatures in the spring. As a result, excessive bolting may occur in response to unusually warm weather in late fall or early winter which induces rapid growth, or in response to unusually cold weather in late winter or spring, which causes vernalization of the large plants. In seasons when both above environmental conditions occur, bolting is commonly excessive, sometimes exceeding 50%, with a significant reduction in yield.

Fields with no bolting usually do not produce maximum possible yields. A low percentage of bolters, about 2 to 5% is an indication that plant development or time of seeding was

TABLE 1
Onion Cultivar, Seedling Dates, and Harvest Dates for Israel (Latitude Ca. 29° 30′ — 33° 30′)

Cultivar	Dates of seeding	Dates of harvest
Ori, Bet Alpha (Arava area)	9/15—9/20	3/5—4/30
Granex	10/5—12/1	5/1—6/10
Grano and Spanish Types	11/1—2/15	5/20—6/30
Egyptian	11/15—2/15	5/20—7/15
Riverside (Ben Shemen)	1/1—3/31	7/1—8/20

TABLE 2
Onion Cultivar, Color, Seeding Dates, and Harvest Dates for Southern New Mexico (Latitude 32 — 33°N)

Cultivar	Skin color	Dates of seeding	Dates of harvest
Granex 33	Y[a]	10/5—10/15	5/25—6/10
Buffalo	Y	9/15—10/5	5/25—6/10
NuMex BRl	Y	9/25—10/5	5/28—6/12
NuMex Sunlite	Y	9/25—10/5	5/28—6/12
Temprana	W	9/20—10/5	6/1—6/15
Texas Early Grano 502	Y	10/5—10/15	6/1—6/15
New Mexico Yellow Grano	Y	10/10—10/15	6/5—6/20
New Mexico White Grano PRR	W	10/10—10/15	6/5—6/15
Texas Grano 1015Y	Y	10/10—10/15	6/5—6/20
NuMex Sundial	Y	10/15—10/20	7/1—7/10
Texspan	Y	1/1—2/1	7/1—7/15
NuMex Suntop	Y	10/15—10/20	7/5—7/15
Ben Shemen	Y	1/1—2/1	7/5—7/20
Pronto S	Y	1/1—2/1	7/12—7/25
Sweet Spanish hybrids	Y + W	1/15—3/1	7/20—8/20
Sweet Spanish OP cvs.	Y + W	1/15—3/1	8/1—9/1

[a] Y = yellow, W = white.

TABLE 3
Onion Cultivar, Color, Seeding Dates, and Harvest Dates in the Lower Rio Grande Valley of Texas (Latitude 26 — 28°N)[a]

Cultivar	Skin[b] color	Dates of seeding	Dates of harvest
Early supreme	W	9/10—9/20	2/20—3/10
Dessex	Y	9/10—9/20	2/20—3/10
Early premium	Y	9/10—9/20	2/20—3/10
Granex 33	Y	9/10—9/20	3/5—3/20
Granex 429	Y	9/15—9/30	3/20—3/25
Crystal wax	W	9/15—9/30	3/20—3/25
Henry's special	Y	9/15—9/30	3/20—3/25
Special 38	Y	9/15—9/30	3/20—3/25
Granex	Y	9/15—9/30	3/25—4/15
Granex	W	9/15—9/30	3/25—4/15
Granex	R	9/15—9/30	3/25—4/15
Eclipse	W	9/15—9/30	3/25—4/15
Contessa	W	9/15—9/30	3/25—4/15
Ringer	Y	9/30—10/15	4/1—4/25
Texas Early Grano 502	Y	9/30—10/15	4/1—4/25
Uno Grande	Y	9/30—10/15	4/1—4/25

TABLE 3 (continued)
Onion Cultivar, Color, Seeding Dates, and Harvest Dates in the Lower Rio Grande Valley of Texas (Latitude 26 — 28°N)[a]

Cultivar	Skin[b] color	Dates of seeding	Dates of harvest
Commander	R	9/30—10/15	4/1—4/25
Texas Grano 1015Y	Y	9/30—10/15	4/1—4/25
Colossal	Y	9/30—10/15	4/1—4/25
Burgundy	R	9/30—10/15	4/25—5/15
Texstar	Y	10/10—10/20	4/25—5/15
Creole	Y	10/10—10/20	5/1—5/20
Texas Grano 1025Y	Y	10/20—10/30	5/20—6/1
New Mexico Grano	Y	10/20—10/30	5/20—6/1
New Mexico Grano	W	10/20—10/30	5/20—6/1
Texspan	Y	10/30—11/15	5/20—6/1
Texas Grano 1105Y	Y	10/30—11/15	5/20—6/1
Paradise	W	10/30—11/15	5/20—6/1

[a] Data provided by Thomas D. Longbrake, Extension Horticulturist, Texas A&M University.
[b] W = white, Y = yellow, and R = red.

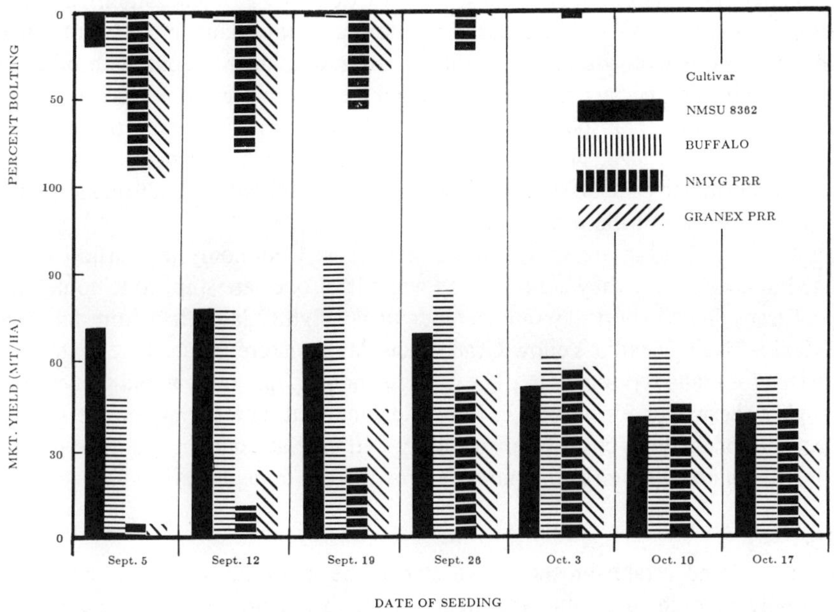

FIGURE 2. The relationship between onion bolting and yield as affected by planting date at Las Cruces, New Mexico, U.S. Unpublished data from Barba, Corgan, and Urquhart, 1986.

optimal. The increased bulb size in such fields will usually more than compensate for the small percentage loss from bolted plants. Of the short day cultivars, the more bolting resistant may be planted earlier, with potential for greater yields.

Most of the crop is seeded from early October to mid-November[7,9,12,13] (Tables 1 to 3). In the warmer winter areas earlier planting is possible, but also later plantings may make sufficient leaf growth before bulbing to produce high yields. Also, the later maturing of the short-day cultivars, or intermediate cultivars, may produce high yields from later plantings.

Harvest (or maturity) date is determined by cultivar, planting date, climatic area, and to some extent by cultural management; and may vary by year, depending on temperatures in late winter or spring. The earliest cultivars in the warmer areas commonly start to mature in late February, and the later (New Mexico Yellow Grano) in the cooler areas about June 10 to 20 (Table 2). Late winter or spring planted intermediate and long-day cultivars extend the harvest season into July and August in some of the subtropical areas.

The length of harvest season for short-day cultivars in the warmer areas of the U.S. may extend 90 or more days from late February to late May (Table 3). In the cooler areas, development of early cultivars is delayed by low temperatures with maturity of granex occurring in late May. In these areas (e.g., southern New Mexico) the maturity season across cultivars is compressed by this delay which is more pronounced on early than later cultivars. The latest planted New Mexico Yellow Grano matures in mid to late June, providing only about a 30-d harvest season for the short day cultivars. Late winter or spring seeding or transplanting of intermediate and long day cultivars extends the length of harvest season to about 90 d (Table 2). In Israel, including all of the production areas, the harvest season across all cultivars extends for more than 150 d, from March 5 to August 20 (Table 1).

B. TRANSPLANTS

Statistics on the fraction of the U.S. short and intermediate day crop grown from transplants are not available, but the percentage varies by area, from relatively large in Georgia, to relatively small in South Texas, New Mexico, and Arizona. Significant acreages are grown from transplants in some areas of California.[9] Transplants provide advantages of a shorter season crop with greater versatility in programming rotations with other crops. They provide an instant, perfect stand of plants, and the shorter season requires less irrigation water and other inputs. Transplanting also accomplishes uniform plant spacing for more uniform bulb shape and size, and a lower percentage of culls. The main disadvantage of transplants is the major cost of stand establishment, and this one disadvantage may more than compensate for all of the advantages.

Plants for transplanting are grown in warmer areas. Commonly the earliest plants are transplanted in areas where they are produced while later ones are shipped to cooler areas.[14] Maturity of transplanted short day onions tends to be slightly later than from fall seeding. In New Mexico, New Mexico Yellow Grano transplants mature in late June, extending the harvest season for grano type cultivars. The Georgia industry has emphasized granex cultivars transplanted in the winter for harvest in May and June, but increasingly growers are direct seeding because of the high cost of transplanting.[12] In the San Joaquin Valley of California, short or intermediate cultivars are commonly grown from transplants.

C. SETS

In Israel if stand establishment is expected to be a serious problem, onion sets are planted, usually a month later than with direct seeding. The onion sets are prepared in spring and stored for late fall planting.

D. FALL ONIONS

A special method for production of onions under decreasing daylength from sets is quite common in Israel.[15] Onion sets, usually of the cultivar Bet Alpha, are harvested in April after seeding in February. The sets are stored and planted in July or early August. Dry or half dry bulbs can be harvested in December for immediate marketing, or they can be stored for limited periods after proper curing.

Daylength in July is between 13 to 14 h, while in December, when the bulbs are harvested, length of day is reduced to about 10 h. Thus,this very early short day cultivar starts bulbing in August and the short days of November and December do not reverse the bulbing process. Lack of irrigation in December and the relatively low temperatures during this month may be contributing factors.

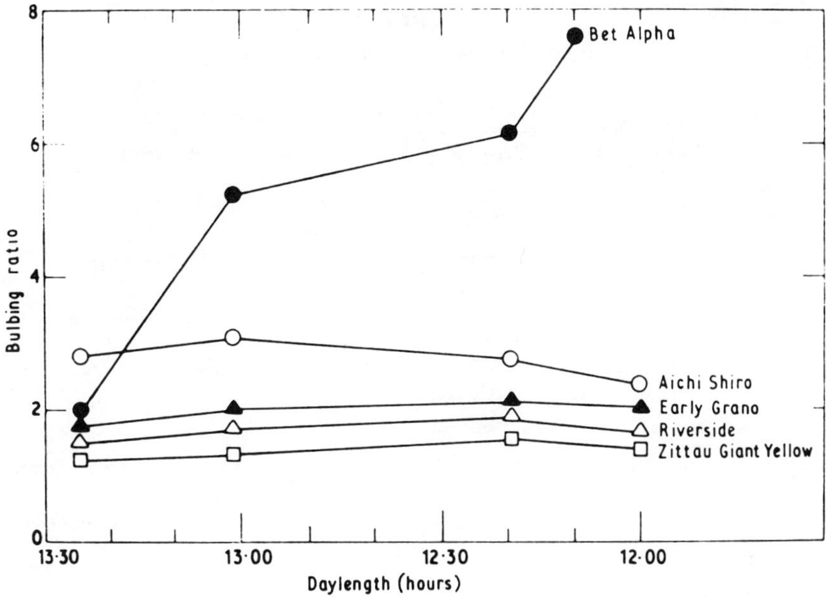

FIGURE 3. Bulbing ratio of 5 onion cultivars grown in the field under decreasing daylength conditions. The seeds were sown on July 12 (daylength 14 h, 1 min) and the experiment was terminated on Sept. 27 (12 h). Each point is the mean of 10 plants. From Kedar, Levy, and Goldschmidt.

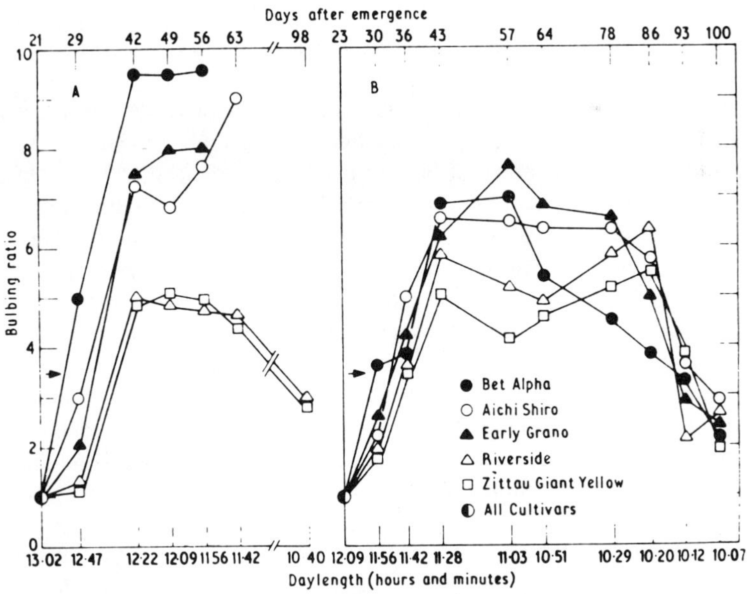

FIGURE 4. Reversibility and bulbing ratio of 5 onion cultivars grown at two dates under long-day conditions (18 h supplementary light by 100 W incandescent lamps, light intensity 18 fc) and transferred to a glasshouse compartment with natural decreasing daylength after reaching a bulbing ratio of 3.5 (indicated by arrows). Each point is the mean of 5 replications, in each replication there were 5 plants. From Kedar, Levy, and Goldschmidt.

TABLE 4
Effect of Light Intensity on Bulbing

Light intensity (foot-candles)	Days-to-bulbing ratio >2		
	Bet Alpha	Early Grano	Riverside
3000	21	21	27
250	35	49	98
100	No bulbing	No bulbing	No bulbing

Note: Plants were grown in a growth chamber under cool white fluorescent 40 W tubes + 75 W incandescent bulbs, at 16 h daylength. (Each number is the mean for 5 — 10 plants). From Kedar, Levy, and Goldschmidt.

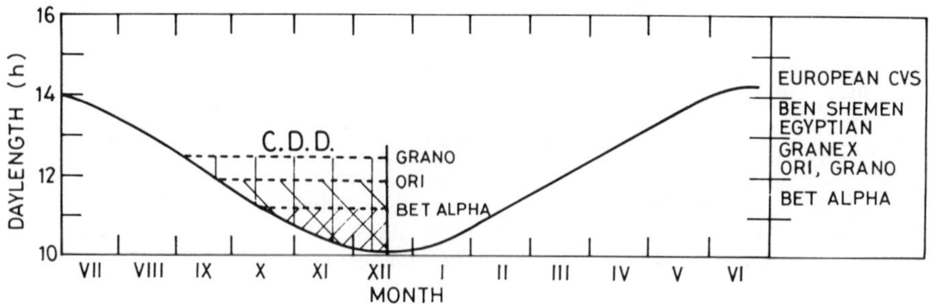

FIGURE 5. Daylength by season in Israel showing adequate daylength for induction of bulbing for various cultivars (right) and CDD for fall onions. European cvs. do not bulb in Israel.

The same production schedule with other cultivars characterized by greater daylength requirements for bulbing do not produce dry bulbs in December[15] (Figure 3). The bulbing rate (ratio between widest diameter of bulb to diameter of neck) of cultivars of grano type, planted from sets in July, decreases during December, and new leaves are produced indicating reversal of the bulbing process. Under certain controlled conditions (Figure 4A), several cultivars reached a bulbing rate of 8 or more. Under shorter daylength (Figure 4B) all cultivars, including Bet Alpha, were reverting to renewed growth.

There seems to be a relationship between high light intensity and early bulbing[15] (Table 4), explaining a somewhat different reaction of onion plants in different experiments. Bulbing under decreasing daylength can be explained by either a minimum number of days within a light period above the critical daylength for a given variety, or by the hypothesis of "critical daylength deficit" (CDD in Figure 5). While CDD for very early cultivars is small enough not to induce reversibility of bulbing, CDD of later cultivars is obviously much greater, and bulbing is reversed.

The principle of fall onions has also been tested in Thailand by the second author. Seeding on July 14, followed by planting of seedlings on August 26 yielded 30 tons/ha of dry or half-dry onions in November. Normal but early seeding and planting dates (Nov. 11 and Oct. 28, respectively) yielded 24 tons/ha during February. Thus, the principle of onion production under decreasing daylength — if properly understood — may be promising for off-season production under varying climates.

IV. STAND ESTABLISHMENT

Onions in the southwestern U.S. are usually seeded or transplanted on standard vegetable beds. Beds are approximately one meter center to center, about 15 cm high, and flat across the top. Top width is usually about 56 cm. Two to eight rows are planted per bed. Either four or six rows are most common for seeded crops. Transplants are hand planted, usually either two or four rows per bed. In some areas transplants are planted in single rows on the level, or on slightly raised beds. Voss suggests use of wider beds for onions where sprinkler irrigation is employed,[9] thereby using the surface area more efficiently. Seeding rates and plant populations are dependent on the size of bulbs desired. Two and one-half to four kilograms of seed per hectare are planted for high yields of small to medium bulbs.

One of the most important innovations in stand establishment in recent years is the use of coated or pelleted seeds combined with special planters for precision seed planting. The main advantage is production of more uniform sized bulbs, and avoidance of misshapen bulbs caused by crowding. Precision planting is most effective and most widely adopted by growers who employ a high level of management to obtain consistently high percentages of germination.

Stand establishment of small-seeded crops is one of the most difficult problems in vegetable production. Problems arise from several sources. Small seed must be planted at shallow depths and moisture must be maintained at these depths until germination. Onions seeded in the fall and winter require from 10 up to 40 d to germinate, sometimes requiring up to 5 irrigations. Rains prior to germination or irrigation water over the seed rows cause soil crusting, restricting seedling emergence. Crusting is particularly severe on alkaline, mineral soils. Furrow irrigation results in concentration of salts at the soil surface over the seed rows. Small amounts of rainfall (1 to 5 mm) may dissolve these salts, killing or injuring emerging seedlings or small plants.

Problems in stand establishment still exist, but progress toward solutions has occurred. The advent of laser leveling provides greater leveling precision, and greater precision in furrow irrigation. The result is less flooding over seeded crops, and much improved plant stands. Sprinkler irrigation for germination is commonly used on onions. Sprinkling may cause soil crusting, but ability to apply small quantities at frequent intervals keeps the crust moist and facilitates seedling emergence. Significantly less water is needed for stand establishment, and sprinkling avoids salt accumulation at the soil surface. Trickle irrigation may be even more effective in alleviating these problems.Research on phosphoric acid and certain organic soil amendments as ''anti-crustants'' is promising as a partial solution to the crusting problem.[16-20] but these soil amendments are not used extensively at this time.

V. FERTILIZATION

Fertilizer needs vary with nutrient availability among soils. Most soils where onions are grown in subtropical climates are deficient in nitrogen and available phosphorus. Soils evolving under conditions of high rainfall may also be deficient in other elements (calcium, magnesium, potassium). Trace element deficiencies are uncommon in U.S. short day onions. Zinc deficiency sometimes occurs on alkaline soils in early spring when soils are still cold, and onions are growing rapidly.

Phosphorus availability is critical for early seedling growth, and response to phosphorus is greatest when soils are cold.[21-23] Relatively slow development of overwintering seedlings dictates a need for available fertilizer phosphorus over an extended period of time. Phosphorus deficiency frequently occurs even after relatively high rates of preplant broadcast phosphorus applications (Figure 6). To avoid this problem, many growers apply very high rates of phosphorus fertilizer before planting, as high as 300 kg P/ha. These are usually broadcast. In many soils, and especially in alkaline soils, phosphorus fertilizers are used more efficiently and maintain available phosphorus longer when applied in concentrated bands below the seed rows. Precision in placement of fertilizer phosphorus on seeded onions may permit the use of lower rates with significant savings in fertilizer costs.

FIGURE 6. The effect of phosphorus fertilization on winter growth of onion. All plants received 100 kg/ha of P preplant incorporated. The +P plants received an additional 200 kg/ha of P sidedressed after germination. Corgan, Unpublished, Las Cruces, New Mexico, U.S., 1987.

Nitrogen is nearly always deficient in soils used for short-day onions. Highest yields are obtained by a continuous supply of N from planting to harvest. The greatest utilization of N by the plants occurs during the periods of most rapid growth,[24,25] but deficiencies that occur in earlier growth stages are translated into significantly lower yields. Excessively high rates of N in latter stages of bulb growth cause delayed maturity,[26] soft bulbs, and generally undesirable characteristics for handling. Ideally, the crop should have adequate N for maximum growth rate until early stages of bulb formation, then gradual depletion of soil N to low levels at time of bulb maturity. Low N during bulb development may result in less total yield, but the alternative of poor handling qualities may be less desirable.

VI. IRRIGATION

Onions, especially from sets, are in Israel sometimes grown without irrigation, depending entirely on winter rains. Yields of grano cultivars in dryland farming are usually 18 to 27 tons, while irrigated fields may yield 36 to 54 tons/ha. Trickle irrigation is quite common while sprinkler irrigation is used at least for germination and stand establishment as well as for incorporation of herbicides, followed by trickle irrigation. Many farmers rely on sprinkling during the entire season. Furrow irrigation has been abandoned because of limitations in availability of water in Israel.

Most short-day onions in the U.S. are grown under supplemental irrigation. In desert regions, nearly all of the water is supplied by irrigation. Irrigation techniques vary by region and individual farmer. The majority of the crop is furrow irrigated. Sprinkler irrigation is used by many growers for stand establishment, but often the crop is finished by furrow application, in part because of concern that high humidities around the plants during and following sprinkling contribute to increased risk of foliage and bulb diseases.[9] Drip or trickle irrigation is not employed extensively on onions in the U.S. at this time.

Evapotranspiration for an entire onion crop is probably less than 100 cm, even in the hot desert regions. However, inefficiencies in application result in significantly higher amounts of water applied. Highest rates are likely used in New Mexico where the crop season may extend from October 1 to about June 1, and most growers employ furrow irrigation for the entire crop. Unpublished estimates are that the average New Mexico grower uses in excess of 150 cm of irrigation water for a fall planted, direct seeded crop. Sprinkler as contrasted to furrow irrigation for germination and early seedling development conserves water, and is commonly used in areas where water costs are relatively high. Trickle irrigation probably offers the greatest water use efficiency.

VII. PEST MANAGEMENT

The most troublesome insect in U.S. short day onions is thrips. The most common species are the onion thrips (*Thrips tabacci* Lind.) and the western flower thrips (*Frankliniella moultoni* Hood).[27] Problems in thrips control arise from several sources. The insect may escape applied pesticides in protected locations between emerging leaves, so that insecticide sprays rarely achieve complete control. A high level of control requires repeated application. In New Mexico five to six applications are commonly applied from February through May. Effectiveness of specific insecticides declines as resistant populations evolve, implying a need for continual research to develop and label new insecticides.

Without question, severe thrips infestation in early plant development is detrimental to yield. Research is needed to determine threshold levels of infestation at various stages of plant development at which insecticide application would give an economic response. Plants with glossy foliage are known to be less attractive to thrips,[28] but the extent of control that can be achieved by incorporating the glossy foliage character into commercial cultivars is unknown.

Pink root disease caused by *Pyrenochaeta terrestris* (Hansen) Gorenz, Walker, and Larson, is one of the most serious disease problems of short day onions. Pink root is most severe in areas where onions have been grown extensively for many years, and the availability of noninfested land is limited. In the U.S. the primary control measures are avoidance of infested fields and cultivar resistance, and/or tolerance. Damage by soil born diseases can be greatly decreased or eliminated by use of the "solarization" method. A detailed description of this new procedure, developed in Israel, is not within the scope of this chapter. In short, the method involves irrigation of the soil, mulching by transparent plastic materials for about 6 to 8 weeks, removal of the cover and seeding or planting. "Solarization" is successful only if done during warm periods with little clouding.

No *Allium cepa* cultivar is known to be completely resistant to *P. terrestris*. However several short day cultivars are sufficiently resistant to be productive on infested soils.[5,7,29] A degree of tolerance to the organism is exhibited by cultivars that have prolific root development during the bulbing phase of growth, when pink root is most severe.[30] Progress in onion breeding for futher improvement in cultivar resistance is promising. Resistance breeding would be more efficient if more were known about the genetics of resistance and host-pathogen interactions, especially the possible existence of different pathotypes, or physiological races of the organism.

Other disease problems are sporadic, or are unique to certain areas of production. Fusarium basal rot caused by *Fusarium oxysporum* f. sp. *cepae* (Hans) Snyder and Hansen is widely distributed throughout the subtropical onion areas. Fields with a long history of onion production often are severely infested. Control is achieved by long term rotation, and

FIGURE 7. Hand harvesting of onions at Hatch, New Mexico, U.S.

avoiding severely infested fields. Possible resistance to basal rot in short day cultivars is not well documented, although resistant long day cultivars are available.[31,32]

Foliage and bulb disorders caused by *Botrytis* species, *Alternaria porri,* (Ell) Cif. or *Peronospora destructor* (Berk) Casp. are most severe in the more humid areas, but even in the desert regions periods of humid weather may cause severe losses.[33] Black mold, caused by *Aspergillus niger* Tiegh, develops on onions near or following maturity and is especially severe on short day cultivars when they are not harvested promptly at maturity. Bulb disorders caused by *Pseudomonas cepacia,* Burkh., are common throughout the subtropical onion areas, and are most serious during periods of rainy weather. The Yellow Dwarf Virus is reported to occur sporadically in California.[9] Smut, caused by *Urocystis colchici* (Schlecht.) Rabenh, is rare in U.S. short day onions. White rot is reported in U.S. long day onions,[34] but is rarely if ever seen in subtropical areas. Voss reports infestations of white rot in a few locations in the San Joaquin valley of California.[9]

VIII. HARVESTING AND HANDLING

Onions are harvested at bulb maturity, when 10 to 100% of the tops have fallen. Most short day onions in the U.S. are harvested by hand. Bulbs are undercut by rod weeders or other implements.[35] Harvest crews trim the roots, clip the tops, drop the bulbs into buckets, and transfer them to burlap sacks for curing (Figure 7). In dryer climates, bulbs are cured in the field for three or more days. They are then bulk loaded by special machinery into dump trucks (Figure 8), hauled to grading sheds (Figure 9) where they are sized, graded, and packed into open mesh bags (Figures 10 and 11). In some areas grading is accomplished in the fields on portable grading machines. In more humid climates, curing may be accomplished in special facilities designed for air (or heated air) movement.[36]

FIGURE 8. Equipment used to bulk load onion from field sacks into trucks for hauling to grading sheds.

FIGURE 9. Onions are unloaded onto equipment for grading and sizing.

FIGURE 10. Onions are inspected for defects.

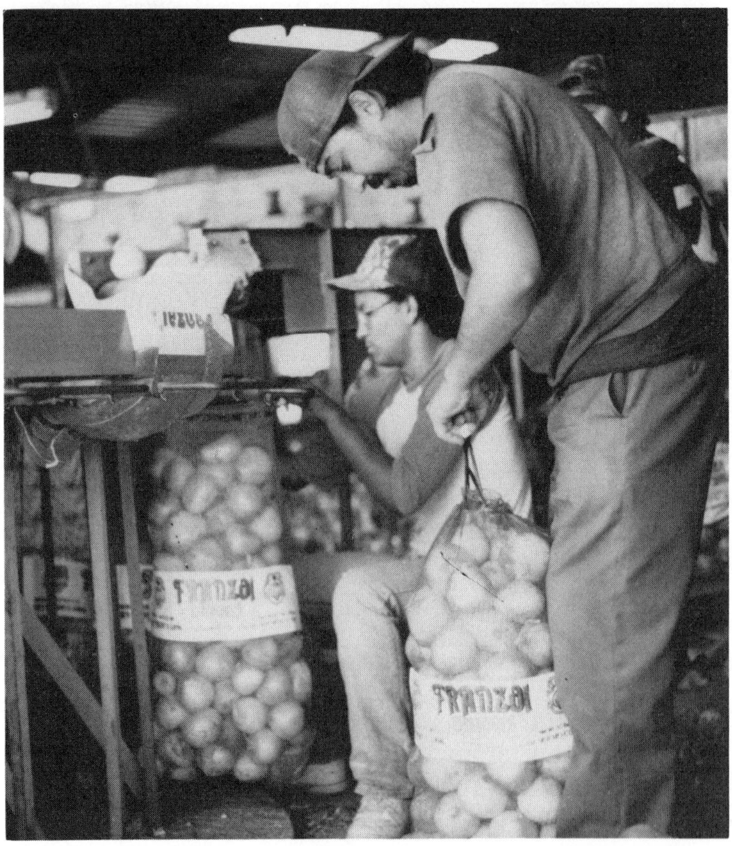

FIGURE 11. Onions for shipment are packaged into 50 lb. mesh bags.

Most onions are sold as 50+ pound sacks, (Figure 11) but 25 lb. sacks are sometimes used. Some are packaged into consumer size packs which are consolidated into larger sacks for shipment. This ''prepackaging'' permits onions to be displayed to the consumer with a grower's or shipper's label and facilitates promotional programs that attempt to promote a special quality, cultivar, or characteristic unique to a particular production area.

Problems, or losses, that occur in harvesting, handling and marketing originate at several points in the system. Most short and many intermediate day cultivars are not long-storage types. As a group they are relatively succulent, low in pungency, and lack the tough dry scales typical of storage type cultivars. The grano and granex type cultivars have very thin dry scales, which are often lost in harvest and grading. The unprotected fleshy scales are easily injured in handling, leading to invasion by microorganisms that cause decay.[37]

Short day onions are usually harvested when temperatures are high, often exceeding 38°C. Under these conditions they are a highly perishable commodity. One of the greatest problems has been that within any production area the cultivars that are grown all mature within a few weeks time. Backlogs of graded onions are often held for several days or even weeks under high temperatures, and in some cases high humidities. Infections by decay causing organisms that occurred in the field or during grading progress very rapidly under these conditions.[37] Also, because of limitations in labor supply, handling facilities and/or markets, bulbs are not always harvested immediately when they mature. After tops have fallen, bulbs in the field are very susceptible to sunburn or heat damage. Also, as onions mature they seem to become more susceptible to infection by a variety of organisms. *Aspergillus niger* develops under the dry outer scales and preconditions the bulbs for invasion by other organisms.

Recent releases of new cultivars by both public and private breeders offer the grower options in maturity leading to an extended harvest season. These and future releases will offer the industry greater flexibility in planning for more orderly harvest and marketing.

IX. WEEDS

Weeds are often very difficult problems in fall and winter planted onions. During early crop development, the winter weeds like various species of mustards are most troublesome. These commonly germinate with the onions,and herbicides that are presently labeled for preplant application are ineffective against many of these species. Some herbicides are effective as post-emergence treatments and are frequently employed for winter weed control. Warm season weeds are most troublesome during latter stages of crop development. Nutsedge is one of the most difficult to control. None of the presently available onion herbicides offers effective nutsedge control.

The most effective weed control programs are those that combine cultivation, judicious use of herbicides, and rotations with other crops which may be maintained free of weed populations.

X. COLD HARDINESS

Overwintering onion seedlings are subject to winter killing in some areas where winter extremes may dip below about −10°C. Such extremes occasionally occur in the production areas of Georgia and New Mexico in the U.S. Early planting with larger overwintering plants minimizes these losses.[38] Many onion cultivars are more cold hardy than those commonly grown in the U.S.[39] Short and intermediate day cultivars of Japanese origin as a group are relatively cold hardy,[39-41] but are not grown extensively in the U.S. Use of more hardy cultivars, or introgression of genes for hardiness into adapted cultivars would reduce losses due to winterkilling, and would extend the areas where onions may be fall-seeded into colder climates.

REFERENCES

1. **USDA,** *Agricultural Statistics,* p. 161, 1985.
2. **Garcia, F. and Fite, A. B.,** Early Grano onion culture, *N. M. Agric. Exp. Stn. Bull.,* 193, 1931.
3. **Hawthorn, L. R.,** Texas Grano onion. *Tex. Agric. Exp. Stn. Prog. Rep.,* 899, 1944.
4. **Enzie, J. V. and Corgan, J. N.,** Temprana, an early maturing white Grano onion, *N. M., Agric. Exp. Stn. Ext. Circ.* 488, 1979.
5. **Pike, L. M. and Leeper, P.,** Five new short day onion varieties for an expanded production season in Texas, *Tex. Agric. Exp. Stn., Misc. Publ.,* 1514, 5, 1982.
6. **Corgan, J. N.,** 'NuMex BR1' onion, *HortScience,* 19, 593, 1984.
7. **Bailey, A. L., and Corgan, J. N.,** Growing onions in New Mexico, *N. M. Agric. Exp. Stn. Ext. Circ.,* 524, 2, 1986.
8. **Bailey, A. L. and Corgan, J. N.,** Onion varieties and classifications, *N. M., Agric, Exp. Stn., Ext. Circ.* 527, 1987.
9. **Voss, R. E.,** Onion production in California, *Univ. Calif. Agric. Ext. Publ.,* 4097, 7, 1979.
10. **Heath, O. V. S.,** Studies on the physiology of the onion plant. I. An investigation of factors concerned in the flowering ("bolting") of onions grown from sets and its prevention. II. Effects of length of day and temperature on onions grown from sets. *Ann. Appl. Biol.* 30, 308, 1943.
11. **Kedar, N.,** What causes bolting in onions?, *Hassadeh* 42, 1, 1961. (Hebrew).
12. **Barber, J. M.,** Onion production (dry), *Coop. Ext. Serv. Univ. Ga. Tifton, Horticulture,* 3, 1984.
13. **Longbrake, T. J., Larson, J., Cotner, S., and Roberts, R.,** Keys to profitable onion production in Texas, *Tex. Agric. Ext. Serv. Leaflet,* 971, 1974.
14. **Jones, H. A. and Mann, L. K.,** *Onions and Their Allies,* Interscience, New York, 1963.
15. **Kedar, N., Levy D., and Goldschmidt, E. E.,** Photoperiodic regulation of bulbing and maturation of Bet Alpha onions (*Allium cepa* L.) under decreasing daylength conditions, *J. Hortic. Sci.,* 50, 373, 1975.
16. **Hemphill, D. D., Jr.,** Phosphoric anticrustant improves seedling emergence and P content, *J. Am. Soc. Hortic. Sci.,* 107, 47, 1982.
17. **Hoyle, B. J.,** Crust control aids seedling emergence, *Calif. Agric., Jan/Feb,* p. 25, 1983.
18. **McGrady, J. J. and Cotter, D. J.,** Anticrustant effects on chile pepper stand and yield, *HortScience,* 19, 408, 1984.
19. **Page, E. R. and Quick, M. J.,** A comparison of the effectiveness of organic polymers as soil anti-crusting agents, *J. Sci. Food. Agric.,* 30, 112, 1979.
20. **Wallace, A. and Wallace, G. A.,** Effect of polymeric soil conditioners on emergence of tomato seedlings, *Soil Sci.,* 141, 321, 1986.
21. **Locascio, S. J. and Warren, G. F.,** Interaction of soil temperature and phosphorus on growth of tomatoes, *Proc. Am. Soc. Hortic. Sci.,* 75, 601, 1960.
22. **Sutton, C. D.,** Effect of low soil temperature on phosphate nutrition of plants — a review, *J. Sci. Food Agric.,* 20, 1, 1969.
23. **Robinson, R. R., Sprague, V. G., and Gross, C. F.,** The relation of temperature and phosphate placement to growth of clover, *Proc. Am. Soil Sci. Soc.,* 23, 225, 1959.
24. **Zinc, F. W.,** Growth and nutrient absorption of green bunching onions, *Proc. Am. Soc. Hortic. Sci.,* 80, 430, 1962.
25. **Zinc, F. W.,** Studies on the growth rate and nutrient absorption of onion, *Hilgardia,* 37, 203, 1966.
26. **Riekels, J. W.,** Nitrogen water relationships of onions grown on organic soils, *J. Am. Soc. Hort. Sci.,* 202, 139, 1977.
27. **Faulkner, L. R.,** Economic thrips of southern New Mexico, *N. M. Agric. Exp. Stn. Bull.,* 387, 1954.
28. **Jones, H. A., Bailey, S. F., and Emsweller, S. L.,** Thrips resistance in the onion, *Hilgardia,* 8, 215, 1934.
29. **Perry, B. A. and Jones, H. A.,** Performance of short-day pink-root-resistant varieties of onions in southern Texas, *Proc. Am. Soc. Hortic. Sci,* 66, 350, 1955.
30. **Levy, D. and Gornik, A.,** Tolerance of onions (*Allium cepa* L.) to the pink root disease caused by *Pyrenochaeta terrestris, Phytoparasitica,* 9, 51, 1981.
31. **Lorbeer, J. W. and Stone, K. W.,** Reaction of onion to fusarium basal rot, *Plant Dis. Rep.,* 49, 522, 1965.
32. **Retig, N., Kurt, A. F., and Gabelman, W. H.,** Greenhouse and field tests for determining the resistance of onion lines to fusarium basal rot, *J. Am. Soc. Hortic. Sci.,* 95, 422, 1970.
33. **Miller, M. E. and Dillon, R. C.,** Survey of bulb diseases on onions in south Texas, *J. Rio Grande Valley Hortic. Soc.,* 33, 25, 1979.
34. **Walker, J. C. and Larson, R. H.,** Onion diseases and their control, *USDA ARS Handbook* No. 208, 1961.
35. **Maw, B.W. and Smittle, D. A.,** Undercutting onions, *HortScience,* 21, 432, 1986.

36. **Smittle, D. A. and Williamson, R. E.,** Onion production and curing in Georgia, *Coastal Plains Exp. Stn., Res. Rep.*, 284, 1978.
37. **Isenberg, F. M.,** The effect of height of fall on onion bruising, *Proc. Am. Soc. Hortic. Sci.*, 66, 331, 1955.
38. **Peffley, E. B., Ortiz, M., and Corgan, J. N.,** A technique for onion cold hardiness evaluation: effects of plant age and size on hardiness, *HortScience*, 16, 773, 1981.
39. **Peffley, E. B.,** Cold Hardiness Characteristics of Overwintering Onions, M. S. Thesis, New Mexico State University, 64, 1981.
40. **Fennell, J. F. M. and Dowker, B. D.,** Screening cultivars for characteristics required in breeding autumn-sown onions, *J. Natl. Inst. Agric. Bot.*, 15, 104, 1979.
41. **Tronickova, E.,** Behavior of short-day onion genotypes under conditions of winter plantation in the CSSR, *Biul. Warzywniczy XXII*, 45.

Chapter 3

CULTURAL SYSTEMS AND AGRONOMIC PRACTICES IN TROPICAL CLIMATES

J. O. Uzo and Lesley Currah

TABLE OF CONTENTS

I. INTRODUCTION

Sturtevant[1] stated that "perhaps onion is indigenous to the area from Palestine to India". From this center of origin it spread along the trade routes to the temperate regions of Europe and North Asia where the long-day onions were selected, and also to North African subtropical regions, where the short day types were derived. Further, these were spread into the sub-Saharan arid and semi-arid tropical parts of Africa. Planting material was then taken to the Caribbean and to Central and South American tropics by settlers from Europe.

Short-day onions are those which bulb and flower in the photoperiods prevailing both in the tropics and the subtropics. Nevertheless, they should not be regarded as true short-day plants. Rather, they are physiologically long-day plants which have a relatively short day-length requirement to induce bulbing, since bulbing still requires a certain minimum or "critical" photoperiod, and the process is faster in photoperiods longer than this minimum. The critical photoperiod for short-day onions ranges from 11 to 13 h of light.

Three schemes run in the 1970s for breeding of onions for tropical conditions have recently been reviewed.[2] Many aspects of crop protection and also storage in tropical conditions have also recently been reviewed.[3]

II. THE TROPICAL ENVIRONMENT IN RELATION TO ONION PRODUCTION

Tropical regions vary greatly in temperature, however, all are similar in having quite a constant day-length of 12 ± 1 h and a wet and a dry season, known in India as Kharif and Rabi.

Short day onions are cultivated throughout the tropics,but commercial production is mostly concentrated either at high altitudes or in semi-arid environments where daily temperatures may range from 15 to 25°C and from 18 to 32°C, respectively. In the lowland humid tropics and in coastal areas, temperatures range from 26 to 32°C. The coastal climate, e.g., in West Africa, is milder but more humid than the hinterland. Onion cultivation in these areas of narrow diurnal temperature range is limited to production for local use only.[4]

Since seasonal changes in photoperiod are rather small, the development of bulbs and inflorescences in tropical onions, is mediated mainly by day/night temperature differences and to a lesser extent by photoperiod.[5] Also, plant age, size, and interplant competition effects may be important for bulbing (see chapter on Physiology of Crop Growth and Bulbing). The differences in onion development and productivity between locations result mainly from differences in the diurnal temperatures. However, light intensity may also play a significant role. The lowland tropics frequently have heavy cloud cover which reduces light energy influx and therefore causes low photosynthetic rates. Consequently, bulbs produced in these areas tend to be small and yields are low. The high relative humidity, also promotes a high incidence of diseases, both in the field and in store.

Low temperatures are usually required to induce flowering in onions. In the tropics, however, vernalization requirements are met by producing seed crops under the wider diurnal range in temperature found at latitudes of some distance from the equator, or in highland areas where cooler temperatures are found. For example, seed crops are produced in Northern Nigeria in Sokoto, latitude 13°N; in Kano, 12°N; in Maiduguri, 11°N; in Niger, 15 to 20°N; in Kenya, 10°S; and in Senegal, 15°N. In addition, newly developed genotypes[6] initiate floral primordia and develop flower heads and seeds even under conditions which normally would not induce blooming (see chapter on Flowering Physiology). Local breeding has also produced genotypes which yield well even in the humid tropical and coastal climate,[7] but seed multiplication is carried out in the neighboring semi-arid areas. Recently, research in Brazil has defined artificial low-temperature treatments of 90 d at 8 to 10°C which are

TABLE 1
Onion Cultivars of Some Tropical Countries

Brazil (Tropical northeast)[16]
Pera IPA-1; Pera IPA-2; Baia Triunfo (these three cultivars are recently developed improved lines); Roxa do Barreiro; Texas Grano 502; Amerela Chata das Canarias
Brunei
Texas Early Grano; Texas Grano; Tropic Ace; Amber Express
Ethiopia[2]
Baladi; Adama Red; Mermiru Brown; Merimu White (the last three cultivars are recent selections, the Merimu types are high in dry matter and suitable for dehydration[2])
Ghana
Bawku (flowers without cold treatment)
Hong Kong
Early Grano; Granex; White Bermuda; Yellow Excel; Tainan nos 1, 2 and 3
Kenya
Bellary Red; Bombay Red; De Wildt; Texas Grano
Niger[10]
Violet de Galmi; Blanc de Soumarana; Blanc de Galmi
Nigeria
Kano Red; Wuyan Makwarwa; Wuyan Bijimi; Gindim Tasa (the last three are local strains of the Red Kano type); Nsukka Red; Texas Early Grano; Crystal White Wax; Lisbon White; Red Creole; Tropical Queen; Bombay Red
Taiwan
Texas Early Grano; Tainan 1 and 2; Yellow Granex; Gold Ring

effective in artificially vernalizing bulbs to allow seed production in hotter areas of the country.[8] A similar system has also been used with success in Zambia.[9] It seems likely that long-term breeding programs in the main onion production areas of the tropics could solve many of the production problems including much needed improvements in storage quality.[2] The work of Nabos[10] in Niger, detailed later, illustrates this. Regardless of their place of origin, onions bred for short-day conditions bulb satisfactorily in the tropics, e.g., *cv.* Texas Early Grano from the U.S.[11] It is, therefore, possible for a seed company located in other latitudes to develop onion cultivars that may outyield local tropical cultivars, as in the example of *cv.* Texas Early Grano.[11] However, the storage qualities of these essentially fresh-market U.S. cultivars are inadequate for long-term storage and transport under tropical conditins. Grano type onions have been bred for quick early bulb production in the southern U.S., and have particularly low dry matter content.

III. ONION PRODUCTION IN VARIOUS PARTS OF THE TROPICS

A. CULTIVARS

The names of some of the cultivars grown in tropical countries are summarized in Table 1.

B. ASPECTS OF PRODUCTION IN VARIOUS TROPICAL COUNTRIES

Practices in a few representative countries are discussed here in some detail.

1. Nigeria

The states of Sokoto, Borno, Kano, Jos, Plateau, and Zaria are the centres of commercial bulb and seed production. The coolest part of the dry season suffices to induce flowering and allow profitable seed yields.

In the northern States of Nigeria, three crops are grown a year, two in the rainy season and the third is an irrigated dry season crop. Main crop onions are grown in the dry season.

These are sown in October and November, transplanted during December and January on flat beds at 15 cm × 20 cm spacing, and irrigated twice weekly. The bulbs are harvested on maturation from March to May, and stored for up to 2 months. The risk of bolting can be minimized by both breeding and proper agrotechniques. Since a tendency to bolting is heritable, selection pressure against flowering can result in nonbolting strains of *cvs.* Kano Red and Bawku.[12] Shading and mulching the seedling nursery results in smaller bulbs, but with virtually no bolting.[12] These effects may be attributed to the low assimilation rates in the shade, which slows growth and development rates. Hence, the plants reach the cool period at a younger physiological age, and are therefore less prone to flower (see chapter on Flowering Physiology).

For the first crop in the rainy season, seed is sown in April and transplanted in June. Plants which have already bulbed precociously are discarded before planting. Bulbs are harvested in August and September. The second rainfed onion crop is sown early in June, transplanted in late July to early August and is harvested in November and December. Pests and diseases cause serious damage to these crops, and bulbs do not keep well in storage.

In the southeastern states of Nigeria the main onion crop is sown by the first week in July and transplanted about the second week of August. The crop grows without irrigation during the late rains and ripens at the onset of the dry season in November and December when there is a paucity of supply from the northern states of Nigeria. The strain "Nsukka Red" is a selection from "Kano Red". At 6° 52′N it flowers after storage at ambient temperatures.[6] In the absence of seed production whole bulbs may be planted and will multiply vegetatively to give 3 to 15 new bulbs per mother bulb.[4]

2. Niger

Onions are a major vegetable crop in Niger.[2,10] They are grown there between 13 and 15°N. The crop is irrigated in the cool dry season from October to December: the mean temperatures ranging from 21 to 24°C in the autumn and winter and rising to 28 to 35°C in the hot season (April to July). The yields are low and amount, on average, to 17.5 t/ha. Rains fall from July to October during which period onion bulbs are stored for replanting for seeds. Yields of *cv.* Violet de Galmi of about 55 t/ha were achieved in some trials[10] and in a storage trial, 73% of bulbs survived 6 months of storage.[10] Local land races from the districts bordering the Niger river and the Maggia valley store well as a result of selection by Niger farmers. Only those bulbs which have stored for 5 months are grown. Vegetative propagation by splitting the bulbs is used to multiply the numbers of bulbs selected for seed.[13] Three major local onion strains were chosen for improvement in the mid 1960s, based on earliness, yield, lack of doubling and skin splitting, shape, color, narrow necks, and resistance to bolting. From these, an early type was selected for its high yield (54 t/ha in some trials), good keeping qualities (80 to 90% of sound bulbs after 6 months in storage), and for their ability to withstand long journeys.[10] The latter two qualities are lacking in "Grano" types which, however, outyielded the local selections in trials, particularly in the case of *cv.* "New Mexico White Grano".

Two local white onion cultivars suitable for dehydration were developed in Niger. These were "Blanc de Soumarana" and "Blanc de Galmi". The former yielded about 45 t/ha and stored for 6 months with 76% surviving. The second was selected from the *cv.* "Violet de Galmi". Its yield was about 54 t/ha and 73% of bulbs survived 6 months of storage.[10]

3. Ethiopia

In Ethiopia onions are being developed as a commercial crop which is grown with irrigation during the dry season from October to May. One growing area is southeast of Addis Ababa near Lake Zwai, at 1000 to 1600 m altitude with mean monthly temperatures of 18.2 to 22.7°C.[2] Another is in the irrigated Awash valley, when onions can be grown during August to November.

New selections derived from Sudanese land races have recently been introduced (Table 1). They are highly pungent, with high soluble solids content and with good storage characteristics. They are susceptible to purple blotch disease (*Alternaria porri* [Ell.] Neerg.) and to downy mildew (*Peronospora destructor* [Berk.] Casp.) (see chapter on Onion Leaf Diseases), thus suffer severe damage when rains arrive unusually early, so new lines are currently being reselected for disease resistance.[14]

4. Brazil

Aguiar et al.[15] reported that after tomatoes and potatoes, onion is the next most important vegetable in Brazil. The main states for onion production are in South and Southeast Brazil, in Rio Grande do Sul, Santa Catarina and Sao Paulo. However, since the 1970s production has been expanded in the tropical northeast region of the Submedio São Francisco valley using irrigation.[16] The successful adaptation of Brazilian cultivars of good storage quality using mass selection has been reported from this region.[17] Techniques for local seed production using artificial vernalization in cold stores like those used for meat storage have been developed.[15] Cold storage of bulbs for approximately 90 d at 8 to 10°C results in economic seed yields.[8] Adoption of similar methods in Zambia with the *cvs.* "Texas Early Grano" and "Pusa Red" has enabled the country to become self-sufficient for onion seed production.[9]

D'Oliveira et al.[11] tested five cultivars: Pera IPA-1, Pera IPA-2, Baia Triunfo, Roxa do Barreiro, and Texas Grano 502 in the São Francisco Valley, and found that late January to early February were the best planting dates. The cultivar Texas Grano 502 proved to be the most stable variety whose yields did not vary significantly with planting date. The low yields obtained from late February to March sowings were probably due to heavy rainfall in March.

IV. CULTURAL PRACTICE

A. SEEDS AND SEED STORAGE

Onion seeds lose their viability shortly after harvest, if they are not dried and stored with care.[18,19] The problem is far more serious during the rainy season than during the dry season. Several factors influence the longevity of onion seeds, the major ones are humidity and temperature in storage. Low humidity of 20 to 25% and low seed moisture content of 6 to 8%, and low storage temperatures result in long storage life[20] (see chapter on Conservation of Genetic Resources).

In the main semi-arid onion producing centres, onion seeds keep fairly well since the relative humidity is usually low. In many parts of the tropics, however, onion seeds must be used immediately in the following season because of lack of modern storage facilities. Simply keeping the onion seeds in an airtight container together with not more than 10% of their volume of a desiccant such as silica gel, as detailed by Grubben[21] can be very effective in extending the life of onion seed in the tropics.

B. SOWING

Percentage and rate of emergence of onions in the tropics, seem to be highest between 20 and 27°C. Those emerging at the lower temperatures seem to be more vigorous than seedlings developed under warmer conditions. Overcrowding in the seed bed may lead to precocious bulbing especially during hot weather.[22]

C. TILLAGE AND FERTILIZING

Because onions have a shallow root system, land cultivation involves rotavation or shallow ploughing to break the soil into a friable texture to a depth of 20 to 30 cm.[4] Prior to transplanting, the field should be properly prepared, principally to avoid weed competition and conserve moisture. A fine tilth improves the physical condition of soils by increasing aeration and the infiltration of water.

Suitable areas for onion growing include the moist areas near irrigated rice fields where the soils retain an adequate amount of water throughout the growing season, without being water-logged during the rainy season. Sticky clay soils must be avoided since they are difficult to work and become easily waterlogged.[4] Excessively dry conditions during bulb development cause the bulbs to split.[10] The best conditions are found in soils rich in humus and in ones derived from alluvial deposits on the banks of rivers.[4]

Recommendations from Nigeria are as follows.[4] The field is ploughed when the land is moist and then harrowed to produce a fine seed bed. If the soil pH is less than 5.5 lime is broadcast at a rate of 1 t/ha. Well-fermented poultry manure should be applied at the rate of 20 t/ha. As an alternative, farm yard manure at a rate of 30 t/ha or compost at 50 t/ha may be used. These should be spread between the planting rows and worked in 2 weeks before transplanting. A compound fertilizer of N:P:K:Mg: 12:12:17:2 + minor elements at 300 kg/ha and lime are added at the time of final tillage. In addition 250 kg/ha of urea is equally divided between the planting date and 2, 4, 6, and 8 weeks later.

Recommendations from Venezuela[23] are to incorporate 800 to 1000 kg/ha of 12-24-12 (N-P_2O_5-K_2O) chemical fertilizers into the soil before making up ridges. The ridge system is used in conjunction with furrow irrigation, either as a system of serpentine furrows within the narrow beds, used where the land has a slight slope or as long furrows where the land is level. Two rows of onions are usually transplanted per ridge. Sprinkler irrigation is used in steep and hilly areas. Other areas of the tropics have specialized local growing systems, often involving division of the fields into small basins which can be flooded at planting, and raised bed systems for growing onions in wet seasons.[22]

D. TRANSPLANTING OF SEEDLINGS

A dry season crop is usually planted on the flat and a rainfed crop on raised beds. Suitable paths are allowed for carrying out of agrotechnical operations like spraying, weeding, fertilizing, and harvesting. The fields are moderately moistened just before transplanting. The premarked beds have rows of 20 to 30 cm apart, and the seedlings are planted at a spacing of 7 to 8 per meter of row. Great care is taken to prevent damage to the leaves during transplanting. After lifting, the roots are carefully spread while planting the seedlings to the same depth as they were in the nursery and pressing the soil. Consequently, the bulbs develop on the soil surface.[4]

Transplanting is successful when done in the later afternoon, as the cool weather of the evening and early morning dew, helps establishment in the field. Mulching with dry grass and irrigation immediately after transplanting is recommended. The mulching helps to conserve soil moisture and keeps down the weeds. As soon as weeds appear they are removed by light hoeing. This is done as often as necessary, since otherwise heavy losses can be expected due to competition for light, water, and nutrients[4] (see chapter on Weed Competition and Herbicides).

E. OPERATIONS DURING BULBING

Bulbs should form above the soil surface to enable maximum aeration, or else they will be more liable to infection with diseases. Frequent hoeing at this stage is good to keep the weed population down and to keep the soil well aerated. During this phase, metabolites are translocated from the green leaves to developing bulbs which rapidly increase in size[4] (see chapter on Crop Growth and Bulbing). The crop is ready for harvest 2 weeks after the shoots fall over. Sometimes fall-over is hastened by twisting and bending over the foliage of individual near-mature plants (Figure 1).

FIGURE 1. A bed of *cv.* "Early Grano" close to maturity. Note the wide spacing and some plant-to-plant variability in the degree of maturity.

F. HARVESTING

After field curing, bulbs are lifted by hand and laid on their sides for 2 days for further curing in the open. If the soil is damp, the onion roots should be pulled free from the soil to prevent rerooting. The leaves should be spread over the bulbs to protect them from sunscald. To avoid disease spread and development, harvesting should be done in dry weather, care being taken not to bruise the bulbs. When dry, the onions are removed to the store for cleaning, and wringing off the shoots and the roots. In many tropical areas strings of onions are formed by plaiting dried leaves together, and the strings can then be stored hanging up in covered sheds or in the home in a naturally well-ventilated state.

Yield varies with variety, season and cultural practices adopted, with ranges from 15 to 55 t/ha in different parts of the tropics. Under experimental conditions yields of 80 to 100 t/ha have been recorded.[10]

A large proportion of the onion crop is often sold during the harvest period and because supply exceeds demand at that time, prices are low. Good storage onions, which can be marketed two to three months later, could improve both supply and returns for farmers. The aim of current breeding schemes in northeast Brazil is to develop locally adapted long storing onions,[17] and such cultivars would be of great benefit in other tropical areas.

V. PESTS AND DISEASES

In semi-arid tropical environments, pests and diseases of the onion are of limited importance on the main season crop. However, a number of pests and diseases have been recorded on rainfed onions and suitable control measures have been developed (see chapter on Onion Leaf Diseases). Inyang[24] reported two pests of the onion from Samaru, Zaria in Nigeria. Crickets cause a lot of damage at the beginning of the rainy season both in the seed-bed and in the field. Control is provided with dieldrin applied to the cricket holes. A nocturnal moth, *Prudenia litura* sometimes lays its eggs on the tubular leaf. The larvae eat their way through the hollow leaves which fall over.[24] Infested leaves and the larvae are manually removed and burnt.

Onion thrips, *Thrips tabaci* may be controlled with methamidophos or quinalphos at recommended doses. Nematodes *Ditylenchus dipsaci* cause bloat and a distorted appearance to infested plants which should be uprooted and burnt. Onions should not be grown on nematode infested soil for at least three years (see also chapters on Insect Pests and on Nematodes Pests of *Allium* Species).

A brief survey of Central and South American advisory literature[23,25,26] shows that onion thrips is the major insect pest in these regions also. It is especially troublesome in hot dry areas. Other notable pests include cutworms (caterpillars of several species of noctuid moths: *Agrotis repleta, Feltia subterranea, Laphynia frugiperda* and *Prodenia eridana*), and a cabbage geometer moth, *Trichoplusia ni*; leaf miners, *Liriomyza* species which attack both onions and garlic; slugs and snails in highland areas; the mite *Aceria tulipae:* a subterranean plant bug, *Cyrtonemus bergi,* and crickets, *Gryllus assimilis* and mole-crickets *Gryllotalpa hexadactyla.* Nematodes of the genera *Ditylenchus, Pratylenchus* and *Meloidogyne* are also encountered.

As regards disease in the American tropics, *Alternaria porri* and *Botrytis* species are damaging everywhere,[23,25,26] while in Brazil the "mal-das-sete-voltas" (seven curls disease) caused by *Colletotrichum gleosporioides* causes serious leaf and bulb damage.[26] Breeding schemes are in progress which are selecting for resistance to this disease.[27,28] Other leaf diseases of importance are downy mildew caused by *Peronospora destructor* an rust (*Puccinia* spp.) (see chapter on Leaf Diseases of Onions). Serious soil-borne diseases are damping-off in seed-beds, caused by various species of *Pythium, Rhizoctonia,* and *Fusarium*; pink root (*Pyrenochaeta terrestris*), which is particularly active at soil temperatures above 30°C; and white rot (*Sclerotium cepivorum*) (see chapter on Root Diseases). In store, *Aspergillus niger* is damaging at high temperatures, as are soft rot bacteria and smudge (*Colletotrichum circinans*)[23,25,26] (see chapter on Storage Diseases of Onions).

General recommendations for the identification and control of pests in the tropics are included in a recent publication,[3] but for specific recommendations which take into account the availability and permitted uses of pesticides it is advisable to consult advisory publications written in the country where onion production takes place.

VI. BULB STORAGE

A. POST-HARVEST DETERIORATION AND STORAGE PRACTICES

The proneness of short day onions to post-harvest deterioration is due to both genetic factors and the technology employed in the storage of the bulbs. Bednarz et al.,[29] in their trials on yield and quality of various short day onion varieties in Northern Nigeria, reported that *cvs.* Texas Early Grano, Yellow Bermuda, Red Creole, and F_1 Granex, were resistant to bolting and gave high yields, but suffered high storage losses as compared with local cultivars. Storage losses in 37 onion cultivars in India where found to be positively correlated with bulb protein content and negatively correlated with ash, potassium, dry matter, total soluble solids, and nonreducing sugar content. The cultivars Seroli-1, Adsuli-1, and Pusa Red gave the lowest storage losses.[30] Another Indian study showed that the high pungency dry season cultivar Ropali was suited for storage and dehydration on the basis of its high dry matter and insoluble solids content.[31]

B. SPROUTING SUPPRESSION

Sprouting and rooting occur in stored onions, shortly after maturation. The rate of sprouting and degree of occurrence depend on both the genotype and the storage conditions.

A wide range of plant growth regulating substances have been tested aiming at regrowth suppression. Application of isopropyl phenyl carbamate (CIPC) plus methyl naphthalene acetate (MNA) entirely suppressed sprouting at 12 to 18°C for 6 months.[32] Maleic hydrazide (MH) is widely used for the control of storage sprouting.[32] Various application rates have been shown to be effective. Randhawa et al.,[33] obtained the least sprouting, rooting, and weight loss when both red and white onions lifted in early June where treated with MH at 2000 ppm + NAA at 2000 ppm and stored in wooden trays till late October. In practice approximately 2.2 l of a 36% a.i. MH preparationn in 225 to 270 l of water per acre is recommended.[32] Various times of application of MH have been recommended, ranging from up to 3 weeks before harvesting to varying stages of top fall.[34-37] Maleic hydrazide must be applied under dry conditions, rain within 24 h of treatment will reduce its effectiveness, and after treatment the crop should not be harvested or treated in any other way for at least 4 d.[32]

C. EFFECTS OF CULTURAL PRACTICES ON STORAGE LIFE

Various cultural practices influence storage ability of onion, though their effects are far from fully understood.[32] Timing and types of various fertilizer application have been reported to have effects on disease incidence, weight loss, and regrowth in storage.[38] Vaughan[39] noted an increased incidence of disease particularly neck rot *Botrytis allii* following excessive nitrogen application and late irrigation. Jones and Mann,[40] concluded that excessive application of nitrogen in the growing season results in the production of bulbs with thick necks which do not store well. In India, however, Wayse[41] found that in addition to yield increases, a low nitrogen application (45 kg/ha) significantly reduced weight losses in storage but had a variable effect on sprouting. The addition of potassium and phosphorus had no effect on the storage life of onions.

There is the danger of excessive neck rot developing when onions are stored with green tops. Hence, Ali and Shabrawy,[42] suggested that bulbs harvested with their tops should be cut to a 1 or 2 cm long top for best storage result. It has also been reported that delayed harvesting increased sprouting in store[43] though in Egypt a decrease in disease severity during storage has been found with delayed harvesting.[42]

Therefore, good cultural practices should be used, which include optimum nitrogen fertilization and avoiding the late irrigation of onion fields. Breeding of bulbs with thin necks, and harvesting ripened firm bulbs with entire skins should be encouraged whenever harvesting is carried out manually. These will help reduce storage losses of onions.

D. EFFECTS OF BULB SIZE AND SKIN COLOR ON STORAGE LIFE

Karmarkar and Joshi,[44] reported that the size of onion bulbs affected both sprouting and water loss during storage. They found that in storage at 11°C, large bulbs sprouted at a faster rate than smaller ones, but that small onions lose weight more rapidly. Saimbhi et al.,[45] found that storage losses (rotting) were greatest in larger bulbs and least in smaller ones. In contrast, Beilinska-Czarnecka et al.,[46] reported that the smallest bulbs were the first to sprout and produce roots and that the largest bulbs were the last. However, in some other investigations, no effect of bulb size on date of commencement of sprouting was found.[43,47,48]

Hurst et al.,[49] reported that white onions showed higher losses than yellow skinned ones. Saxena et al.[50] reported that red cultivars have a higher storage potential in comparison with yellow and white (Table 2) under conditions of Guyana though *cv.* White Creole proved to be an exception.

Emphasis should therefore be placed on breeding the medium size, red and yellow cultivars which can store with only 6 to 9% losses for a 6 months.[49]

TABLE 2
Losses of Onions in Storage in Guyana[50]

Skin color group	Number of cvs in each group	Days in storage: 21	44	55	82
		Mean bulb loss in storage (%)			
'White Creole' (control)	1	11	36	46	57
White (excluding 'White Creole')	5	33	64	75	86
Yellow	5	16	58	72	89
Red	3	0	30	42	62

E. DRYING AND CURING

When onions are not properly dried and cured to remove excessive moisture from the outer skin and neck of the bulbs, infection by pathogens is expected. Curing, also aids the development of skin colors.

A number of methods have been employed for onion bulb curing, all of which are based on exposure to high temperature for periods of time ranging up to 16 d.[32] Curing can be carried out in the field or done artificially. Under tropical conditions, curing is readily carried out in the field. Gobhurdun,[51] in Mauritius reported that field curing of onions *cv.* Local Red for 21 d extended the storage life from 3 months (control) to 5 months and this was further prolonged to 7 months when artificially cured at 37°C for 7 d.

Recommended times and temperatures for artificial curing range from 16 h at 46°C to 14 to 17 d at 36 to 37°C with relative humidity between 70 and 80%. For good skin retention, the RH must not fall below 65% and color is enhanced by a period at a temperature of about 21°C. The higher the temperature above this level the shorter the period required for a given color to be reached.[52] In New Zealand, Harrow and Harris,[53] reported the maximum safe temperature for artificial curing to be about 38°C.

F. STORAGE AND TEMPERATURE AND HUMIDITY

Storage temperature and humidity affect sprouting and rooting, loss in weight, respiration rate, the incidence and severity of rots and many other qualities of stored bulbs. Selection of suitable storage temperature and humidity can preserve the bulbs for months. The storage of onion bulbs under tropical conditions has been reviewed by Thompson et al.[32]

Loss in weight of bulbs increases with increasing storage temperature.[44,49] Karmarkar and Joshi,[44] reported an increase in weight losses of bulbs with increase in temperature from 0 to 10°C followed by a decrease in losses up to 27°C and further increase at still higher temperatures (Figure 2). This effect is probably as a result of an increase in sprouting between 0 and 15°C (Figure 3)[44,49] combined with increasing respiration of onions at higher temperatures.[32] However, the respiration rate of onion rises more slowly as temperature rises than is the case with other crops.[54] (see chapter on Bulb Dormancy and Storage Physiology). This may contribute, in part, to the feasibility of storing onion bulbs at high temperatures. Yamaguchi et al.[55] found a similar percentage of sound bulbs resulting from storage at 0 to 7.5°C to that from 25 to 30°C after 4 months storage. At temperatures between 7.5 and 25.5°C they recorded a significant reduction in the percentage of sound bulbs (Figure 4).

G. STORAGE STRUCTURES

The storage structures can influence the post-harvest deterioration of onions. Storage structures around the world vary from simple shelters to insulated buildings with automatically controlled heating and cooling.

Bulbs can be harvested with tops which are sometimes tied or plaited together into

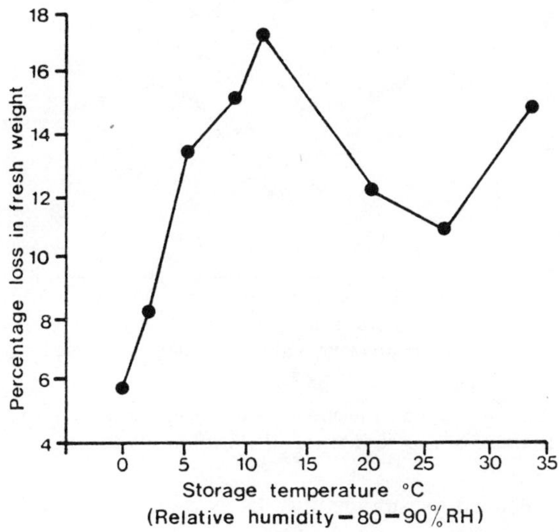

FIGURE 2. Effects of temperature on the percentage of fresh weight loss of bulbs after few months of storage at 80 to 90% RH.[44]

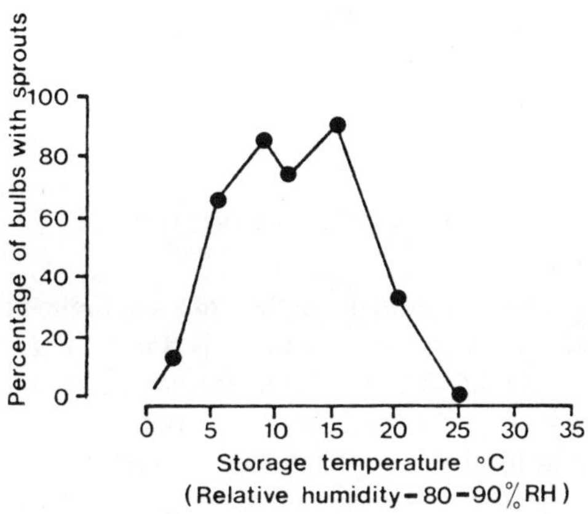

FIGURE 3. Effects of temperature on the percentage of bulbs sprouting after four months storage at 80 to 90% RH.[44]

bundles and then hung up. This is done in Japan, where onions are stored in bundles hung on bamboo poles in ventilated shelters to protect them from sun and rain. About 7.5 t/ha of onions can be stored in a hut of area 4.1 m^2.[32] In India and Sudan onions are bulk-stored in special houses with conical thatched roofs; the walls are made of bamboo canes which are well spaced to ensure good air circulation.[40,56]

The bulbs can also be stored in crates or bags in well-ventilated houses. The crates are stacked on slatted floors in storage houses built so that air can circulate from below the floor, upward through the crates and out through vents in the roof.

H. HANDLING

Poor handling of onions before or during storage can increease bulb losses. Damage due to poor handling is most commonly manifested as bruises caused during harvest. The bruises cause increased respiration rates, and infestation with diseases and pests, thus accelerating weight loss and they may facilitate infection by disease causing organisms.[57]

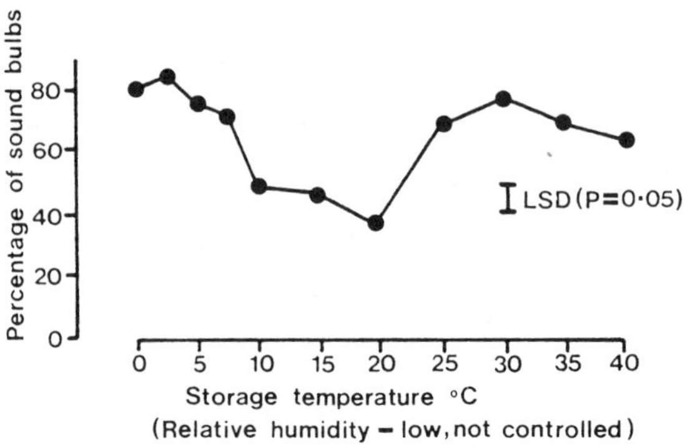

FIGURE 4. Effects of temperatures on the percentage of sound bulbs after four months of storage (RH not controlled).[55]

Therefore, bulbs intended for storage must be as free as possible from cuts and bruises and must be handled with maximum care. Major post-harvest losses in onions are mostly due to sprouting and rots (see chapter on Storage Diseases of Onions).

Both variety improvement and better storage techniques are ways of controlling the magnitude of storage losses in short-day onions. At present improved storage technology tends to be the most effective procedure.

The bulbs should first be well cured, then stored at low temperatures around 0°C; or, if not available, high temperatures of about 30°C with relative humidity of between 65 to 80% RH should be used. This will help to reduce losses during storage. Diseased bulbs should be sorted out before storage.

VII. CONCLUSIONS

Scientific information on onion growing in the tropics is limited and the existing sources of information come from widely scattered sources. It is difficult to obtain an accurate idea of the available literature relating to research on tropical onions both because research on the crop is at a comparatively early stage in many tropical countries, and because many publications appear only in local language journals of limited circulation. The abstracting journals *Horticultural Abstracts* (CAB International, Wallingford UK) and *Abstracts of Tropical Agriculture* (KIT, Amsterdam) perform a valuable service in abstracting from the tropical onion literature, as do the *Review of Applied Entomology* and the *Review of Applied Mycology* (CAB International). It would be useful if research workers could periodically submit critical reviews of their work and that of their local colleagues to journals of wide international circulation. This would not only help the dissemination of material to onion growers and breeders in many countries, but it would promote the cause of onion research with funding agencies internationally.

REFERENCES

1. **Sturtevant, E. L.,** Notes on edible plants, *N. Y. Agric. Exp. Stn. Rep.,* New York, 1919.
2. **Currah, L.,** Review of three onion improvement schemes in the tropics, *Trop. Agric. (Trinidad),* 62, 131, 1985.
3. **Anon.,** Pest Control in Tropical Onions, *Trop. Develop. Res. Inst.,* London, 1983.

4. **Uzo, J. O.,** How to grow onions in Eastern Nigeria, *Agric. Info. Sec. Min. Agric.*, Enugu, Nigeria, 1964.
5. **Abdalla, A. A.,** Effect of temperature and photoperiod on bulbing of the common onion (*Allium cepa* L.) under arid tropical conditions of the Sudan, *Exp. Agric.*, 3, 137, 1967.
6. **Uzo, J. O.,** Induction of biennial reproductive development in short day onion cultivars in a low latitude tropical environment, *Gartenbauwissenschaft*, 48, 149, 1983.
7. **Eavis, B. W. and Jeffers, W. D.,** Onions — from a possibility to a reality in Barbados, in *Proceedings of the 7th Meeting Caribbean Food Crops Society*, Martinique-Guadeloupe, 182, 1969.
8. **Aguiar, P. A. A.,** Período de vernalização dos bulbes de cebola para producao de sementes, no nordeste do Brasil, *Pesqui. Agropecu. Bras.*, 9, 197, 1984.
9. **Mathai, P. J.,** personal communication, 1986.
10. **Nabos, J.,** L'Amélioration de l'oignon (*Allium cepa* L) au Niger, *Agron. Trop.*, 31, 387, 1976.
11. **D'Oliveira, L. O. B., Aguiar, P. A. A., and Assuncao, M. V.,** Competição de cultivares de cebola no Submedio São Francisco visando producao de bulbos, *Pesqui. Agropecu. Bras.*, 19, 193, 1984.
12. **Uzo, J. O.,** Studies on bolting phenomenon in a short day onion cultivar, *Ghana J. Agric. Sci.*, 11, 49, 1978.
13. **Turner, M.,** personal communication, 1983.
14. **Jackson, T. H.,** personal communication, 1982.
15. **Aguiar, P. A., D'Oliveira, L. O. B., and Assuncão, M. V.,** Vernalizacão de bulbos na producão de sementes de Cebola na região do submedio São Francisco, *Pesqui. Agropecu. Bras.*, 18, 741, 1983.
16. **D'Oliveria, L. O. B. and Lima, V. de, P. M. S.,** A cultura da cebola no Nordeste, *Fortaleza., CE, BNB-ETENE*, Brazil, p. 98, 1979.
17. **Melo, P. C. T., Costa, C. P., Wanderley, L. J. G., Menezes, D., and Queiroz, M. A.,** Selecão massal estratificada em duas populacãos de cebola (*Allium cepa* L.) Baia Periforme no vale do Submedio São Francisco, *Pesqui. Agropecu. Pernamb.*, 2, 95, 1978.
18. **Beattie, J. H. and Boswell, V. R.,** Longevity of onion seeds in relation to storage conditions, *Proc. Am. Soc. Hortic. Sci.*, 35, 553, 1937.
19. **Barton, L. V.,** Storage of vegetable seeds, *Contrib. Boyce Thompson Inst.*, 7, 323, 1935.
20. **Biasutti, O. E.,** The storage of seeds for maintenance of viability, *Bulletin 43 Commonwealth Agric. Bureau. Farnham*, UK, 1957.
21. **Grubben, G. J. H.,** Vegetable seeds for the tropics. *Bulletin 301, Dept. Agric. Res., Royal Tropical Institute, Amsterdam*, p. 40, 1978.
22. **Currah, L.,** personal communication, 1987.
23. **Ayala, H., Medrano, C., Rodriguez, P. J., Cermelli, P. J., Morales, F., Renaud, J., Flores, A., Auba de Garcia, J. and Gouveia, J.,** Amarilidaceas cebolla y ajo, Fundacion Servicio para el Agricultur, Cagua, Venezuela, p. 95, 1975.
24. **Inyang, A. O.,** Onion cultivation in northern Nigeria, *Samaru Agric. Newsl.*, 8, 60, 1966.
25. **Medina, J. N.,** Cultivo de la cebolla, *Camaro del Agro Guatemala*, 2, 16, 1980.
26. **Rêgo, G. M., Menezes, J. E., Castro, M. E. B., Lima, M. F. B. F., and Silva, J. B. T.,** Eds., *Cebola: Resumos Informativos, Embrapa/Dept. of Information and Documentation, Brasilia*, p. 313, 1980.
27. **Melo, I. S. and Costa, C. P.,** Estimativas de paramétros geneticos em cebola (*Allium cepa* L.), populacões Pira Rosa e Pira Tropical, para o carater resistência a *Colletotrichum gloeosporioides*, Penz sensu Arx, 1957, *Summa Phytopathol.*, 9, 207, 1984.
28. **Melo, I. S. and Costa, C. P.,** Selecão massa em cebola (*Allium cepa* L.), populacão "Pira Ouro", para resistência a *Collectotrichum gloeosporioides*, Penz sensu Arx, 1957, *Summa Phytopathol.*, 9, 214, 1984.
29. **Bednarz, F., Kadam, K., and Dosunmu, G. O.,** Yield and quality of various short day onion varieties (*Allium cepa:* L.), in *Northern Nigeria in Relation to Date of Sowing, in Nortson 8th Annual Conference/ 10th Anniversary of Nihort*, Ibadan, 1986, 1.
30. **Patil, R. S. and Kale, P. N.,** Correlation studies on chemical composition and keeping quality of some onion cultivars, *J. Maharashtra Agric. Univ. (India)*, 10, 154, 1985.
31. **Maini, S. B., Diwan, B., and Anand, J. C.,** Storage behaviour and drying characteristics of commercial cultivars of onion, *J. Food Sci. Technol. (India)*, 21, 417, 1984.
32. **Thompson, A. K., Booth, R. H., and Proctor, F. J.,** Onion storage in the tropics, *Trop. Sci.*, 14, 19, 1972.
33. **Randhawa, K. S., Arora, S. K., Mohan, S., and Gupta, A. K.,** Effects of different chemicals on storage of onions under ordinary conditions, *J. Res. Punjab Agric. Univ.*, 22, 261, 1985.
34. **Isenberg, F. M., Jensen, C. O., and Odland, M. L.,** Effect of Maleic Hydrazide on the respiration of mature onion bulbs, *Science*, 120, 464, 1954.
35. **Wittwer, S. H. and Sharma, R. C.,** The control of storage sprouting in onions by pre-harvest foliage sprays of Maleic hydrazide, *Science*, 112, 597, 1950.
36. **Randhawa, K. S. and Nandpuri, K. S.,** Effect of plant growth regulators on sprouting of onion under ordinary storage condition, *Indian J. Agron*, 11, 238, 1966.

37. **Tzeng, T. C.,** Studies on the factors influencing the quality and yield of onions, *Taiwan Agric. Q.,* 10, 97, 1974.
38. **Proctor, F. J., Goodlife, J. P., and Coursey, D. G.,** Post-harvest losses of vegetables and their control in the tropics, in *Vegetable Productivity,* Spedding, C. R. W., Ed., Macmillan, London, 1981, 139.
39. **Vaughan, E. K.,** Influence of growing, curing and storage practices on development of neck rot in onions, *Phytopathology,* 50, 87, 1960.
40. **Jones, H. A. and Mann, L. K.,** *Onions and their Allies,* Leonard Hill Ltd., London, 1963, 286.
41. **Wayse, S. B.,** Effect of N, P and K on yield and keeping quality of onion bulbs, *Indian J. Agron.,* 12, 379, 1967.
42. **Ali, A. A. and Shabrawy, A. M.,** Effect of some cultural practices and some chemicals on the control of neck rot disease caused by *Botrytis allii* during storage and in the field for seed onion production in A.R.E., *Agric. Res. Rev. Plant Pathol.,* 57, 103, 1979.
43. **Ward, C. M.,** The effect of bulb size on the storage life of onions, *Ann. Appl. Biol.,* 91, 113, 1979.
44. **Karmarker, D. V. and Joshi, B. M.,** Investigations on the storage of onions, *Indian J. Agric. Sci.,* 11, 82, 1941.
45. **Saimbhi, M. S. and Randhawa, K. S.,** Losses in white onion variety Punjab-48 under ordinary storage conditions as influenced by bulb size, *J. Res. India,* 19, 188, 1982.
46. **Bielinska-Czarnecka M., Kepkowa, A., Kielak, E., and Zdanowska, E.,** Influence of size of onion bulb *cv.* "Czerniakowska" on its dormancy, sprouting and rooting, *Acta Agrobotanica Warsaw,* 34, 129, 1981.
47. **Kunkel, R.,** The effect of various levels of N and K on the keeping quality of onions, *Proc. Am. Soc. Hortic. Sci.,* 50, 361, 1947.
48. **Kato, T.,** Physiological studies on the bulbing and dormancy on onion plants. V. Relations between dormancy and organic constituents of bulbs, *J. Jpn. Soc. Hortic. Sci.,* 35, 143, 1966.
49. **Hurst, W. C., Shewfelt, R. L., and Schuler, G. A.,** Shelf life and quality changes in summer storage of onions *(Allium cepa), J. Food Sci.,* 50, 761, 1985.
50. **Saxena, G. K., Halsey, L. H., Gull, D. D., and Persuad, N.,** Evaluation of carrot and onion *cvs.* for commercial production in Guyana, *Sci. Hortic.,* 2, 257, 1974.
51. **Gobhurdun, S.,** Aspects of warm temperature storage of onions, *Rev. Agric. Sucr. Ile Maurice,* 57, 157, 1978.
52. **Bleasdale, J. K. A., Tucker, W. G., and Hough, K. H.,** Bulk storage of onion, onion skin colour and keeping quality, *Ann. Rep. Natl. Veg. Res. Stn. 1969,* 78, 1970.
53. **Harrow, K. M. and Harris, S.,** Artificial curing of onion for control of neck-rot, *N Z. J. Agric. Res.,* 12, 592, 1969.
54. **Lutz, J. M. and Hardenburg, R. E.,** The commercial storage of fruit, vegetables, florist and nursery stock, in *USDA Agriculture Handbook,* Vol. 66, U.S. Government Printing Office, Washington, D.C., 1968.
55. **Yamaguchi, M., Pratt, K. H., and Morris, L. L.,** Effect of storage temperature on keeping quality and composition of onion bulbs and on subsequent darkening of dehydrated flakes, *Proc. Am. Soc. Hortic. Sci.,* 69, 421, 1957.
56. **Sulafa, K. M., Habish, A. H., Abdalla, A. A., and Adlan, A. B.,** Problems of onion in the Sudan, *Trop. Sci.,* 15, 319, 1973.
57. **Green, J. H.,** Suggestions for improved storage of onions in Northern Nigeria, *Samaru Agric. Newsl.,* 14, 56, 1972.

Chapter 4

WEED COMPETITION AND WEED CONTROL IN *ALLIUM* CROPS

Baruch Rubin

TABLE OF CONTENTS

I. INTRODUCTION

Competition between weeds and a crop is of major importance in determining the economics of weed control, thus affecting directly the income of the farmer. As long as 40 years ago, Comin[1] commented that "onion weeding is the most laborious and most expensive operation connected with growing the crop "weed control . . . represents the greatest single cost in the production of onions, and frequently runs to 15 to 30% of the total costs".

Several growth factors, e.g., light, CO_2, oxygen, water, and mineral nutrients are required for the development of plants. When any or all of the above essentials, are in short supply to a population of plants, competition occurs. The stronger competitor, weed or crop, inhibits the growth of its neighboring plant. In their early stages of development, alliaceous crops have both a slow growth rate and thin foliage. Therefore these plants are considerably weaker competitors than most crops.

Weed competition at the early stages of growth results in severe yield losses. Weed competition adversely affects both total yield and the marketable share of the yield, namely the percentage of large or medium sized bulbs. Late infestation with weeds may not directly affect the yield, but may interfere with mechanical harvesting. Additionally, it may produce seeds which will add to the seed bank in the soil, thus increasing infestation in future crops. The introduction of both herbicidal and cultural techniques and particularly the selective herbicides, have dramatically reduced the cost of weed control. However, although essential, weed-free management of the onion crop remains difficult to achieve.

Chemical weed control started in the late nineteen forties. Various salts and dilute acids were applied postemergence, but these did not discriminate sufficiently between weeds and crop. The introduction of pre- and post-emergence herbicides, with greater selectivity, resulted in a considerable reduction in costly hand labor, decreased weed competition, and increased yields.

II. WEED COMPETITION AND YIELD LOSS

A. GENERAL ASPECTS OF WEED-CROP COMPETITION

Most arable lands contain large numbers of viable seeds of annual and perennial weeds, part of which are stimulated to germinate with every cultivation. This results in a flush of weed seedlings, which compete with the emerging crop.[2] In the semi-arid zone, new flushes of weed seedlings may appear after each irrigation or rain. Thus, in these areas, weeds may infest the fields not only after sowing but throughout the growing season.

Allium crops, due to their slow germination, morphology (their upright narrow leaves), and slow growth rate are very sensitive to early-season weed competition.These crops grow all over the world, being exposed to a variety of climatic conditions. Thus, the time from sowing to emergence varies from several days to several weeks. In England, onions are sown in March, when the soil is still cold. Therefore, germination is rather slow and 50% emergence requires between 5 to 6 weeks, reaching the two-leaf stage by the end of May. During this period, the fast emerging weeds grow rapidly accumulating dry weight per unit–area 20 times that of the crop.[2,3] On the other hand, in the Mediterranean countries and in California, onions and garlic are sown mainly in the fall. The time required for both crop germination and emergence, is much shorter than that in England. However, the gradual decrease in temperatures with the onset of winter slows down the growth rate of the crop, exposing it to severe competition from winter annual weeds.

As mentioned above, weed competition affects both total yield and its quality. As weed density increases, and/or the duration of competition extends during the critical period of the crop growth (see Section II.B), the proportion of small bulbs is increased.[4-9] Yield reduction, due to weed competition, is accompanied both by a delay of fall-over of the tops,

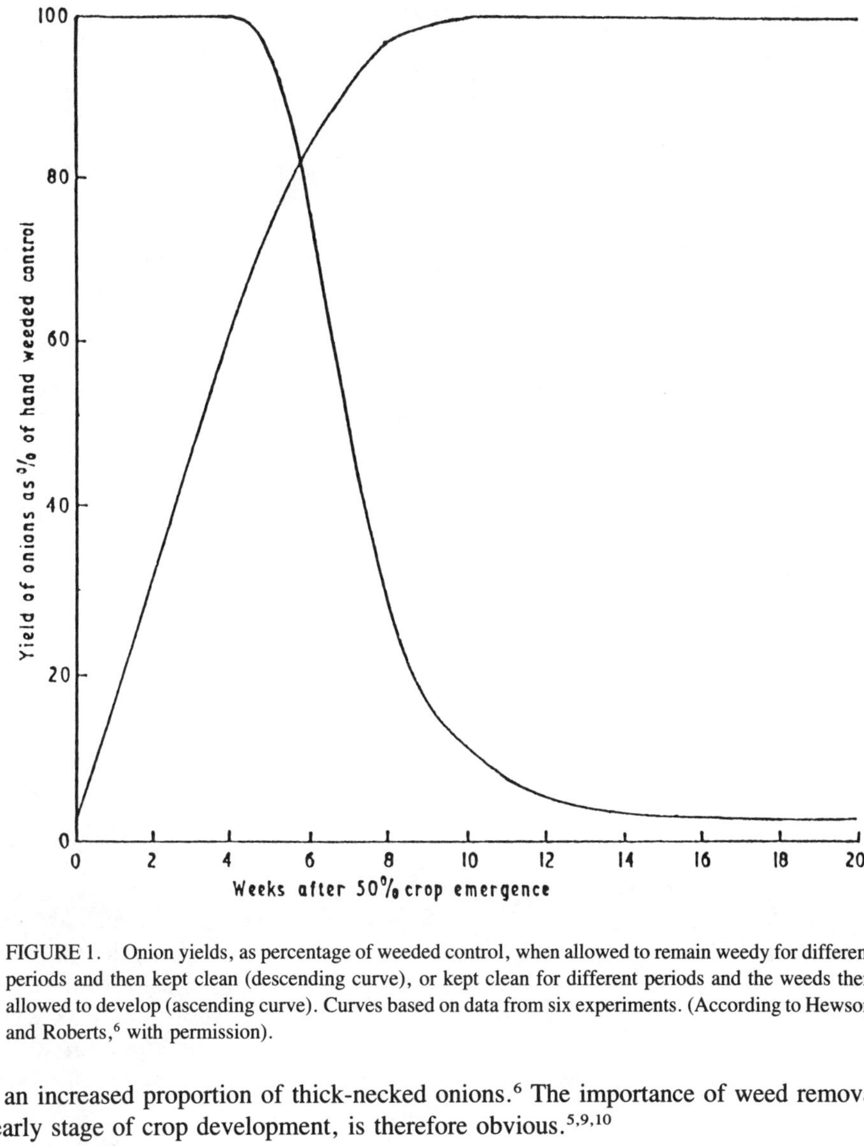

FIGURE 1. Onion yields, as percentage of weeded control, when allowed to remain weedy for different periods and then kept clean (descending curve), or kept clean for different periods and the weeds then allowed to develop (ascending curve). Curves based on data from six experiments. (According to Hewson and Roberts,[6] with permission).

and an increased proportion of thick-necked onions.[6] The importance of weed removal, at an early stage of crop development, is therefore obvious.[5,9,10]

B. THE CRITICAL PERIOD FOR WEED COMPETITION

Bleasdale[5] showed that weeds, present even for a short period early in the growth of onions, significantly reduce the final yield. The ''critical period'' of competition has been defined by various researchers as the period before and after which the presence of weeds has little effect on the final yield.[2,6,9,15] Hewson and Roberts[6] studied in England the effect of weed removal at different stages of crop development on the yield of onions. They reported that the critical period for onions, sown in March, occurred during the development of the third true leaf (Figure 1). At earlier stages, i.e., from the first 5 to 5.5 weeks after 50% emergence, weed infestation at densities in the range of 150 to 850/m^2 caused no damage, provided the crop was kept weed-free thereafter until the third true leaf developed. Also, weed infestation at a later stages did not bring about any yield reduction. If weeds were allowed to remain during this critical period, yield was reduced. They therefore concluded that the critical period lasts two weeks. The lack of damage by late infestation was attributed to the fact that the main spring peak of weed germination had passed, and in the absence of soil disturbance, only a few weeds emerged.

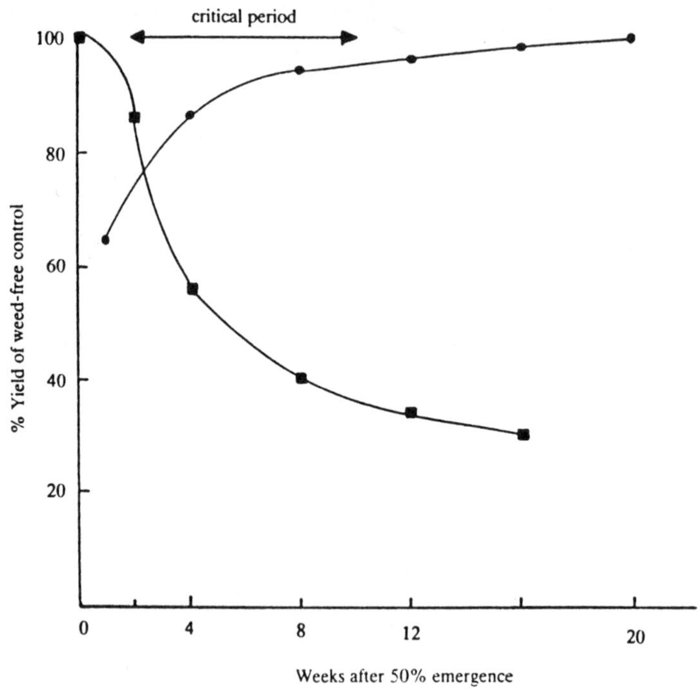

FIGURE 2. Illustration of the critical period of growth of the onion crop in relation to weed competition. Crop under competition then subsequently kept weed free. (■—■) Crop kept weed-free then competition established. (●—●) (According to Thomas and Wright,[9] with permission).

Similar experiments, conducted in England under comparable conditions, yielded somewhat different results.[9] Onions were sown at 68 seeds/m^2 to give a plant stand of 42 plants/m^2, together with ryegrass or turnip, as a substitute for the natural weed flora, at a plant stand of 200 to 250 plants/m^2. The latter were present for 2, 4, 8, 12, and 16 weeks, after 50% emergence. This competition resulted in 14, 44, and 60 to 70% yield reductions for the 8, 12, and 16 weeks treatment, respectively. When the field was kept clean for 1, 4, and 8 weeks, before weed competition was established, yield reduction was 36, 14, and 6%, respectively. Weeds present after 12 to 16 weeks resulted in negligible yield losses. It was estimated that the critical competition period lasts 8 weeks, occurring between the second to the 10th week after crop emergence (Figure 2). This critical period is fourfold longer than that reported by Hewson and Roberts,[6] and occurred chronologically earlier, but at a similar developmental stage of the crop (at 2 to 4 leaves). The critical competition period seems to occur at the same date every year. This suggests some photoperiodic involvement in the onset of this period.[9]

Shadbolt and Holm,[4] in Wisconsin, concluded that the critical stage in the competition between weeds and row crops, including onions, is from the time of crop emergence and up to 4 weeks later. Wicks et al.[7] found that competition by a mixture of annual broadleaf and grass weeds, for 2, 4, 6, or 8 weeks after onion emergence, reduced yield by 20, 20, 40, and 65%, respectively. Plots which were kept clean until emergence or 2, 4, 5, 8, 10, or 12 weeks later, suffered yield losses of 100, 99, 87, 75, 46, 25, and 5%, respectively. It was concluded that onion fields need to be kept weed-free for 12 weeks after emergence, due to their both, lack of vigorous crop foliage, and an inability to recover from competition.

Williams et al.,[14] in Corvallis, Oregon reported 100% yield loss with weed competition throughout the growing season. The yield was equally and significantly reduced by weeds present during the first 5 weeks, or by those present later in the season. They concluded that weeds should be controlled during the entire season to prevent onion yield loss.

In many slow growing or non-competitive crops, transplants are used to shorten the critical competition period.[11] However, transplanted onions should be kept weed-free during the first third of their growth. Unrestricted weed growth reduced bulb yield by 26 to 48% in transplanted onions, in Gezira, Sudan. One hand weeding, commencing 42 days after transplanting, resulted in losses of only 15% of the yield from clean-weeded control plots.[16] Paller et al.,[17] in the Philippines, reported that the critical period in 48-d-old transplanted onions began only 2 weeks after transplanting and lasted for 5 to 6 weeks, or longer.

William and Warren[13] studied competition between the perennial weed purple nutsedge *(Cyperus rotundus)* and garlic, in Vicosa, Brazil. This weed is worldwide one of the most troublesome weeds.[18] Nutsedge emerges before garlic and the crop's vertically growing leaves never form a competitive canopy. When the weed, at a stand of 1600 shoots/m^2, competed with garlic for the entire growth season, yield losses amounted to 89%. The critical period, however, occurred between 3 and 13 weeks after planting.[13] It was, therefore, concluded that the first weeding should take place within 3 weeks of field planting. Any delay results in yield loss. The crop must be maintained weed free for 13 weeks or until bulbing occurs, whichever is sooner.

Roberts,[2,8] stated that competition begins when weed weight starts to increase rapidly. Once this occurs, the effects on the crop become severe. Thus, the critical period reflects the development pattern of both crop and weeds. The onset and duration of this period is affected by the composition and density of the weed population, by crop management, and by crop density.[6] Differences in these factors may explain the differences in onset and duration of the critical period reported by different researchers. The critical period of weed competition, in onions and in similar slow growing crops, is a practical rather than a basic concept. It indicates the optimum time for weed removal using cultural and/or chemical measures.

C. EFFECTS OF WEED POPULATION DENSITY

Weed population density is likely to affect the time at which irreversible effects of competition become apparent.[5,19] Shadbolt and Holm,[4] studied the effect of an early weed competition, in onions, using a natural weed density of approximately 200 plants/m^2, and, by careful thinning, reduced it to 15, 30, and 50% of the normal stand. These were allowed to compete with the crop for 4, 5, and 6 weeks after emergence. Four weeks of competition, with a 50% weed stand, caused more than 40% yield reduction. An additional week of competition, with 15, 30, and 50% weed stand, resulted in 48, 68, and 76% yield reduction, respectively. Six weeks of competition at 15% of the weed stand, resulted in approximately 90% yield reduction. They concluded that, within certain limits, actual weed population density is not important when yield is considered, since strong shading and a large weed biomass is produced from both, low and high populations. Hewson and Roberts[6] reported that the time at which competition begins was considerably affected by weed density, from 0 to 150 plants/m^2; but further increase in weed population, up to 850 plants/m^2, had only a little additional effect. It is clear that a few vigorous and tall weeds can cause as much damage to the crop as larger populations of smaller plants.

D. EFFECTS OF THE TYPE OF WEED

Weed species differ greatly in their competitive ability. This is determined mainly by the time of their emergence, morphology, and growth rate. It is only logical to assume that early emerging, fast growing weeds which form a dense leaf canopy cause more damage and advance the onset-and increase the duration of the critical period. Indeed, interference studies showed that the presence of London rocket (*Sisymbrium irio*), a tall growing plant, for 4, 8, 12, and 16 weeks after onion planting reduced yields by 20, 23, 65, and 92%, respectively. In comparison, the presence for the same periods, of ridgeseed spurge (*Euphorbia glyptosperma*), a prostrate and creeping weed, did not affect yields.[20] Purple nutsedge, a fast emerging and rapidly growing weed, covers the field, even before the crop emerges, resulting in early and long competition and an almost complete crop loss.[13]

Similarly, the infestation of onion fields with the vigorous and tall-growing redroot pigweed (*Amaranthus retroflexus*), results in severe competition and yield reduction.[4,7,14] This is the most prevalent weed in Nebraska and part of Massachusetts, where it infests approximately 90% of all onion fields.[7] Putnam and his co-workers[19] in Michigan, studied the effect of common purslane (*Portulaca oleracea*), a common inhabitant of vegetable crops in the U.S., and elsewhere. Onion yield losses occurred if competition at 450 plants/m^2 lasted for 20 days or longer after seeding. At weed densities of 50 plants/m^2, yield losses were significant only with competitions of 30 d or more after seeding. Paller et al.,[17] reported the onset of the critical period as early as 2 weeks after the transplanting of 48-d-old onion seedlings, when plots were predominantly infested by the fast growing weeds, barnyardgrass (*Echinochloa colonum*) and horse purslane (*Trianthema portulacasrtum*). In California, it is recommended that neither onion nor garlic be planted in fields infested with fast growing annual grasses, such as barnyardgrass, and/or with some perennial weeds.[21,22]

The most troublesome winter annuals in alliaceous crops in California are: London rocket, mustard (*Brassica* spp.), pineappleweed (*Matricaria matricarioides*), common groundsel (*Senecio vulgaris*), shepherdspurse (*Capsella bursa-pastoris*), little mallow (*Malva parviflora*), annual sowthistle (*Sonchus oleraceus*), and volunteer cereals. In the summer, in California, the most damaging weeds are barnyardgrass, common lambsquarters (*Chenopodium album*), pigweed, Russian thistle (*Salsola iberica*), and little mallow.[21,22]

Wicks et al.,[7] in Nebraska, showed a significant negative correlation between the yield of weeds and that of the onions. The strongest correlation was found with red root pigweed, proso millet (*Panicum miliaceum*), green foxtail (*Setaria viridis*), and stinkgrass (*Eragrostis cilianensis*), which compete with the crop throughout the season. Kochia (*Kochia scoparia*) yield, although present only early in the season, was also highly negatively correlated with onions yields.

Hewson and Roberts,[6] reported the onset of the critical period, when the fresh weight of the weeds; prostrate knotweed (*Polygonum aviculare*), annual bluegrass (*Poa annua*), and chickweed (*Stellaria media*), began to increase rapidly 5 to 6 weeks after 50% crop emergence. This belated competition is understandable, since at Wellesbourne, England, these weeds have a slow initial growth rate, thus, delaying the onset of the critical period.

Late infestation of onions, with late germinating, but fast growing weeds, may also significantly reduce onion yields by interfering with mechanical harvesting. In Israel, late infestation with hairy fleabane (*Conyza bonariensis*), and horseweed (*C. canadensis*), significantly reduced marketable onion yields.[23]

In addition, some weeds, e.g., common purslane and purple nutsedge, may exert part of their adverse effect by producing allelopathic compounds, which further reduce the growth and vigor of the crop.[13,19] Purple nutsedge also provides a favorable environment for certain onion-damaging insects and diseases.[13]

E. EFFECT OF CROP PLANT DENSITY

Crop density also affects the onset of the critical period, although in a slow growing crop, such as onion, this may not be very important. Hewson and Roberts,[6] reported that, in fields with a crop density of 191 plants/m^2, no critical period was observed, and late establishing weeds were suppressed. This stand, however, is twice the recommended plant density in onion fields, and is not commercially acceptable. Williams et al.,[14] studied the effects of row spacings on weed competition. Row spacing of 12.7, 25, and 38 cm (7.6 cm

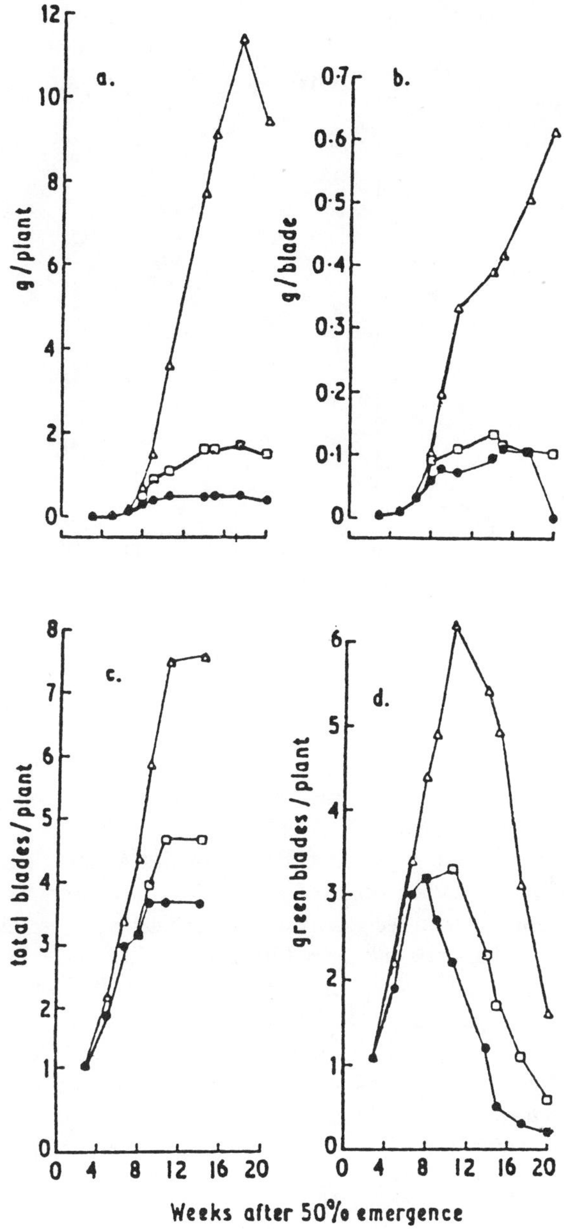

FIGURE 3. Effect of weed competition on the growth characteristics of onion: (a) mean plant weight; (b) mean green leaf-blade dry weight; (c) production of foliage leaves; (d) numbers of green leaves. △, Onions in the absence of weeds; ●, weeds present throughout; □, weeds present for the first 7.5 weeks only. (Adopted from Hewson and Roberts,[6] with permission).

space within the row) resulted in crop densities of 103, 51, and 34 plants/m^2, respectively. Fields were cleared of weeds either early or late in the season, representing late or early infestation, respectively. No yield was obtained from plots with full season weed competition at any tested spacing. When the weeds were present for part of the growing season, (early or late infestation), total yield losses were significantly lower at the highest onion density, as compared with those from medium and low crop densities.

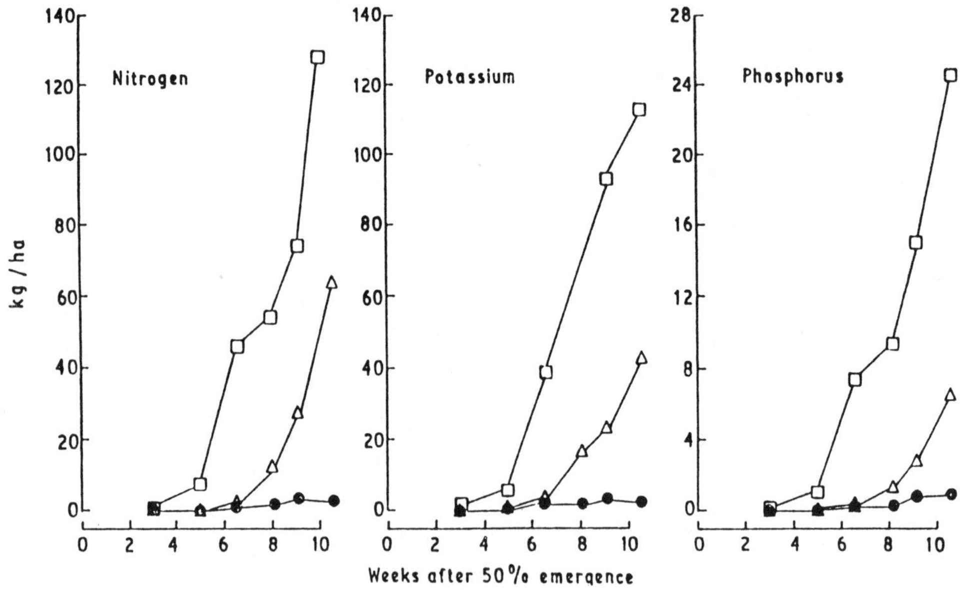

FIGURE 4. Removal of major nutrients from the soil by onions and weeds. △, Onions in the absence of weeds; ●, onions in the presence of weeds; □, weeds (According to Hewson and Roberts,[2] with permission).

F. COMPETITION FOR LIGHT

Competition for light, in the early stages of onion development, is of major importance, because of the small, cylindrical and erect leaves, and the low growth rate of the crop. Competing weeds reduce as much as 85% of the light intensity at crop level. The degree of shading imposed on the crop is closely correlated with the biomass of weeds present. In one experiment maximum shading occurred about 25 d after crop emergence, just before the weeds reached their maximal weight.[4] Common purslane, 15 cm high, reduced at soil level the penetration of blue, red, and far red light by 97, 94, and 98%, respectively. High densities of common purslane, present for a 30-d period, reduced both onion leaf length and number of leaves per plant by 59%.[19]

Similarly, under continuous competition with annual weeds (at 160 to 230/m^2), the photosynthetic capacity of the crop was drastically reduced, due to a reduction in the number of functional leaves, their size and chlorophyll content (Figure 3). These leaves were also shorter lived.[3] Severe competition for light, between garlic and purple nutsedge has been reported. It was suggested that the marked reduction in light intensity is the most important factor affecting the yield.

Under strong competition, bulbing occurs earlier in the season and at a younger physiological age (lower leaf number), as compared to weed-free conditions.[3,4,9] Hence, only small bulbs are produced, and no increase in dry weight or bulb size is observed, even if weeds are removed.[3] Thomas and Wright,[9] suggested that under weedy conditions, the plant is induced to produce higher levels of endogenous growth substances which promote bulbing. This, however, could also be explained by the effect of a higher far-red:red ratio in the radiation under a weed leaf canopy (see chapter on Physiology of Crop Growth and Bulbing).

G. COMPETITION FOR WATER AND NUTRIENTS

Competition between weeds and onions for nutrients, was studied by Hewson and Roberts,[3] (Figure 4). Nutrients accumulation by weeds started earlier than in the onions and, by the end of the first half of the growing season, the nutrient content in the weeds was much higher than that in the crop (calculated on the basis of the total biomass of weeds and crop present). At this time, the weeds had removed half of the nitrogen and a third of the potassium applied in the base fertilizer dressing, whereas the crop under competition contained negligible amounts of these nutrients (Figure 4).

Similar results were obtained for purple nutsedge,[13] and for common purslane.[19] The latter removed as much as 100 kg/ha of potassium from the onion field.

Competition for water is another important factor, particularly in arid and semi-arid zones. However, no attempts have been made to directly determine the effect of weeds on water availability, since in all experiments reported, the crops were regularly irrigated or rain-fed.[7,13,17,19]

III. METHODS OF WEED CONTROL

The primary goal of a weed control program is to reduce as much as possible the weed stand, with minimal crop injury.[2,4] Several approaches and methods are outlined below. However, the effectiveness of each technique used separately is limited, since no single method can economically control enough weed species under all conditions. Thus, only a rational combination of several management approaches can provide adequate and economical weed control.

A. CULTURAL CONTROL

Growers should be aware of the field history so as to minimize problems with volunteer crops (e.g., potatoes), weeds and herbicide residues left over from the preceding seasons.[21,22,24,25] This may be achieved by crop rotation, a commonly used practice for sanitation purposes in onion fields. Crop rotation, combined with tillage and chemical measures, may provide a solution to some problems of weed control, especially those which are difficult to control in onions. Kurtz,[21,22] stated that crop rotaton is extremely important, both in terms of which crops and weeds species to avoid preceding the onion and garlic crops, and which crops should not follow garlic and onions. Since grass control in onion fields is quite difficult, cereal fields which cannot be adequately prepared (e.g., by burning of stubble, preplant irrigation, and deep tillage) should be avoided. Similarly fields with known infestation by parasitic weeds (e.g., *Cuscuta* spp.), annual grasses like barnyardgrass, and/or perennial weeds such as field bindweed (*Convolvulus arvensis*), Johnsongrass (*Sorghum halepense*), yellow nutsedge (*Cyperus esculentus*), and quackgrass (*Agropyron repnes*) should be avoided.

Herbicide residues, from the previous season, should be considered when a field is selected. Certain herbicides, used on the previous crop (e.g., atrazine in corn; see Table 1 for list of chemical names), may persist in the soil and damage onion seedlings.

Cultural practices, which favor the initial growth of onion plants rather than that of weeds, are of primary importance in successful weed management.[26] Well-prepared seedbeds, without established weeds, should facilitate seed drilling and thus result in better crop germination, emergence, and initial stand. It also facilitates later weed control operations, since the rapid emergence results in seedlings that are all at a similar stage of subsequent development. The timing of application of post-emergence herbicides depends critically on the growth stage of the crop (see Section III.C.3.a). Hence, it is essential that all crop plants are at a uniform growth stage, if these herbicides are to be used successfully. Campbell and Anderson,[27] reported that seed production from planted bulbs was significantly lower in no-tillage plots compared to those with conventional tillage.

Increased density may, in some crops, reduce weed competition. However, as discussed above (see Section II.E) these effects are small and are not economically worthwhile with onions. On the other hand, postponing drilling for 2 to 3 weeks after seedbed preparation, often permits the emergence of a flush of weeds (the "stale seed bed" technique). These

TABLE 1
List of Common and Chemical Names of Herbicides

Common name	Chemical name
Alachlor	2-Chloro-*N*-(2,6-diethylphenyl)-*N*-(methoxymethyl)acetamide.
Alloxydim	Methyl 4-hydroxy-6, 6-dimethyl-2-oxo-3-{l-[(2-propenyloxy) imino]butyl}-3-=cyclohexene-carboxylate
Atrazine	2-Chloro-4-(ethylamino)-6-(isoprpylamino)-*s*-triazine
Bensulide	*O,O*-bis(1-Methylethyl) *S*-{2-[(phenylsulfonyl)amino]ethyl} phosphorodithioate
Bentazon	3-(1-Methylethyl)-(1*H*)-2,1,3-benzothiadiazin-4(3*H*)-one 2,2-dioxide
Bromoxynil	3,5-Dibromo-4-hydroxybenzonitrile
CDAA[a]	2-Chloro-*N,N*-di-prophenylacetamide
Chlorbufam	1-Methylprop-2-ynyl 3-chlorocarbanilate
Chloroxuron	*N'*-[4-(4-Chlorophenoxy)phenyl]-*N,N*-dimethylurea
Chlorpropham[b]	1-Methylethyl 3-chlorophenylcarbamate
Clopyralid	3,6-Dichloro-2-pyridinecarboxylic acid
DCPA[c]	Dimethyl 2,3,5,6-tetrachloro-1,4-benzenedicarboxylate
Diclofop	(q)-2-[4-(2,4-Dichlorophenoxy)phenoxy]propanoic acid
Dinoseb	2-(1-Methylpropyl)-4,6-dinitrophenol
Diquat	6,7-Dihydrodipyrido[1,2-:2',1'-*c*]pyrazinediium ion
Ethofumesate	(±)-2-Ethoxy-2,3-dihydro-3,3-dimethyl-5-benzofuranyl methansolfunate
Fenoxaprop	(±)-2-{4-[(6-Chloro-2-benzoxazolyl)oxy]phenoxy}propanoic acid
Fluchloralin	*N*-(2-Chlorethyl)-2,6-dinitro-*N*-propyl-4-(trifluoromethyl) benzenamine
Glyphosate	*N*-(Phosphonomethyl)glycine
Haloxyfop	2-{4-[[3-Chloro-5(trifluoromethyl)-2-pyridinyl]oxy]phenoxy}propanoic acid
Ioxynil	4-Hydroxy-3,5-diiodobenzonitrile
Linuron	3-(3,4-Dichlorophenyl)-1-methoxy-1-methylurea
Methabenzthiazuron	1-Benzothiazol-2-yl-1,3-dimethylurea
Methazole	2-(3,4-Dichlorophenyl)-4-methyl-1,2,4-oxadiazolidine-3,5-dione
Metolachlor	2-Chloro-*N*-(2-ethyl-6-methylphenyl)-*N*-(2-methoxy-1-methylethyl) acetamide
Nitrofen	2,4-Dichloro-1-(4-nitrophenoxy)benzene
Oxadiazon	3-[2,4-Dichloro-5-(1-methylethoxy)phenyl]-5-(1,1-dimethylethyl)-1,3,4-oxadiazol-2-(3*H*)-one
Oxyfluorfen	2-Chloro-1-(3-ethoxy-4-nitrophenoxy)-4-(trifluoromethyl) benzene
Paraquat	1,1'-Dimethyl-4,4'-bipyridinium ion
Pendimethalin	*N*-(1-Ethylpropyl)3,4-dimethyl-2,6-dinitrobenzenamine
Phenisopham	Isopropyl 3-[ethyl(phenyl)carbamoyloxy]carbanilate
Propachlor	2-Chlor-*N*-(1-methylethyl)-*N*-phenylacetamide
Propham[d]	1-Methylethyl phenylcarbamate
Pyrazon[e]	5-Amino-4-chloro-2-phenyl-3(2*H*)-pyridazinone
Sethoxydim	2-[1-(Ethoxyimino)butyl]-5-[2-(ethylthio)propyl]-3-hydroxy-2-cyclohexen-1-one

[a] Allidochlor.
[b] CIPC.
[c] Chlorthal-dimethyl.
[d] IPC.
[e] Chloridazon.

can easily be controlled by a non-residual herbicide, such as paraquat or glyphosate. Additional treatment, with contact herbicide, can be applied just prior to the crop emergence, controlling another flush of emerged weeds, thus allowing the crop a few more weeks with little or no weed competition.

Solar heating of the soil was developed, in Israel, as another cultural approach to reduce weed infestation in onions, and other crops.[28] The method is based on mulching or tarping wet soil with transparent polyethylene sheets, in the hot season, thus heating the soil to a level lethal to various soil pests.[28,29] The efficiency of this technique, for weed control, has been studied extensively.[30-34] Soil solarization of prepared beds for 4 to 6 weeks before sowing significantly reduced the infestation with annual broad-leaf and grass weeds, but was much less effective against perennial weeds (Rubin and Eviatar, unpublished). Annual broadleaf weeds with hard seeds, such as sweetclover (*Melilotus* spp.), milkvetch (*Astragalus* spp.) and mallow, are heat resistant and are insensitive to soil solarization.[30,31]

B. MECHANICAL CONTROL

Cultivation is another means of controlling weeds in alliaceous crops, provided the field layout is suitable.[21,22,24] Straight beds, crop rows of uniform width and proper interow spacing are required. Early in the season, when the crop plants are still small, both beds and furrows can be cultivated. However, as plants increase in size, cultivation is restricted to furrows only. Cultivation is commonly combined with chemical treatments and hand weeding, i.e., when the crop cover is substantial, the furrows can still be cultivated, but beds are treated with herbicides. Herbicides are also applied when cultivation alone is not sufficient to free the crop from weeds.[35]

Where and when chemical weed control programs are not available, or their use is too expensive, weed populations can be reduced by cultivation, so that hand weeding becomes easier and quicker. In California, mechanical undercutting, prior to harvest, is a common method for controlling weeds established in garlic and onions fields.[21,22] This technique is used primarily to reduce the weed population established between the last irrigation and harvest, thereby facilitating mechanical harvest.

In some countries, hand weeding is the principle means used for weed control.[25,36] In Sudan, a single hand weeding, 42 d after transplanting onions, decreased both weed population and yield loss to acceptable levels, but the labor requirement, for this weeding, was 127 manhours/ha, and the weeds tended to re-invade the plots.[16] Hand weeding was inefficient in decreasing the seed population of palmer amaranth (*Amaranthus palmeri*) and of common purslane in onions.[37] However, the damaging effect of weed re-infestation, following hand weeding, cannot be tolerated, as the crop must be weed-free at the end of the growing season, when mechanical harvest procedures are employed.[38] Hand weeding may be very expensive, costing 10 to 100 times more than an effective herbicide program giving similar yields.[12,38] Moreover, in many developed countries, hand weeding is not practical as the man power is not available. Thus, hand weeding can only be a supplementary method for the advanced grower, when herbicides and/or cultivation cannot be applied. The proper timing of hand weeding is important in order to reduce both weed competition and the yield reduction which results from the accidental removal of young seedlings or bulbs during weeding.[13,21,22]

Regardless of the high cost, hand weeding is still the only method permitted for some leafy alliaceous crops, such as chives, because of the dangers of the residual effect to the consumer (see chapter on Chives).

C. CHEMICAL WEED CONTROL

The use of herbicides for weed control in alliaceous crops started in the mid-forties of the 20th century, when dilute sulfuric acid and several inorganic salts were used for post emergence treatments.[1] In the early 1950s, the cost of hand weeding increased, hence, the use of more selective herbicides became more popular.[39] Weed control with selective compounds in alliaceous crops, was readily accepted because of the availability of a number of chemicals providing economical control of a wide range of weed species, thus permitting flexibility in application time as well as being relatively cheap.[38] Currently, the introduction and registration of new herbicides is limited, due to legal, economical and scientific difficulties.[40]

Both the activity and selectivity of herbicides are greatly influenced by many interacting variables, including plant age of both crop and weeds, soil type and moisture content, weather conditions, method and time of application, and previous herbicidal treatment. These subjects have been recently reviewed,[41-44] and will not be discussed in this chapter.

Herbicides can be applied at several stages in the development of onions, garlic and leeks, as will be discussed later. Only a few of the available methods are necessary in a given field provided the farmer knows both which weeds infest his field and their response to the available herbicides.

1. Presowing Application of Herbicides

Chemical control of existing weeds is desired when beds are prepared early in the season, and the soil is not disturbed before sowing or planting. Nonselective and nonresidual herbicides can then be used without endangering the safety of the crop. Paraquat, diquat, a mixture of both, and/or glyphosate, are commonly used for this purpose. Paraquat and diquat act rapidly on young seedlings of annual weeds, while glyphosate provides good control of both annual and perennial weeds.

Preplant incorporation (PPI) of selective herbicides is usually done during seedbed preparation, using a rotary tiller.[45,46] The herbicides are broadcast, but are only effectively incorporated on the cultivated beds, thus providing only poor weed control in the furrows. The main advantage of mechanical soil incorporation of herbicides occurs when insoluble and/or volatile compounds are used. Field and laboratory experiments have shown that, while PPI treatment may improve the efficacy of certain herbicides, it can cause severe damage to the crop with others.[47] In Punjab, India, preplant incorporation of fluchloralin to onion bulbs, grown for seed production, resulted in excellent weed control and led to a marked improvement in plant height and seed yield over the unweeded control.[35] It was shown that PPI application of bensulide and propachlor provided better broad leaf and grass weed control, with no damage to the onion yield, as compared to non-incorporated treatments. However, soil-incorporation of DCPA (chlorthal-dimethyl) and chlorpropham (CIPC) resulted in poorer weed control and reduced yield.[45,46]

2. Preemergence Application of Herbicides

Preemergence application of herbicides can be done at any time between sowing and crop emergence. The major advantages of preemergence treatments are that they can be applied as part of the sowing or planting operation, prior to weed development, or can be delayed and applied just before crop emergence. In temperate areas, germination of spring-sown onions is slow, and crop emergence requires between 5 and 6 weeks. Under these conditions, the application of a nonresidual foliar herbicides, such as paraquat, plus a soil acting chemical, can be delayed to just before emergence, thus controlling the first flush of fast germinating weeds. Such a management frees the crop from early competition.[2,8] The timing of these treatments is extremely critical, as, any crop emerged will either be damaged or eliminated, if contacted by the herbicide.[22]

Application of soil acting preemergence and pretransplanting herbicides is of great importance in the overall weed control program. The aim of these treatments is to control weeds which germinate and emerge during the early growth of the crop. An optimal preemergence treatment is expected to control both broad leaf and monocot weeds, for the entire crop growth without damaging the crop. There are several herbicides which in combination, have proved efficient for this purpose in onions. Unfortunately, none of them provide complete weed control and additional measures have to be taken.

Propham (IPC) and chlorpropham (CIPC) were the first selective pre–emergence herbicides introduced for onions.[39,48] These herbicides are more effective against grasses than against dicots. Chlorpropham is more persistent in the soil than propham and thus its herbicidal activity lasts for a longer period, and it soon became the main preemergence herbicide for onions, garlic, and chives.[21,22,25,39,45,46] However, it is rarely used nowadays. CDAA (allidochlor), DCPA and propachlor alone, or in combination, are commonly used worldwide as preemergence treatments in direct seeded or set planted onions, garlic and leeks. When applied to mineral or organic soils these herbicides, do not adversely affect the crop, but provide acceptable control of several annual weeds. In spite of their limited killing spectrum, these herbicides are recommended in order to reduce weed competition during the critical early stage of crop establishment. The three herbicides differ in performance with respect to climatic conditions, range of species controlled and persistence.[46,49]

DCPA applied preemergence, does not adequately control either grasses nor several important winter weeds, including greater ammi (*Ammi Hmajus*) toothpick ammi (*A. visnaga*), and wild carrot (*Daucus* spp.).[50,51] In addition, some weeds of the Compositae family, such as pineappleweed, hairy galinsoga (*Galinsoga ciliata*), annual sowthistle, ragweed (*Parthenium hysterophorus*), and of the Cruciferae family, e.g., London rocket and mustard are not sufficiently controlled.[20,50,52] Babiker and Ahmed,[16] reported, from Sudan, that in transplanted onions, DCPA was very effective on grasses (90 to 100% control), but was less effective on broad-leaf weeds (38 to 85% control). Wilson[53] reported recently that when applied preemergence, this herbicide provides on the average, 70% of annual weed control. DCPA was less selective between onions and weeds on very light soils.[54] Sanok et al.[52] reported that propachlor provided fair to poor control of grass weeds, but excellent control of galinsoga, lambsquarter, and pineappleweed. The use of CDAA as a herbicide was recently restricted in most countries, and therefore will not be discussed here.[55,56] In the U.S. and in Europe, pendimethalin, applied preemergence or at onion emergence, is quite safe and provides good control of several broadleaf weeds, but not of redroot pigweed,[54,56-58] whereas PPI treatment with this herbicide resulted in severe crop damage.[59] In Israel also, such early treatments resulted in severe damage to onions.[60] It is much safer when used at or before planting of sets or seedlings.[27,57]

Ethofumesate, applied preemergence in drilled onions, provides excellent full season control of annual grasses, including volunteer wheat and barley, as well as some broadleaf weeds.[61-64] No visible damage to the crop was observed at 0.8 to 1.0 kg/ha,[62-64] but at higher rates, a significant reduction in both crop stand and yield was observed.[61,65] Leaf curling and twisting is observed, with concomitant peeling of the outer transparent membrane of the leaves (Rubin and Adler, unpublished). PPI application of ethofumesate (2.2 and 4.5 kg/ha), before planting onion bulbs for seed production, resulted in unexpectedly low seed yields. The only indications of phytotoxicity were a reduction in flowers per umbel numbers and a delay in maturity.[27]

The adverse effects of ethofumesate on the crop become evident under unfavorable environmental conditions, or when certain postemergence herbicides are applied.[63,66,67] Ethofumesate applied preemergence, impairs both the quantity and the quality of the leaf epicuticular wax deposition in various plants, including onions.[66-69] Photomicrographs of ethofumesate treated and nontreated onion leaves, taken with the scanning electron microscope (SEM), have shown that the herbicide changes the wax crystallization on the external layer of both the cotyledon and leaf.[66] Similar effects were observed in different onion cultivars (Figure 5). Thus, by affecting the leaves' epicuticular wax deposition, ethofumesate treatment in onions increases the spray retention on and the penetration into the leaf of postemergence applied chemicals.[66,67] The implications of this phenomenon on the response of onions to subsequent herbicide treatments is discussed later (see Section III.3.)

All preemergence treatments, registered for use in seeded onions, may also be used pretransplanting or preemergence following set or bulb planting. Moreover, due to the advanced growth stage of transplants, sets, and bulbs, lower sensitivity to soil applied herbicide is observed. This permits the use of higher rates of various herbicide without damaging the crop.[16,27,35,54]

3. Postemergence Application of Selective Herbicides

Only a limited number of selective preplanting and pre-emergence herbicides are avail-

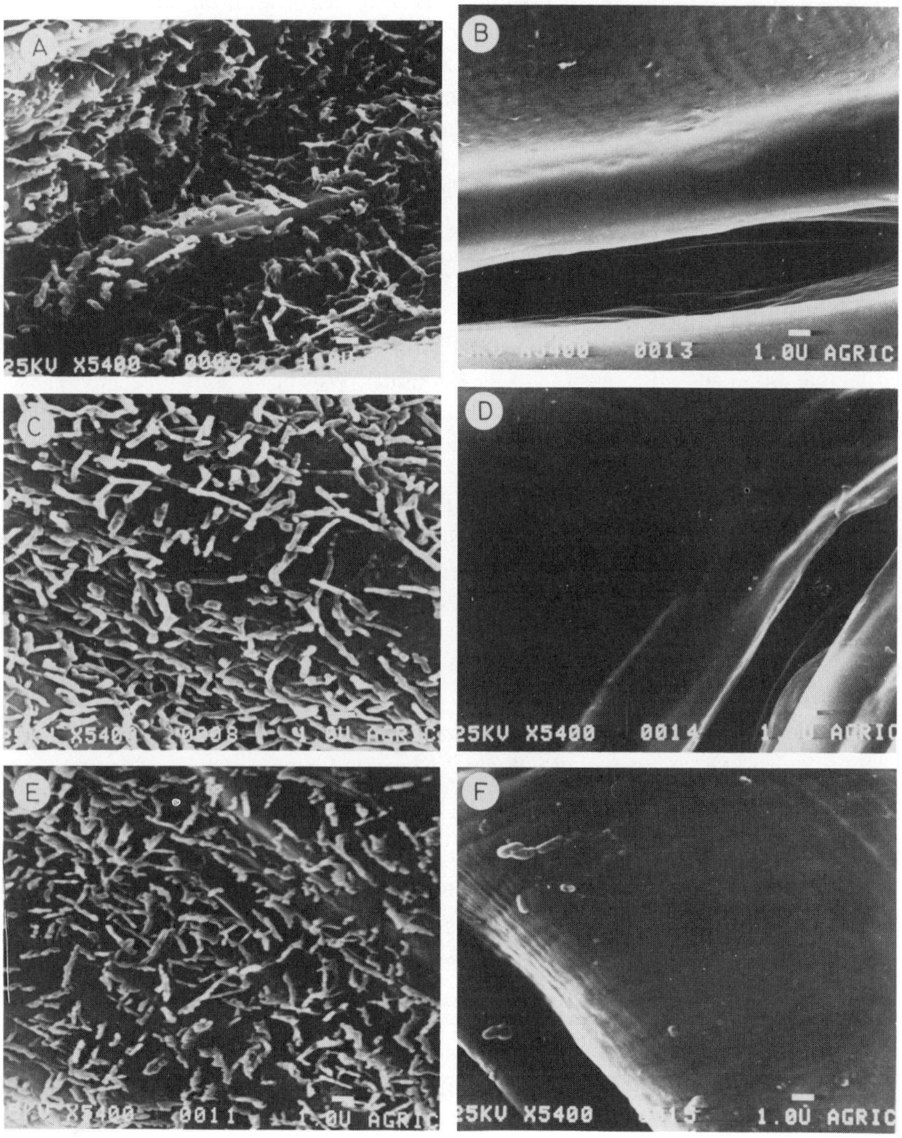

FIGURE 5. SEM micrographs of leaf surface of onion cultivars, with (right), and without (left), preemergence treatment with ethofumesate at 0.8 kg/ha. *cv.* Ori; pictures a, b, *cv.* Grano 3; pictures c, d, *cv.* Ben Shemen, pictures e, f. (According to Adler[65]).

able for onions. None controls all weeds during the entire season. Hence, additional selective post-emergence herbicides are needed, and more so in onions grown from seeds than from transplants, sets or bulbs. Weed competition should be minimized early in the development of the crop. However, young seedlings of both onions and weeds are more susceptible to herbicides than older plants. Therefore, post-emergence treatment is given as soon as the crop has acquired acceptable tolerance to these chemicals.[20,50,55,61,67,70] However, when weed infestation endangers the crop, herbicides are applied even when the onion seedlings suffer some damage, thus allowing reasonable development and acceptable yield, rather than a total yield loss.[21,22]

a. Mode of Selectivity

The response of plants to foliar-applied herbicides is greatly affected by the amount of

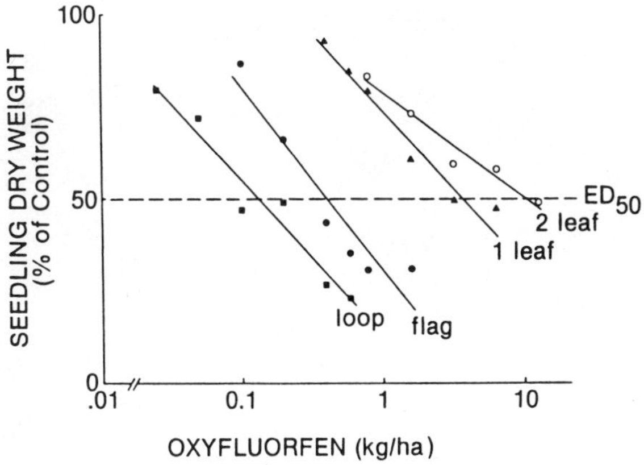

FIGURE 6. Response of onion seedlings, to varying application rates of oxyfluorfen, during early seedling development. (According to Akey and Souza Machado,[55] with permission).

chemical absorbed by the leaves. This is determined by the ability of leaves to intercept and retain the herbicide spray.[67] These traits depend on leaf morphology and surface wetability (see Bucovac,[71] for an in depth review).

Both onions and weeds become more tolerant to foliar applied methazole with age. In onions, the amount of structural crystalline epicuticular wax varies among cultivars,[72] and increases with successive leaves. This results in a larger contact angle between spray droplets and the leaf surface. Hence, a lower spray retention, per unit leaf dry-weight, is observed. This contributes to the increased tolerance to methazole with age. Onion leaves retain smaller amounts of spray, per unit dry-weight, than most weeds, and are thus, more tolerant to the herbicide.[67,73]

Onion tolerance to post–emergence application of oxyfluorfen increases more than 70-fold, from the late loop- to the two fully expended leaf stage of development (Figure 6). This also is attributed to a progressive decrease in spray retention, per unit dry-weight, as a result of a corresponding increase in epicuticular wax with plants' age (Figure 7).[55]

The tolerance of onion seedlings to various foliarly applied herbicides is reduced following preemergence treatment of ethofumesate (Table 2). This adverse effect of ethofumesate is evident with postemergence application of herbicides which interfere with various processes of photosynthesis, but not with pendimethalin which impairs cell division.[66,67]

b. Postemergence Management

Nitrofen has been used world wide for postemergence control of both annual broadleaf and some grass weeds in onion.[35,52,61,74-76] With the removal of this chemical from the market because of toxicological problems, suitable alternative herbicides were sought.

Putnam et al.,[77] reported better control, without yield loss, of common purslane, common chickweed, lady's thumb (*Polygonum persicaria*), and redroot pigweed, with oxyfluorfen (0.28 kg/ha) than with nitrofen (1.2 kg/ha), when applied at the two to three leaf stage. Oxyfluorfen was examined in the U.S. in three north central states, as a replacement for nitrofen on onions grown in muck soils. It was found to be safe to the crop, when applied after the second leaf has developed.[70,78] Oxyfluorfen at 0.3 kg/ha, applied at the two true leaf stage to Yellow Sweet Spanish onions, provided better weed control than that applied at the one true leaf stage of onion growth. Additionally, a combination of oxyfluorfen with bromoxynil resulted in a better weed control than either one applied alone.[53] In contrast, more than 40% yield reduction, following application of 0.25 kg/ha oxyfluorfen at the two-leaf stage,[79] or before four leaves were developed (Rubin and Adler, unpublished). Tem-

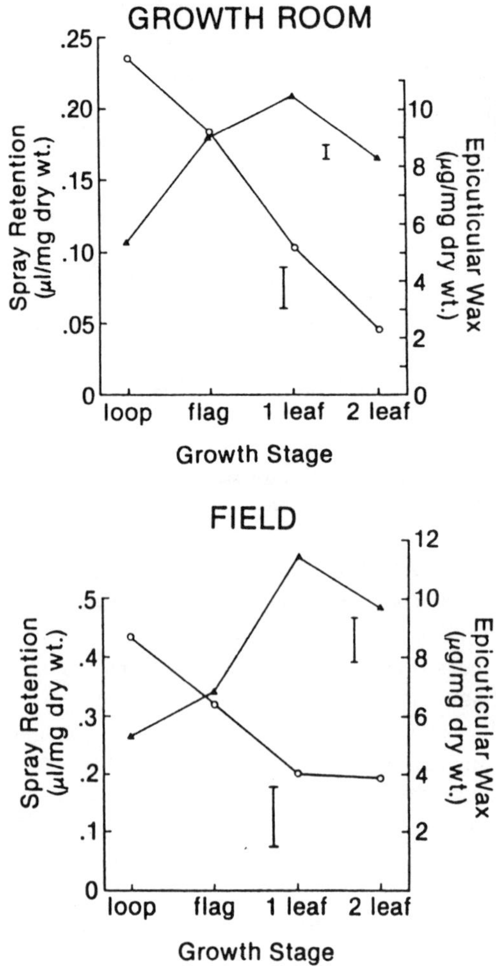

FIGURE 7. Effect of plant age on spray retention and epicuticular wax development of leaves of onion seedlings, under growth room conditions (upper), and under field conditions (lower). Spray retention, (○—○—○); epicuticular wax, (▲—▲). (According to Akey and Souza Machado,[55] with permission).

porary leaf burn was observed when this herbicide was applied at the recommended rate of 0.37 kg/ha to onions at the four leaf stage,[80] or when applied at 0.7 kg/ha to garlic field at 2 or 29 d after planting.[81] Oxyfluorfen effectively controls a wide range of broadleaf weeds, including some perennials such as field bindweed, and annual grasses such as hood canarygrass (*Phalaris paradoxa*) and wild oats (*Avena sterilis*).[82] This herbicide is highly selective when applied on transplanted onions, even at much higher rates,[16,83] and can be applied safely via sprinkler irrigation.[84]

Oxadiazon is at present one of the most commonly used post-emergence herbicide.[35,60,83,85,86] Starting as early as the first true leaf, the chemical is applied at low rates (0.2 kg/ha), either alone or in combination with pendimethalin, without damage to the crop. At the two fully developed leaves stage, higher rates (up to 0.6 kg/ha) are applied with excellent selectivity to the crop.[64] In bulb planted onions in Punjab, India, oxadiazon at 2.0 to 3.0 kg/ha alone, in combination, or in sequence with other herbicides, provided excellent weed control with no damage to the crop.[35] Similarly, oxadiazon, applied at transplanting,

TABLE 2
Effect of Pre-Emergence Treatment with Ethofumesate, (0.8 kg/ha) on the Response of Onion Seedlings of *cv.* Grano 3, to Post-Emergence Treatment with Five Herbicides, Applied at the First Leaf Stage of Development

Treatment		Post-emergence herbicide (kg/ha)				
Pre-emerg.	Post-emerg.	Oxadiazon (0.2)	Bentazon (0.4)	Oxyfluorfen (1.75)	Pendimethalin (0.75)	Phenisopham (0.75)
		Shoot fresh weight (mg/plant)				
−	−	169	179	116	130	182
+	−	165	180	100	118	176
−	+	164	175	113	148	144
+	+	64**	84**	60**	95NS	57**

Note: Statistical analyses were carried out for each post-emergence herbicide separately. Plants were harvested 7 d after foliar application. (Modified according to Rubin et al.)[66]

(**) = significant interaction at $p < 0.01$ respectively. NS = not significant.

or 7 to 14 d later, gave excellent and persistent weed control (75 to 99%). Application of oxadiazon, 14 d prior to planting, resulted in less persistent activity.[16] Oxadiazon is also highly selective pre-emergence herbicide in garlic.[26,87] At post–emergence treatments, this herbicide has an excellent activity in controlling existing weed seedlings, and also has soil activity, thus reducing further infestation of late germinating weeds.

Pendimethalin is widely used in Israel and other countries, as a post-emergence herbicide. It can be applied safely to the crop, at the first true leaf stage, or after transplanting.[16,56,60,83] Earlier application to seeded onion may, however, result in severe stand reduction. Hence, application at the flag leaf stage, resulted in 50% stand reduction as compared to the untreated plots, when a hail storm had destroyed onion foliage, and the recovering seedlings had to re-emerge through a soil layer containing the herbicide.[60] The herbicide effectively controls young (up to two leaf stage) broadleaf weeds, such as lambsquarter, prostrate spurge, and common purslane, but not mallow, redroot pigweed, velvetleaf (*Abutilon theophrasti*)[56] several Umbelliferae, and fleabane.[60] Pendimethalin activity against grasses and broadleaf weeds with more than two leaves, is improved when applied with oxadiazon, oxyfluorfen, or bentazon.

The inadequate control of purple and yellow nutsedge by registered chemicals, is one of the major factors preventing the increase in production of onions in various parts of the world.[13,88] Postemergence treatments with metolachlor, firstly at the two-leaf stage, and followed by a second application 4 weeks later (6 to 8 leaf stage), provided full season control of yellow nutsedge.[88,89] This herbicide, as well as alachlor, which also controls yellow nutsedge, may be included in an onion weed management program.[78] Bentazon also controls this troublesome weed, when applied at high rates. However, onions at early stages are quite sensitive to rates above 0.5 to 1.0 kg/ha of bentazon.[90] Thus, the herbicide can be used against nutsedge in seeded onions, as preharvest treatment only.

One of the major weeds, in many vegetable crops in the Mediterranean region, is fleabane, the seeds of which are easily dispersed by wind. This weed not only competes with the crop, but interferes also with the application of insecticides and fungicides to onion foliage, as well as with mechanical harvest operations. Very poor control is achieved by the available herbicides such as DCPA, and oxadiazon. Post–emergence application of clopyralid (0.04 to 0.06 kg/ha), at the three leaf stage of onions, resulted in very good control of the weed. Only light and temporary crop growth inhibition, and some leaf twisting, were observed, without any yield loss, as compared to the hand weeded control.[23] Neary and Majek[91] reported recently that no crop injury was observed when clopyralid was applied postemergence to *Allium* spp. at a rate (0.42 kg/ha) much higher than that reported from Israel. At 0.21 and 0.42 kg/ha, this herbicide controlled common lambsquarter, common ragweed (*Ambrosia artemisiifolia*), and common cocklbur (*Xanthium strumarium*), but did not control velvetleaf, smooth pigweed (*Amaranthus hybridus*), shepherdspurse, and annual grasses.

The use of other post–emergence herbicides in onions, garlic and leeks, for control of broadleaf weeds, e.g., dinoseb, methabenzthiazuron, bentazon, bromoxynil, ioxynil, linuron, chloridazon (pyrazon), chlorbufam, chloroxuron, and diluted sulfuric acid, is limited to low rates, specific conditions, specific weeds, or to certain regions.[21,22,54,87,92,93.] Usually, these compounds are given as a single application, due to possible crop damage. Under certain circumstances, these chemicals are used only in order to salvage the crop from severe weed infestation, following failures of pre-emergence treatments.

In recent years, repeated applications of oxadiazon in Israel, or ioxynil and ioxynil plus linuron, in England, at one eighth to one quarter the normally recommended dose rates, have been widely used, particularly on muck soils. These treatments commence as early as the one leaf stage, are repeated at the 1.5 and 2 leaf stages and give successful weed control without apparent crop damage.[94] A similar approach was tested with bentazon,[95] but was less successful.

Grass weeds, particularly the perennials, present serious problems to the growth and mechanical harvest of onions and garlic.[21,22] Recently, several new grass herbicides with outstanding selectivity in *Allium* crops, were introduced, e.g., diclofop, alloxydim-sodium, sethoxydim, fluazifop, and haloxyfop. Although differing in the range of grass species controlled, and in the optimal rate required, these herbicides, when applied post-emergence, provide economical and efficient control of annual and perennial grasses.[96-102]

Timing of application is very important, as the purpose of using these herbicides is, primarily, to free the crop from grass weed competition, as early as possible. Since the soil activity of these grass killers is poor, too early an application may result in control of the existing grasses, but not of those emerging later in the season. Thus, one should consider using more than one application, or to split the recommended dose to two applications.[103] The addition of adjuvants to some of these herbicides improved their activity.[100,102] Tank mixture of grass-killers with broad-leaf herbicides should be avoided, as the phytotoxicity of the mixture to the crop may thus increased.[104-106]

In New York State, some onion growers have been interplanting barley on organic soils, between onion rows, in order to reduce wind erosion. Soil particles, when moved by wind, can cause severe abrasion of the young onion seedlings, and these losses are often large enough to necessitate replanting. The barley plants are removed, when protection is no longer needed, by mechanical means or by nonselective herbicides. Application of fluazifop, at the appropriate time, reduces the competition imposed by the barley on the onion seedlings, without reducing the protection against wind damage.[101]

IV. CONCLUSIONS

It is evident, that by competing for light, nutrients, and water, weeds adversely affect the growth of alliaceous crops, particularly at the early growth stages. In order to prevent yield losses, the crop has to be free from weed competition, at least during the critical period. The onset and duration of this period vary in different countries, and depend on the environmental condition prevailing at the time of crop sowing or planting, on crop density, on type and density of weeds infesting the field, and on their growth habit. Although weeds, present beyond the critical period, have no direct effect on the crop yield they may indirectly reduce it, by their interference in the mechanical harvest process.

The major purpose of the available weed control methods is to reduce the competition imposed on the crop by the weeds during the critical period. Combining more than one method is necessary in order to achieve this goal.

Rational combination of cultural practices such as field selection, crop rotation, and seedbed preparation and cultivation, with careful selection and integration of proper herbicide program, may help the grower in his endless struggle with the weeds, and reduce costs.

In spite of the advanced technology achieved in the development of selective herbicides in agriculture, there is no single chemical that, by itself, can successfully provide perfect weed control in alliaceous crops. Properly timed weed control program as a part of integrated pest management, is still necessary, and must include presowing preparation, a pre- and post-emergence herbicide program, together with cultivation and hand weeding.

REFERENCES

1. **Comin, D.,** *Onion Production,* Orange Judd Publishing Company, New York, 1946, 100.
2. **Roberts, H. A.,** Weeds and the onion crop, *J.R. Hortic. Soc.,* 98, 230, 1973.
3. **Hewson, R. T. and Roberts, H. A.,** Some effects of weed competition on the growth of onions, *J. Hortic. Sci.,* 48, 51, 1973.
4. **Shadbolt, C. A. and Holm, L. G.,** Some quantitative aspects of weed competition in vegetable crops, *Weeds,* 4, 111, 1956.
5. **Bleasdale, J. K. A.,** The yield of onions and red beet as affected by weeds, *J. Hortic. Sci.,* 34, 7, 1959.
6. **Hewson, R. T. and Roberts, H. A.,** The effect of seed removal at different times on the yield of bulb onions, *J. Hortic. Sci.,* 46, 471, 1971.
7. **Wicks, G. A., Johnston, D. N., Nuland, D. S., and Kinbacher, E. J.,** Competition between annual weeds and Sweet Spanish onions, *Weed Sci.,* 21, 436, 1973.
8. **Roberts, H. A.,** Weed competition in vegetable crops, *Ann. Appl. Biol.,* 83, 321, 1976.
9. **Thomas, M. N. and Wright, C. J.,** A study of the factors affecting the onset of the critical period of weed competition in the onion crop *(Allium cepa* L.), *Sci. Hortic.,* 35, 94, 1984.
10. **Van Heemst, H. D. J.,** The influence of weed competition on crop yield, *Agric. Systems,* 18, 81, 1985.
11. **Zimdahl, R. L.,** *Weed-Crop Competition A Review,* International Plant Protection Center, Corvallis, Oregon, 1980, 196.
12. **Nylund, R. E., Nelson, D. C., and Dinkel, D. H.,** Comparative costs of weeding onions by hand or with monuron, CIPC, and CDAA, *Weeds,* 6, 304, 1958.
13. **William, R. D. and Warren, G. F.,** Competition between purple nutsedge and vegetables, *Weed Sci.,* 23, 317, 1975.
14. **Williams, C. F., Crabtree, G., Mack, H. J., and Laws, W. D.,** Effect of spacing on weed competition in sweet corn, snap beans, and onions, *J. Am. Soc. Hortic. Sci.,* 98, 526, 1973.
15. **Nieto, J., Brando, M. A., and Gonzalez, J. T.,** Critical periods of the crop growth cycle for competition from weeds, *PANS,* 14, 159, 1968.
16. **Babiker, A. G. T. and Ahmed, M. K.,** Chemical weed control in transplanted onion (*Allium cepa* L.) in the Sudan Gezira, *Weed Res.,* 26, 133, 1986.
17. **Paller, E. C., Jr., Guantes, M. M., Soriano, J. M., and Vega, M. R.,** Duration of weed competition and weed control and yield. II. Transplanted onions, *Philippines Agric.,* 55, 221, 1971.
18. **Rubin, B. and Kigel, J.,** *Cyperus rotundus* and *C. esculentus,* in *CRC Handbook of Flowering,* Vol. 2, Halevy, A. H. Ed., CRC Press, Boca Raton, FL, 1985, 402.
19. **Putnam, R. A., Werner, G. M., and Boldt, P. F.,** Interference between common purslane and onion on organic soil, *Proc. North Central Weed Control Conf.,* 33, 171, 1978.
20. **Menges, R. M. and Tamez, S.,** Response of onion *(Allium cepa)* to annual weeds and postemergence herbicides, *Weed Sci.,* 29, 74, 1981.
21. **Kurtz, E. A.,** Garlic (*Allium sativum),* in *Principles of Weed Control in California,* California Weed Conference, Thomson Publ., Fresno, CA, 1985, 299.
22. **Kurtz, E. A.,** Onions *(Allium cepa),* in *Principles of Weed Control in California,* California Weed Conference, Thomson Publ., Fresno, CA, 1985, 311.
23. **Graph, S., Kleifeld, Y., Herzlinger, G., Bargutti, A., and Keren, Y.,** Control of *Conyza* in onion in the Jordan Valley, *Phytoparasitica,* 13, 255, 1985.

24. **Voss, R. E.,** *Onion Production in California,* Div. Agric. Sci. University of Calif., California, 1979, 29.
25. **Jones, H. A. and Mann, L. K.,** *Onions and Their Allies,* Interscience Publ., New York, 1963. 170.
26. **Rahman, A., James, T. K., and Mortimer, J.,** Weed control and use of herbicides in garlic, *Proc. N. Z. Weed Pest Control Conf.,* 142, 1985.
27. **Campbell, W. F. and Anderson, J. L.,** Effects of no-tillage and herbicides on carrot and onion seed production, *HortScience,* 15, 662, 1980.
28. **Katan, J., Greenberg, A., Alon, H., and Grinstein, A.,** solar heating by polyethylene mulching for the control of diseases caused by soil-borne pathogens, *Phytopathology,* 66, 683, 1976.
29. **Rabinowitch, H. D., Katan, J., and Rotem, I.,** The response of onions to solar heating, agricultural practices and pink-root disease, *Scientia Hortic.,* 15, 331, 1981.
30. **Rubin, B. and Benjamin, A.,** Solar heating of the soil: effect on weed control and on soil-incorporated herbicides, *Weed Sci.,* 31, 819, 1983.
31. **Rubin, B. and Benjamin, A.,** Solar heating of the soil: involvement of environmental factors in the weed control process, *Weed Sci.,* 32, 138, 1984.
32. **Horowitz, M., Regev, Y., and Herzlinger, G.,** Solarization for weed control, *Weed Sci.,* 31, 170, 1983.
33. **Eagly, G. H.,** Reduction of weed seed populations by solar irradiation of soil. *Abs. Weed Sci. Soc. Amer.,* 175, 1981.
34. **Jacobson, R., Greenberg, A., Katan, J., Levy, M., and Alon, H.,** Control of Egyptian broomrape *(Orobanche eagyptiaca)* and other weeds by means of solar heating of the soil by polyethylene mulching, *Weed Sci.,* 28, 312, 1980.
35. **Sandhu, K. S. and Randhawa, K. S.,** Chemical weed control in onion seed crop in Punjab, India, *Trop. Pest Manag.,* 26, 41, 1980.
36. **Rana, M. K., Singh, K. P., and Pandey, U. K.,** Economic yield of onion in Haryana, *Indian J. Agric. Sci.,* 55, 21, 1985.
37. **Menges, R. M.,** Weed seed population dynamics during six years of weed management systems in crop rotations on irrigated soil, *Weed Sci.,* 35, 328, 1987.
38. **Majek, B. A.,** Cost and effectiveness of hand weeding and cultivation compared to labeled and experimental herbicide programs, *Proc. Northeast Weed Sci. Soc.,* 39, 124, 1985.
39. **Lachman, W. H.,** Weeding set onions with chloro IPC, *Proc. Am. Soc. Hortic. Sci.,* 61, 496, 1953.
40. **Binning, L. K.,** Registration and availability of herbicides for minor crops, *Proc. North Central Weed Control Conf.,* 31, 116, 1976.
41. **Ashton, F. M. and Crafts, A. S.,** *Mode of Action of Herbicides,* John Wiley & Sons, New York, 1981, 525.
42. **Duke, S. D.,** *Weed Physiology, II. Herbicide Physiology,* CRC Press, Boca Raton, FL, 1985, 257.
43. **Fedtke, C.,** *Biochemistry and Physiology of Herbicide Action,* Springer-Verlag, New York, 1982, 202.
44. **Hatzios, K. K. and Penner, D.,** *Metabolism of Herbicides in Higher Plants,* Burgess Publishing, Minneapolis, MN, 1982, 142.
45. **Menges, R. M. and Hubbard, J. L.,** Influence of climate and soil incorporation on herbicides in onions, *Proc. Am. Soc. Hortic. Sci.,* 92, 498, 1968.
46. **Menges, R. M. and Tamez, S.,** Effect of soil incorporation on selectivity, movement, and persistence of herbicides in onion plantings, *J. Am. Soc. Hortic. Sci.,* 98, 390, 1973.
47. **Alspaugh, K. R., Abernathy, J. R., Lipe, W. N., and Gipson, J. R.,** Response of onion varieties and seed placement to herbicides, *Proc. South. Weed Sci. Soc.,* 36, 211, 1983.
48. **Davis, G. N.,** *Onion Production in California,* Div. Agric. Sci. University of California, 1957, 12.
49. **Binning, L. K., Hughes, R. L., Michaelis, B. A., and Bewick, T. A.,** Weed management in onion, *Res. Rep. North Central Weed Control Conf.,* 43, 15, 1986.
50. **Dubitzki, E.,** Post-emergence weed control in onion, *Proc. 5th Conf. Weed Sci. Soc. Israel,* 120, 1973.
51. **Luz, E., El-Al, G., and Alon, Y.,** Effect of mechanical incorporation on DCPA activity, *Proc. 4th Conf. Weed Sci. Soc. Israel,* 2, 212, 1970.
52. **Sanok, W. J., Selleck, G. W., and Creighton, J. F.,** Herbicide trial in onions-a three-year summary, *Proc. Northeastern Weed Sci. Soc.,* 34, 407, 1980.
53. **Wilson, R. G.,** New developments in onion weed control, *Abs. Weed Sci. Soc. Amer.,* 24, 1988.
54. **Sinha, T. D. and Lagoke, S. T. O.,** Pre-transplant herbicides for weed control in irrigated onion *(Allium cepa* L.) in northern Nigeria, *Crop Prot.,* 2, 455, 1983.
55. **Akey, W. C. and Souza Machado, V.,** Response of onions *(Allium cepa)* to oxyfluorfen during early seedling development, *Can. J. Plant Sci.,* 65, 357, 1985.
56. **Warholic, D. T., Ellerbrock, L. A., and Bellinder, R. R.,** Control of weeds in muck onions and diethatyl-ethyl, metolachlor and pendimethalin, *Proc. Northeastern Weed Sci. Soc.,* 41, 120, 1987.
57. **Anon.,** *Stomp Herbicide,* American Cyanamid Company, New Jersey, 32, 1981.
58. **Dobrzanski, A. and Szwonek, E.,** Effect of herbicides on the uptake of mineral elements by onions, *Biul. Warzywniczy,* 27, 381, 1984.

59. **Esau, R. and Rumney, V.,** Weed control in vegetables crops, *Ann. Rep. Alberta Hortic. Res. Center,* Brooks, Alberta, Canada, 69, 1985.
60. **Sofer, S. and Gardji, G.,** Selective weed control in onion, 1978-80, *Phytoparasitica,* 10, 269, 1982.
61. **Boldt, P. F. and Putnam, A. R.,** Onion and weed response to several newer herbicides, *Proc. North Central Weed Control Conf.,* 31, 107, 1976.
62. **Lifshitz, N.,** Pre-emergence treatments with ethofumesate for the control of grasses and other weeds in seeded onions, *Phytoparasitica,* 4, 155, 1976.
63. **Sofer, S.,** Evaluation of pre-emergence treatments with ethofumesate and chlorthal-dimethyl followed by post-emergence application with oxadiazon in onions, *Phytoparasitica,* 4, 155, 1976.
64. **Eshel, Y., Ohali, I., and Baduiah, G.,** Selectivity of ethofumesate to onion, *Phytoparasitica,* 7, 150, 1979.
65. **Adler, U.,** Effect of Pre-Emergence Treatment with Ethofumesate on the Selective Response of Onion *(Allium cepa L.)* to Herbicide: Possible Mechanism, M.Sc. thesis, Hebrew University of Jerusalem, Rehovot, 1983, 37.
66. **Rubin, B., Adler, U., Varsano, R., and Rabinowitch, H. D.,** Effect of ethofumesate on the epicuticular waxes of onion leaves, and on the response of plants to foliage-applied herbicides, *Ann. Appl. Biol.,* 108, 365, 1986.
67. **Verity, J., Walker, A., and Drennan, D. S. H.,** Aspect of the selective phytotoxicity of methazole. I. Measurements of species response, spray retention and leaf surface characteristics, *Weed Res.,* 21, 243, 1981a.
68. **Leavitt, J. R. C., Duncan, D. N., Penner, D., and Meggitt, W. F.,** Inhibition of epicuticular wax deposition on cabbage by ethofumesate, *Plant Physiol.,* 61, 1034, 1987.
69. **Duncan, D. N., Meggitt, W. F., and Penner, D.,** Basis for increased activity from herbicide combinations with ethofumesate applied on sugarbeet *(Beta vulgaris), Weed Sci.,* 30, 195, 1982.
70. **Boldt, P. F., Binning, L., and Putnam, A. R.,** Onion weed control with oxyfluorfen in Minnesota, Wisconsin and Michigan, *Proc. North Central Weed Control Conf.,* 36, 62, 1981.
71. **Bukovac, M. J.,** Herbicide entry into plants, in *Herbicides Physiology Biochemistry Ecology,* Vol. 1, Audus, L. J., Ed., Academic Press, London, 1976, 335.
72. **Ferreira, P. V. and Costa, C. P.,** Effects of foliar wax on the reaction of onion varieties *(Allium cepa* L.) to post-emergence herbicides, *Planta Daninha,* 5, 29, 1982.
73. **Verity, J., Walker, A., and Drennan, D. S. H.,** Aspects of the selective phytotoxicity of methazole. III. Behaviour in plants following foliar treatment, *Weed Res.,* 21, 317, 1981.
74. **Gornick, A. and Zur, Y.,** Use of nitrofen to control weeds in transplanted onions, *Phytoparasitica,* 1, 75, 1973.
75. **Kempen, H. M.,** Research results on post-emergence herbicides in onions, *Proc. West. Soc. Weed Sci.,* 28, 51, 1975.
76. **Lang, A., Hoyle, B., and Tyler, K.,** Early postemergence annual weed control in direct seeded onions, *West Soc. Weed Sci. Res. Prog. Rep.,* 91, 1973.
77. **Putnam, A. R., Riley, J. E., and Chase, W. R.,** Onion and weed response to postemergence applications of oxyfluorfen, *Proc. North Central Weed Control Conf.,* 35, 95, 1980.
78. **Binning, L. K., Michaelis, B., and Hughes, R. L.,** Post emergence herbicides for onion weed control, *Proc. North Central Weed Control Conf.,* 37, 94, 1982.
79. **Souza Machado, V. and Kitchen, D.,** Onion I weed control, *Res. Rep. Eastern Sect. Exp. Comm. Weeds,* 378, 1982.
80. **Abada, M.,** The effect of wetting on the activity of oxyfluorfen sprayed on onions, *Phytoparasitica,* 13, 255, 1985.
81. **Nortje, P. F. and Henrico, P. J.,** Evaluation of herbicides for long-term weed control in garlic *(Allium sativum* L.), *South Afr. J. Plant Soil.,* 2, 85, 1985.
82. **Abada, M., Ziplievitch, Y., and Gornik, A.,** Tests with oxyfluorfen for the control of annual weeds in onions, *Phytoparasitica,* 10, 269, 1982.
83. **Malik, Y. S., Singh, K. P., and Pandita, M. L.,** Chemical weed control in onion *(Allium cepa* L.), *Indian J. Weed Sci.,* 14, 24, 1982.
84. **West, L. D., Hildreth, R. C., Jehle, M. F., and Schlesselman, J. T.,** Application of oxyfluorfen in irrigation water, *Proc. West. Soc. Weed Sci.,* 93, 1984.
85. **Boshvitz, Y.,** Lasting effect of oxadiazon in the soil and its influence on different crops in rotation, *Phytoparasitica,* 1, 84, 1973.
86. **Sharma, A. B., Shreehari, G., Tiwari, J. P., and Nair, P. K. R.,** Effect of oxadiazon and mulching on weed control in garlic, *Indian J. Weed Sci.,* 16, 194, 1984.
87. **Sharma, A. B., Patel, R. K., and Tuvari, J. P.,** Chemical weed control in garlic, *Indian J. Weed Sci.,* 15, 17, 1983.
88. **Senesac, A. F. and Sieczka, J. B.,** Yellow nutsedge control in onions grown on mineral soil, *Proc. Northeastern Weed Sci. Soc.,* 41, 121, 1987.

89. **Powell, A. W. and Hoxworth, G.,** Yellow nutsedge control in onion with metolachlor, *West. Soc. Weed Sci. Res. Prog. Rep.,* 131, 1987.
90. **Lifshitz, N., Gur, Y., Shlevin, E., and Alon, Y.,** The use of bentazon for the control of *Cyperus rotundus* and *Tibulus terrestris* in peanuts and of *Ammi* species in onions, *Phytoparasitica,* 4, 150, 1976.
91. **Neary, P. A. and Majek, B. A.,** Clopyralid for broadleaf weed control in cole crops, cucurbit, and onion species, *Abs. Annu. Meet. Weed Sci. Soc. Am.,* 24, 1988.
92. **Scheffer, J. J. C. and Wilson, G. J.,** Early season weed control in onions, *Proc. 35th N.Z. Weed Pest Control Conf.,* 35, 304, 1982.
93. **Wilson, G. J. and Scheffer, J. J. C.,** Pre- and early post-emergence weed control in onions, *Proc. 6th Aust. Weed Control Conf.,* 1, 221, 1981.
94. **Anon.,** Getting pre-emergence control right is key to onion low-dose technique, *Arable Farming,* 11, 87, 1984.
95. **Lifshitz, N.,** Combination of pre- and post-emergence herbicide treatments in seeded onions, *Hassadeh,* 57, 437, 1977.
96. **Boshvitz, Y.,** Application of dichlofop methyl for grass control in onion, *Phytoparasitica,* 7, 149, 1979.
97. **Friedland, S.,** Control of gramineous weeds in garlic, *Phytoparasitica,* 10, 270, 1982.
98. **Binning, L. K.,** Efficacy and crop safety of postemergence grass herbicides-vegetables, *Proc. North Central Weed Control Conf.,* 149, 1984.
99. **Hartley, M. J.,** Selective grass weed control, *Proc. N. Z. Weed Pest Control Conf.,* 183, 1984.
100. **Johnson, J. R. and Hopen, H. J.,** Response of tomato *(Lycopersicon esculentum)* and onion *(Allium cepa)* to several postemergence grass herbicides, *Weed Sci.,* 32, 168, 1984.
101. **Lanterman, W. S., Warholic, D. T., Ellerbrock, L. A., and Stachowski, P. J.,** Effects of fluazifop and barley windbreak competition on seeded onions in organic soil, *Proc. Northeast. Weed Sci. Soc.,* 39, 184, 1985.
102. **Lunsford, J.,** Effect of herbicide rate, additives, growth stage and competitive crop on the control of Texas panicum with fluazifop-butyl, *Proc. South. Weed Sci. Soc.,* 36, 149, 1983.
103. **Watkins, S. D. and Hargrave, M. R.,** Control of annual grasses in onions with fluazifop-butyl, *Proc. West. Soc. Weed Sci.,* 70, 1984.
104. **Colwell, S. G.,** The response of seedling onions to fluazifop-butyl and bromoxynil, *Proc. West. Soc. Weed Sci.,* 161, 1984.
105. **Giannopolites, K. N.,** New possibilities for post-emergence weed control in seed onions, *Zizaniologia,* 1, 97, 1983.
106. **Hartzler, R. G. and Taber, H. G.,** Effect of oxyfluorfen and sethoxy dim combinations on onion yield, *Proc. North Central Weed Control Conf.,* 38, 72, 1983.

Chapter 5

MYCORRHIZAL ASSOCIATIONS AND THEIR SIGNIFICANCE

David P. Stribley

TABLE OF CONTENTS

I. INTRODUCTION

In their acclaimed book *Fleas, Flukes and Cuckoos* Rothschild and Clay[1] vividly portrayed the vertebrate animal as host to an impressive range of small invertebrate animals whose activities range from the parasitic to the benign. Likewise, the microbiologist views the vascular plant not as an unsullied hierarchy of cells but rather as a composite structure that is invaded by a wide variety of microbes. One of the most important plant-microbe relationships is that formed between roots and certain soil fungi, in which the fungus forms an intimate and stable association. In the plant kingdom as a whole these dual organisms are very diverse in structure[2] but are known by the blanket term "mycorrhiza" (from the Greek: "fungus root"). The purpose of the present chapter is to make the reader aware of the existence and great physiological significance of mycorrhizae in *Allium*.

Species of *Allium* have been widely used in research on mycorrhizae, not because of an interest in *Allium* physiology per se, but because plants of this genus are ideal for laboratory studies on mycorrhizae, for they grow relatively slowly, possess a simple root structure and are very physiologically responsive to colonization by mycorrhizal fungi. The happy consequence of this is that we know more about effects of mycorrhizae on *Allium* than about their effects on any other genus (except perhaps for *Trifolium*). This means, however, that a thorough review of mycorrhizae of *Allium* could cover most of what is known about the biology and function of mycorrhizae, and as such would be inappropriate in the present context. The scope of this essay will thus be restricted to a description of the salient features of the biology of mycorrhizae in *Allium*, a brief review of their physiological effects and some speculation about the possible commercial application of mycorrhizal inoculants.

II. GENERAL BIOLOGY OF *ALLIUM* MYCORRHIZAE

A. WHAT ARE VESICULAR-ARBUSCULAR MYCORRHIZAE?

Alliums possess but one type of mycorrhiza: vesicular-arbuscular (VA) mycorrhiza. The fungal partner is a species of the *Endogonaceae* and these fungi are ubiquitous in soils from an extremely wide range of habitats from deserts to the tundra. Inoculum is present variously as spores, infected fragments of roots and hyphae. The invading fungus is present both within the root tissues, where it is restricted to the cortex (Figure 1a), and is continuous with an external mycelium (Figure 2) which ramifies in the surrounding soil and bears asexual chlamydospores. This external mycelium can extend up to 8 cm from roots of *Allium*,[3] but it should be noted that the distribution of hyphal length, as a function of distance from the root, is entirely unknown. The internal phase needs special staining and clearing techniques[4,5] before it can be seen easily. In *Allium* it ramifies almost exclusively in the intercellular spaces with each space being occupied by several hyphae (Figure 1a) so that a squash of the root shows an obviously parallel disposition of hyphae (Figure 1b). This internal phase forms two unique structures, vesicles and arbuscules, after which this type of mycorrhiza is named. Vesicles are intercalary or terminal swellings, rich in oil droplets, whose function is unknown. Arbuscules are formed by invagination of the host cells by the fungus which branches dichotomously and remains outside the plasmalemma of the host. The host cell responds by a massive hypertrophy of cytoplasm, which increases in volume some 20- to 30-fold.[6] The arbuscule is an evanescent structure with a life span of only several days and several arbuscules may form in one cell at a time. The analogy of arbuscules with haustoria of biotrophic parasitic fungi, such as mildews, is obvious and suggests that they are the main site of nutrient interchange between host and fungus, but this has not yet been proven.[7] For a detailed and excellent account of the morphology of VA mycorrhizas the reader is referred to the paper of Bonfante-Fasolo.[8]

The fungal partner of VA mycorrhizae has not yet been cultured axenically.[9] Only

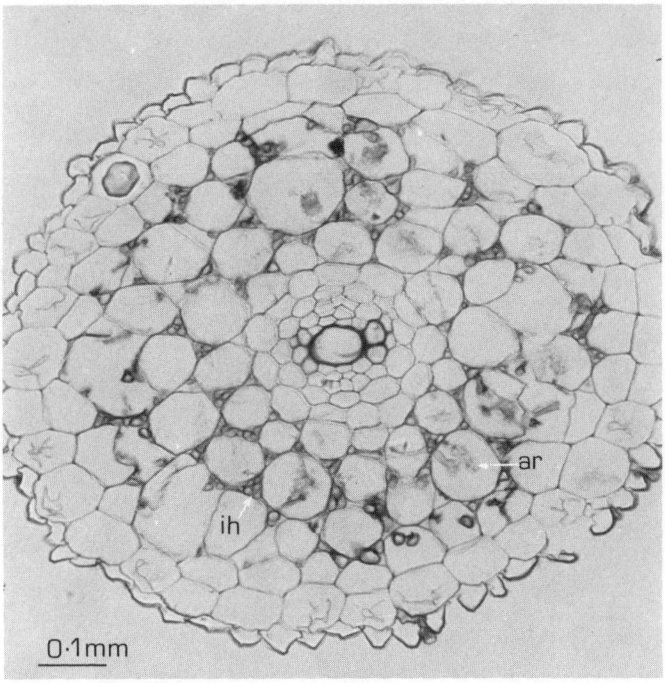

a

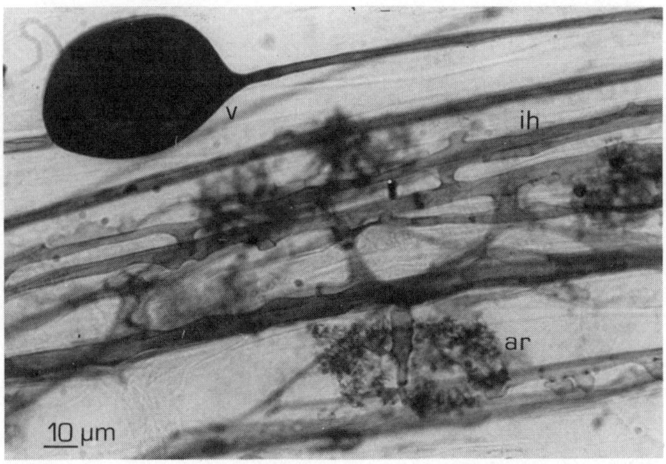

b

FIGURE 1. (a) Transverse section of a main axis of an adventitious root of leek (*Allium porrum*) colonized by the VA mycorrhizal fungus *Glomus mosseae*. Note that the internal hyphae (ih) occupy the intercellular spaces and that the fungus is restricted to the cortex. Cortical cells are penetrated by fungal arbuscules (ar). (b) Squash of a sample of the same root which has been cleared with KOH and stained with Trypan Blue. Note vesicles (v) and arbuscules (ar) and parallel disposition of the internal hyphae (ih). Photographs kindly donated by Dr. F. Amijee.

limited growth of hyphae from germinated chlamydospores has been obtained, but it is notable that such growth on water agar can be markedly stimulated by addition of reduced sulfur compounds.[10]

VA mycorrhizae are present in almost all species of plants, with the notable exception of the *Cruciferae* and *Chenopodiaceae*, most species of which seem to be inherently immune

FIGURE 2. External mycelium of a leek (*Allium porrum*) *Glomus mosseae* mycorrhiza. Photograph kindly donated by Dr. F. Amijee.

to colonization. They are also absent from forest trees of the *Cupuliferae* which bear a type of mycorrhiza known as ectomycorrhiza (or sheathing mycorrhiza) where the root is externally invested by the fungus.[2] It is a reasonable proposition that VA mycorrhizae constitute the most important symbiosis on earth, both in terms of absolute mass, and also of ecological and economic importance.

B. OCCURRENCE IN *ALLIUM*

The fungal partner of VA mycorrhizae has a very low host specificity. Species of *Allium* are readily colonized and thus it seems that under natural conditions VA mycorrhizae are widely present in *Allium*, but the degree of colonization of the host root will depend on the particular environment (see Section II.C.3).

Mycorrhizae have been recorded in all important crop species of *Allium*, e.g., *A. cepa, A. porrum, A. sativum, A. schoenoprasum,* etc. Harley and Harley[11] have compiled a bibliography of references to the occurrence of mycorrhizae both in the British flora and in Europe and this includes a useful section on *Allium*.

In an historical context it is interesting to recall that one of the earliest observations of what we now know as VA mycorrhiza was made in a species of *Allium (A. scorodoprasum)* as early as 1884 by Mollberg.[12]

C. COLONIZATION OF ROOTS

1. Importance

As implied above, VA mycorrhizae are formed *de novo* by colonization of the developing root system by hyphae from existing mycorrhizae or propagules of VA mycorrhizal fungi present in soil; the infection is not systemic. The most important physiological effect of VA mycorrhizae on the host is to increase the rate of uptake of phosphorus (P) from a P-deficient soil (see Section III.A.1), and, while many physiological attributes of the fungus may influence this process, there is strong evidence that the speed of colonization of a developing root system is a very large determinant of the phosphorus uptake response.[13,14] For annual and biennial species such as alliums this rate of colonization is of particular importance. The elucidation of processes which control the colonization of developing root systems by VA mycorrhizal fungi is therefore of paramount importance in understanding the response of the hosts, and considerable research effort has been put into it. Here alliums have been the host of choice and it is particularly interesting to note that as long ago as 1904 Gallaud[15] pointed out that onion roots were excellent material for the study of mycorrhizal colonization.

2. Infection Processes

Hyphae from a propagule of a VA mycorrhizal fungus appear to show little chemotropism when approaching a root. At the point of contact between hypha and root an appressorium forms and this may be preceded by branching of the hypha to form a fan-like structure.[8] The invading hyphae proceed mainly intracellularly and spread distally and proximally through the root cortex via the intercellular spaces. Cox and Sanders[6] considered that a discrete length of infection of about 5 mm was formed from a single entry point and they termed this "the infection unit". The time sequence of formation of fungal structures of an infection unit in leek has been clearly described by Brundrett et al.[15a] Further entry points are formed from growth of the external mycelium, and on current evidence this seems to be the mechanism by which the fungus extends linearly along the cortex, but the histological details have never been fully examined. Amijee et al.[16] showed that if inoculum was placed in a discrete layer, then the cortex of leek roots that penetrated this layer became invaded circumferentially by the fungus. It rapidly attained a maximum density within the cortex and then spread distally and proximally as a unit of constant density, termed a "segment", which was sharply demarcated by the advancing "fronts".[17] When two opposing fronts (initiated by separate bands of inoculum) met, there was no increase in hyphal density, implying either that opposing hyphae had anastomosed or that they overlapped each at a reduced density. Root-to-root infection undoubtedly does occur but there is evidence that this process is slow compared to linear spread.

The susceptibility of roots of leek to colonization by VA mycorrhizal fungi in relation to distance from the root tip was studied by Hepper.[18] Seedlings were grown for up to 32 d on partly sterilized soil and transplanted to soil heavily infested with inoculum of *Glomus mosseae*. All parts of the root, except the extreme tip, and basal regions of adventitious roots, become more or less equally colonized. However Amijee,[19] using an apparatus in which the entire length of the root of leek could be simultaneously challenged with inoculum, showed that the probability of colonization of a main axis of a root was greatest at about 12 cm from the tip. Thus, there was a definite "window of probability" for primary colonization.

It must be emphasized that most studies on VA mycorrhizae record the length of infected root or the ratio of the length of infected root:total length of root ("fractional infection"), to express the degree of colonization, and hence internal biomass of the fungus is rarely measured. Becker and Gerdemann[20] noted that mycorrhizal colonization of onion roots caused a yellow discoloration, and they suggested that the formation of this pigment could be used as a basis for estimating colonization. However, this method had already been shown, as long ago as 1924,[21] to be erroneous because the yellow pigmentation is not confined to the colonized zone. Methods based on assay of chitin as a unique fungal component have been developed to measure fungal mass and look promising; these show that in a typical mycorrhiza the fungus accounts for about 10% (dry weight basis) of the mass of the host root.[22]

Recent techniques involving identification of isoenzyme banding patterns revealed by polyacrylamide gel electrophoresis now make it possible to identify the fungal endophyte present in a VA mycorrhiza.[23] These techniques are being used to study competition between different species of VA mycorrhizal fungi colonizing roots of leeks.[24]

3. Environment and Colonization

The rate of colonization of roots by VA mycorrhizal fungi is greatly influenced by many factors of the chemical and physical environment of the plant. Of the chemical factors it is the concentration of P in soil which has the most marked and well documented effects. It has been known for at least 20 years that high concentrations of soil P inhibit mycorrhizal colonization. Mosse[25] extensively investigated effects of P on mycorrhizae in *A. cepa* and showed that when more than about 250 mg P/kg were added, mycorrhizae of transplanted infected plants tended to die out and the morphology of the infection became abnormal, for formation of arbuscules was inhibited and hyphae were attenuated. Certain morphologically distinct cells in the subepidermal layer only were infected. Mosse suggested that the antomical evidence implied that the decline of infection was a result of high concentration of P in the host cell. This was later confirmed in an elegant experiment by Sanders,[26] who showed that injection of hollow leaves of *A. cepa* with phosphate solution strongly inhibited the progress of colonization. It has recently been shown in *A. cepa* and *A. porrum* that the relationship between applied soil P and total length of infected root follows a step function; at a critical concentration of soil P the length of infected root is sharply decreased.[19,27] Amijee et al.[28] have shown that, at this critical concentration of P, for *A. porrum* infected from a band of placed inoculum of *Glomus mosseae,* there is a sharp increase in the delay in initial infection of the root, an approximate halving of the rate of linear extension of the infection segment, and a reduction in the internal mass of fungus. Further investigation of the effect of P should be most rewarding for it may elucidate the nature of the host control over the mycorrhizal endophyte.

Effects of fungicides are of interest to the student of mycorrhizae in *Allium* crops. Soil drenches of benomyl and thiophanate-methyl reduced P uptake by inoculated onion plants grown in irradiated soil, but were ineffective for plants on unsterile soil.[29] The failure of fungicide in the latter soil was attributed to breakdown by the general microbiological population. Jabaji-Hare and Kendrick[30] tested the two systemic anti-oomycete fungicides fosetyl (Al salt) and metalaxyl on leek. Both colonization and growth of inoculated seedlings were reduced by soil drenches of metalaxyl, but were increased by foliar application of fosetyl. Effects of fungicides on mycorrhizal colonization of *Allium* crops do not seem to have been tested in the field at realistic rates of application.

Of the physical conditions that affect colonization, effects of temperature and light have been tested, but little is known about effects of water potential of the soil. Furlan and Fortin[31] found that under low irradiance colonization of onion by *Gigaspora calospora* was increased. Hayman,[22] working with the same host, found that effects of light on colonization by *Glomus mosseae* were mainly qualitative, in that arbuscules were generally smaller at lower irradiance.

At low temperature the longitudinal spread of *Glomus caledonium* in roots of onion was reduced and the mycelium became much less branched and bore fewer arbuscules.[33] It is important to note when considering temperature that most laboratory studies on the physiology of the mycorrhizal association are done at unrealistically high temperatures. In northern temperate latitudes, for example, soil temperature is in the range of 0 to 15°C whereas many experiments on mycorrhizae use soil temperatures in excess of 20°C. Further investigation into temperature effects and especially its interaction with phosphate nutrition in *Allium* crops are urgently needed.

4. Dynamics

In many studies on mycorrhizae the extent of colonization is expressed as fractional infection, and often this is recorded at a single arbitrary date of harvest only. This is unfortunate because it overlooks the highly dynamic process of colonization in which the developing root system is colonized by an endophyte which is itself growing within the root. Tinker[34] has argued that mathematical modeling is a powerful tool for analysis of such dynamic systems. Mathematical models of colonization processes in onion and leek have been developed at laboratories in Leeds U.K. and at Rothamsted Experimental Station U.K. The aims of these models are to elucidate processes that govern colonization, and particularly to distinguish between effects on the host and on the fungus.

Sanders and Sheikh[35] modeled colonization of onion roots by assuming that the rate of increase of entry points of the VA mycorrhizal fungus with time was exponential. They did not attempt to distinguish between primary infection (i.e., from soil-borne propagules) and secondary infection (infection arising from an already infected piece of root).[36] They also assumed that the rate of longitudinal growth of the fungus within the cortex was constant. Using this model they were able to predict with some success the growth response of the host to colonization on a P-deficient soil (see Section III.A.1). Buwalda et al.[17] considered the special case of roots that had become infected by passing through a layer of banded inoculum. They derived a model which was formally similar to the logistic equation but which, although convenient for preliminary analyses, lacked mechanistic content.[34] In a later more mechanistic model[37] it was shown that segments of infection extended down the root at a constant rate over long periods of time and that this rate of extension was approximately half that of the roots themselves. This model only considered non-branched adventitious roots but it has been extended to include branched roots by Amijee et al.[28]

It must be emphasized that all current models of mycorrhizal colonization are expressed in terms of length of infected root only and take no account of internal biomass of the endophyte.

III. EFFECTS ON HOST PHYSIOLOGY

Two important points must be noted. First, it is essential to realize that nearly all experiments on effects of VA mycorrhizae on the physiology of their hosts have been done on sterilized or partly sterilized media. This is because of the need to eliminate indigenous mycorrhizal fungi in order to grow uninfected control plants. Soils thus treated differ chem-

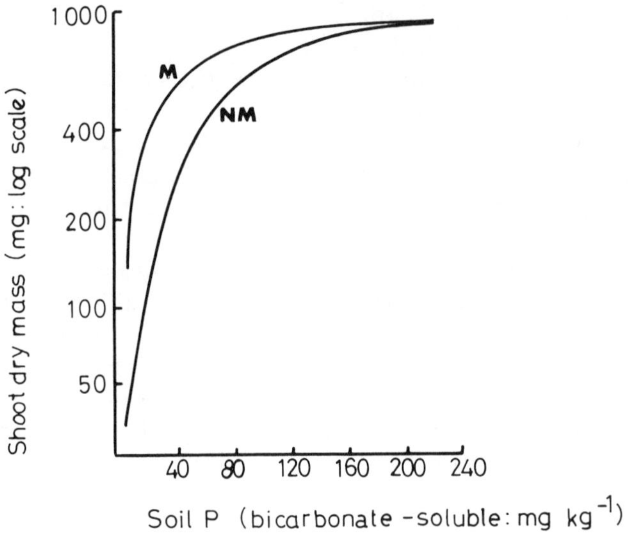

FIGURE 3. Relationship between yield of shoot dry matter of leek (*Allium porrum*), colonized (M) or not (NM) by the VA mycorrhizal fungus *Glomus mosseae*, grown in the glasshouse on γ-irradiated clay-loam soil supplemented with added P to give a wide range of soil bicarbonate-soluble P. Figure adapted from Stribley et al.[48]

ically and biologically in many important ways to untreated soil.[38,39] Great caution therefore must be exercised in extrapolating results of such experiments to the field situation.[40] For example, Warnock et al.[41] found that *Collembola* could graze the external mycelium of leek, thereby greatly reducing the efficacy of mycorrhiza in improving P uptake. Such fungivorous fauna may be important in modifying the physiology of mycorrhiza in the field.[42] However, most worrying are the recent reports that VA mycorrhizae may reduce manganese toxicity.[43] Since soluble manganese is released in large quantities by soil sterilization[38,39] many experiments in the literature may have been confounded by Mn/VA mycorrhiza interactions. This is a topic worthy of careful future consideration.

Second, I have tried to avoid using the term "symbiosis" to describe the VA mycorrhizal association because, although acceptable when used in the wide sense of de Bary to mean a prolonged and intimate association between two partners, it implies in the minds of many a *mutualistic* relationship with each partner benefitting from the association. There is no general agreement at present on the preferred use of "symbiosis".[44] Indeed it is important to emphasize that the physiological significance of VA mycorrhizae may vary from mutualism, to parasitism, to benign association, depending on the environmental circumstances.

A. MINERAL UPTAKE

1. Phosphorus

There is general agreement that the most important physiological effect of VA mycorrhizae is to increase the rate of uptake of P by the host from P-deficient soils.[2] It has been shown repeatedly in pot studies and field studies, both on sterilized and nonsterilized soil, that inoculation with VA mycorrhizal fungi can cause spectacular increases in dry weight of *Allium* species, particularly in onion and leek.[45] It is not immediately clear why inoculation of unsterile soils gives a response, for here the roots are normally found to be colonized by indigenous VA mycorrhizal fungi.[46,47] It may be that the inoculant fungus is inherently more effective than are the indigenous fungi, but more probably it is because inoculation increases the inoculum density and hence the rate of colonization of the root system. Figure 3 shows a typical experiment with leek. The magnitude of the mycorrhizal effect clearly diminishes with increased addition of P to the medium. This is almost certainly due, in part, to the elimination of infection by P (see Section II.C.3) and partly due to the progressive decline in the response of the host to added P. Under the ideal conditions of the pot experiment described in Figure 3 the maximum effect of mycorrhizal inoculation was equivalent to about 75 mg/kg of added P,[48] illustrating the very great responsiveness of *Allium* to VA mycorrhizae. Complete response curves for *Allium* with and without VA mycorrhizae (as in Figure 3) are not available from field studies, so it is not yet possible to infer how effective mycorrhizae are under field conditions. It is instructive, however, to consider the field trial of Yost and Fox[49] who compared plants naturally infected by indigenous mycorrhizal fungi in the soil to those growing on plots sterilized by methyl bromide. Here the ''mycorrhizal effect'' was equivalent to about 0.03 to 0.04 mg P/ml of soil solution, that is about equivalent to an additional 250 kg/ha P.

There is now wide agreement about the underlying mechanism of the effect of VA mycorrhizae and, as this has been discussed in detail in several texts,[2,7] it will be described only briefly here. Phosphorus is present in soil solution at very low concentrations (in the order of 15 to 20 μM in agricultural soils[50]) and mass flow to the plant root is unable to meet the plant demand. Phosphorus diffuses very slowly in soil and diffusion to the root becomes the rate limiting step. Thus, around the absorbing roots a depletion zone quickly develops and the rate of absorption by the root becomes very slow. The external hyphae of VA mycorrhizae extend beyond this depletion zone, absorb phosphorus from the surrounding soil[51] and translocate it to the plant (probably in the form of polyphosphate granules)[52] where it is transferred to the host. This hypothesis was propounded in seminal papers by Sanders and Tinker[53-55] who showed that nonmycorrhizal roots of onion were absorbing P at the maximum rate that could be supported by the diffusive processes of the soil. They found that the inflow of P (flux - mass per unit time - per unit length of root) to mycorrhizal roots was about four times that of nonmycorrhizal roots. They suggested that this extra inflow could only be accounted for by uptake and translocation of P by fungal hyphae of the mycorrhiza. In the 15 years or so since this hypothesis was suggested it has never been seriously challenged. It needs to be refined to include, for example, effects of overlap of depletion zones of individual hyphae, and of maturation of hyphae. However, the recent finding by Jurinak et al.,[56] that external hyphae of VA mycorrhizal *Pascopyrum smithii* can secrete oxalate and thereby increase the solubility of Ca apatite, suggests that the hypothesis may be too simple. Studies which have compared the specific activity of P in tissues of mycorrhizal and nonmycorrhizal plants, grown on soils labeled with ^{32}P, have shown that mycorrhizal plants did not have access to nonlabile P.[57] However, such studies are not capable of testing whether mycorrhizae could alter the equilibrium between solid phase and solution P.[57,58]

The large responsiveness of P uptake in *Allium* plants to mycorrhizal colonization is probably largely attributable to the low efficiency with which their roots take up soil P. This, in turn, is probably related to a low specific root length[59] and to paucity of root hairs.[60] At present, however, it is not possible to explore this problem more deeply because the attributes of the root system which make it efficient in P uptake are even now incompletely understood.[61]

2. Other Elements

Rhodes and Gerdemann[3] demonstrated that external hyphae of onion mycorrhizae were able to take up and translocate sulfur to the host but under natural conditions this is unlikely to be an important phenomenon because mass flow is usually sufficient to supply plants with enough sulfate.

Potassium uptake was increased by mycorrhizal colonization of leek but the internal concentration expressed on a cell water basis was unchanged.[62] It seems that here the improved uptake of potassium was a secondary effect of better P nutrition and growth in the mycorrhizal plant, and that the plant took up more K to maintain a constant vacuolar concentration of this important osmoticum.

Again, increased uptake of nitrogen by mycorrhizal onion plants seemed to be a consequence of improved P nutrition and there was no evidence for a direct physiological effect of mycorrhizae.[63] Smith et al.[64] showed that the fungal partner of the *G. mosseae*/onion association possessed glutamine synthetase activity. They suggested that VA mycorrhizal fungi can assimilate ammonium via this enzyme and that this ability may be important in increasing uptake of nitrogen concomitantly with P.

Little is known about uptake of trace elements by VA mycorrhizae. Tinker and Gildon[65] pointed out that metals such as copper, zinc, and manganese have low mobility in soil and thus VA mycorrhizae may enhance their rate of uptake. Gildon and Tinker[66] unequivocally demonstrated that VA mycorrhizae could increase uptake of copper by leeks from deficient soil, and that this effect was not a simple consequence of improved phosphorus nutrition. The role of mycorrhizae in uptake of trace elements by *Allium* crops deserves further study.

B. CARBON PHYSIOLOGY

It is widely agreed, and there is evidence from experiments with ^{14}C,[67,68] that the VA mycorrhizal endophyte obtains its carbon supply from its host. However, neither the chemical form transferred nor its site of transfer are known. The fungal endophyte is rich in lipids[69] and these may be an important sink for assimilated carbon.[68] The putative "carbon demand" by the fungus has led to speculation that there is a real cost in terms of carbon to the host. Thus, Stribley et al.[70] suggested that the unexpectedly high concentrations of internal P they observed in mycorrhizal leeks resulted from carbon drain by the mycorrhizae preventing the host attaining the potential dry weight that should have been realized through improved P nutrition. Mycorrhizal colonization has been shown to reduce yield in plants of onion[71] and leek[19] well supplied with P. It is difficult to envisage that carbon demand by the endophyte is responsible, because at such high P supply infection is normally severely reduced. Mosse[25] demonstrated, however, that if infected plants are transplanted to soil high in P, excess uptake of P may occur, and she suggested that this P toxicity may reduce yield. The report that mycorrhizal onions take up more P than do nonmycorrhizal plants, even where extra P does not increase growth (luxury uptake),[63] supports this contention.

Vascular plants appear to lose a great deal of carbon (up to 30% or more of the total fixed C) as exudation from roots.[72] This might imply that the mycorrhizal endophyte simply lives gratuitously on carbon that the plant would lose normally and therefore does not constitute a drain. However, Snellgrove et al.[73] demonstrated that mycorrhizal plants of leek lost 7% more fixed carbon through below-ground respiration than did nonmycorrhizal plants. The latter were given extra P so that they grew at a similar rate to the mycorrhizal plants. The increased below-ground respiration loss was met by increased downward translocation from the shoot. The leaves of mycorrhizal plants had a greater specific leaf area (leaf area:leaf dry weight ratio). In a subsequent experiment mycorrhizal plants showed a greater rate of net photosynthesis and rate of photosynthesis per unit leaf area, together with an increased specific leaf area and a greater stomatal conductance.[74] The conclusion was that the carbon physiology of mycorrhizal plants was quite different to that of nonmycorrhizal plants of similar growth rate, and that these physiological differences could not be related in any simple way to internal concentrations of P. Further evidence that mycorrhizal colonization affects carbon metabolism in ways that P nutrition does not is provided by the observation that mycorrhizal roots of leek accumulate soluble carbohydrates,[19] and have enhanced levels of invertase activity of host cells.[62]

A pertinent question is whether mycorrhizal plants can compensate for the extra below ground loss of carbon, i.e., is there a *mechanistic* link between increased photosynthesis and increased carbon loss by mycorrhizae? Such compensation has been clearly demonstrated for shoots of leek infected by leek rust, where net assimilation of uninfected leaves was greatly increased by infection of other leaves.[75] However, in VA mycorrhizal plants the confounding effects of improved P uptake make it extremely difficult to test for any such compensation.

C. OTHER EFFECTS

It is essential to ask: are all effects of VA mycorrhizae upon their host plants attributable to improved P nutrition alone? Many experiments have explored this question, by using nonmycorrhizal plants given extra P so that they grew similarly to mycorrhizal plants. Such "matched" plants were used in the experiments cited above which purported to show that there are unique effects of VA mycorrhizae on the carbon physiology of leek. However, matters are not so simple. Snellgrove and colleagues[62,76,77,78] have shown that in leek seedlings, early colonization (when fractional infection is as low as 3%) triggers a very rapid increase in uptake of P, leading to high internal concentrations of this element. There is an accompanying increase in fresh weight:dry weight ratio, but acceleration in dry weight accumulation is delayed, so that percent P and relative growth rate are out of phase for some time. These early and large perturbations of physiology in mycorrhizal plants mean that, although it may be possible to produce nonmycorrhizal controls of roughly similar growth rate to mycorrhizal plants, the two sets of plants are not strictly comparable at any time and have had different physiological histories.

Many studies have reported that VA mycorrhizae can alter the water relations of the host. In particular it is often observed that mycorrhizal colonization lowers root hydraulic conductivity. Such work has been carefully reviewed by Nelsen[79] who concluded that most mycorrhizal effects were explicable in terms of improved phosphorus nutrition. The suggestion that external mycelium may conduct water directly along hyphae to the host remains highly controversial. Sanders and Tinker[53] calculated that the flux of water via hyphae is very low but, by contrast, Hardie[80] showed that severing external mycelium of leek seedlings reduced their rate of transpiration. Nelsen and Safir,[81] however, reported an apparently uniquely mycorrhizal effect on drought resistance in onion. Nonmycorrhizal plants were given additional phosphate, so that they grew at the same rate as did mycorrhizal plants in well watered soils. Half of the plants were subjected to repeated cycles of drought by withholding water and rewatering when the soil moisture had dropped to a predetermined potential. Leaf water potentials were unaffected by mycorrhizal colonization, but during the stress treatment the mycorrhizal plants grew better and contained higher concentrations of P than did nonmycorrhizal plants. Nelsen and Safir concluded that mycorrhizae enabled the plants to continue to take up phosphorus from dry soils, even though under these conditions the diffusion coefficient of phosphorus in the soil was sharply reduced. It is difficult to see a mechanism that could explain this phenomenon. Undoubtedly, diffusive resistance to movement of phosphorus to the plant root will be greatly increased by reduction in soil water content, and conceivably VA mycorrhizae may become even more important to the host under these conditions.[40] Dunham and Nye[82] showed, however, that the rate limiting step in uptake of phosphate from dry soils by roots of onion was the root absorbing power and that diffusion and mass flow were well able to satisfy the greatly reduced demand by the plant. It is by no means certain that uptake properties of mycorrhizae are similar to those of nonmycorrhizal roots and the important data of Nelsen and Safir deserve further investigation.

Mycorrhizal onions appear to be more resistant to the detrimental effects of saline soil than do nonmycorrhizal plants, but this effect again seems to be attributable to improved P nutrition.[83]

There is considerable literature on effects of VA mycorrhizae on resistance of roots to disease caused by various agents. The evidence however is very conflicting and depends on complex circumstances, such as the timing of inoculation with mycorrhizal fungi in relation to the timing of infestation by the pathogen. For example, MacGuidwin et al.[85] showed that there was no direct competition between the nematode *Meloidogyne hapla* and the VA mycorrhizal fungus *G. fasciculatum* in onion. Inoculation with the VA mycorrhizal fungus reduced the numbers of nematodes penetrating the roots, but this effect was ascribed to an indirect effect of improved P uptake. The data showed that there were complex interactions between inoculation, sampling schedule and duration of treatments.

IV. SIGNIFICANCE IN THE FIELD

It is difficult to assess the contribution made by natural VA mycorrhizae to total P uptake by crops of *Allium* in the field. No technique is yet available to eliminate natural infections without altering other soil properties. The correct tools, such as genotypes of hosts that are immune to infection, or fungicides specific to VA mycorrhizal fungi, are urgently needed. Yost and Fox[49] compared the growth and P concentration of various crops grown on a tropical oxisol with and without fumigation at various levels of soil P. Plants of onion on untreated soil maintained higher internal P concentrations than did the nonmycorrhizal plants on the treated soils, up to a soil solution concentration of 0.8 mg P/ml. This result suggests that natural mycorrhizae contribute greatly to P uptake by field-grown onions. Similarly, Plenchette et al.[86] showed that fumigation with methyl bromide severely reduced yield of field-grown onion whereas, by contrast, oat and wheat were unaffected. A subsequent experiment in pots[87] showed that the stunting could be corrected by the reintroduction of natural inoculum.

Future experimentation on the role of VA mycorrhizae in the field could make use of the fact that soil drenches of the fungicide benlate rapidly reduce mycorrhizal colonization and physiological functions.[29,88] An alternative approach may be to infer the functioning of mycorrhizae by using models of solute movements in the soil-root system.[89] The approach of Sanders and Tinker has already been discussed. A similar method was used by Barraclough,[90] who used the model of Baldwin et al.[91] to calculate the concentration of nutrients in the soil solution necessary to support the observed rates of uptake through the processes of diffusion. A large discrepancy between these two quantities would imply mediation by mycorrhizae.

V. POTENTIAL FOR PRACTICAL APPLICATION

The marked effects of VA mycorrhizae on their host plants, particularly in relation to P nutrition, have led to intense speculation and discussion about the potential commercial application of inoculum of VA mycorrhizal fungi.[92,93] Field experiments with onions and leek on P-deficient soils show that spectacular increases in growth can follow artificial inoculation.[94-96] It is thus undoubted that VA mycorrhizae can substitute for P fertilizer. However, *Allium* crops are very profitable; although exact figures for profits are difficult to obtain, the value of the crop may be gauged from the gross margins which are about £2200 per ha in the U.K.[97] Fertilizer accounts for only about 5% of the total variable costs of such an onion crop[98] so there is little saving to be made on this. Superphosphate has been used for over 140 years; it is highly effective and there are ample reserves of phosphate rock. All these considerations make it very unlikely that VA mycorrhizal fungi have any future as a "biological fertilizer" for *Allium* crops. However, if VA mycorrhizae could improve the consistency of yield of *Allium* crops, or increase drought tolerance, for example, and thereby increase yield even only slightly, then artificial inoculation would become a commercial proposition. Snellgrove et al.[45] found that onion plants raised in peat modules and inoculated with *Glomus mosseae* yielded more fresh weight of bulbs than did seedlings raised under normal commercial conditions. This arose through an unexpected reduction in number of plants that failed to bulb (''thicknecks'') on irrigated plots. The mechanism of this effect is unknown.

A potential, but as yet unexplored, area for application of inoculum of VA mycorrhizal fungi is in organic agriculture. In this type of husbandry, the market for which is expanding rapidly, there is strict control over the forms of fertilizer used, with emphasis on recycling of plant nutrients. The phosphorus fertilizers permitted by the British Organic Standards Committee, for example, are poorly soluble organic (e.g., bonemeal, seaweed) or inorganic (e.g., basic slag, rock phosphate) forms.[98a] Here, VA mycorrhizae would be essential for adequate growth of *Allium* crops. Carefully selected strains of inoculant fungi, efficient in uptake of phosphorus, could be used to great advantage.

Another great problem in the potential use of inoculum of VA mycorrhizal fungi is the technology of production and application of inoculum. Much of the work until now has used inoculum from stock plants growing in open pots, with its associated problems of quality control and susceptibility to ingress of pathogens. The situation has now greatly improved with the recent introduction of granular inocula which can be surface sterilized.[99,100]

VI. CONCLUSIONS

Two decades of research on VA mycorrhizae, mainly in pot studies under controlled conditions, have shown that these associations have profound effects on the physiology of their hosts, particularly in *Allium*. It is a matter of great concern to mycorrhizologists that these findings have made little impression on the study of plant physiology as a whole. There is now an urgent need to extend experimentation to the field situation. The possibility that mycorrhizae have effects on the host not wholly ascribable to improved uptake of P, in particular the suggestion that mycorrhizae can improve drought tolerance, should be rigorously tested in relation to the possible commercial applications of inoculum of VA mycorrhizal fungi.

REFERENCES

1. **Rothschild, M. and Clay, T.,** *Fleas, Flukes and Cuckoos,* Collins, London, 1952.
2. **Harley, J. L. and Smith, S. E.,** *Mycorrhizal Symbiosis,* Academic Press, London 1983.
3. **Rhodes, L. H. and Gerdemann, J. W.,** Hyphal translocation and uptake of sulphur by vesicular-arbuscular mycorrhizae of onion, *Soil Biol. Biochem.,* 70, 355, 1978.
4. **Phillips, J. M. and Hayman, D. S.,** Improved procedures for clearing roots and staining parasitic and vesicular-arbuscular mycorrhizal fungi for rapid assessment of infection, *Trans. Br. Mycol. Soc.,* 55, 158, 1970.
5. **Brundrett, M. C., Piché, Y., and Peterson, R. L.,** A new method for observing the morphology of vesicular-arbuscular mycorrhizae, *Can. J. Bot.,* 62, 2128, 1984.
6. **Cox, G. and Sanders, F. E.,** Ultrastructure of the host-fungus interface in a vesicular-arbuscular mycorrhiza, *New Phytol.,* 73, 901, 1974.
7. **Clarkson, D. T.,** Factors affecting mineral nutrient acquisition by plants, *Ann. Rev. Plant Physiol.,* 36, 77, 1985.
8. **Bonfante-Fasolo, P.,** Anatomy and morphology of VA mycorrhizae, in *VA Mycorrhiza,* Powell, C. Ll. and Bagyaraj, D. J., Eds., CRC Press, Boca Raton, FL, 1984, chap. 2.
9. **Hepper, C. M.,** Isolation and culture of VA mycorrhizal (VAM) fungi, in *VA Mycorrhiza,* Powell, C.Ll. and Bagyaraj, D. J., Eds., CRC Press, Boca Raton, FL, 1984, chap. 5.
10. **Hepper, C. M.,** Inorganic sulphur nutrition of the vesicular-arbuscular mycorrhizal fungus *Glomus caledonium, Soil Biol. Biochem.,* 16, 669, 1984.

11. **Harley, J. L. and Harley, E. L.,** A check-list of mycorrhiza in the British flora, *New Phytol. (Suppl.)*, 105, 1, 1987.
12. **Trappe, J. M. and Berch, S. M.,** The prehistory of mycorrhizae: A. B. Frank's predecessors, in *Proceedings of the 6th North American Conference on Mycorrhizae,* Molina, R., Ed., Forest Research Laboratory, Corvallis, OR, 1985, 2.
13. **Menge, J. A., Tinker, P. B., Stribley, D. P., and Snellgrove, R.,** Inoculum potential — its role in early infection and mycorrhizal efficiency, in *Proceedings of the 6th North American Conference on Mycorrhizae,* Molina, R., Ed., Forest Research Laboratory, Carvallis, OR, 1985, 394.
14. **Sanders, F. E., Tinker, P. B., Black, R. L. B., and Palmerley, S. M.,** The development of endomycorrhizal root systems. I. Spread of infection and growth-promoting effects with four species of vesicular-arbuscular endophyte, *New Phytol.*, 78, 257, 1977.
15. **Gallaud, I.,** Etudes sur les Mycorhizes Endotrophes, Ph.D. Thesis, Faculté des Sciences de Paris, University of Paris, 1904.

15a. **Brundrett, M. C., Piché, Y., and Peterson, R. L.,** A developmental study of the early stages in vesicular-arbuscular mycorrhiza formation, *Can. J. Bot.*, 63, 184, 1985.

16. **Amijee, F., Stribley, D. P., and Tinker, P. B.,** The development of endomycorrhizal root systems. VI. The relationship between intensity of infection and development of infection in young leek roots, *New Phytol.*, 102, 293, 1986.
17. **Buwalda, J. G., Ross, G. J. S., Stribley, D. P., and Tinker, P. B.,** The development of endomycorrhizal root systems. III. The mathematical representation of the spread of vesicular-arbuscular mycorrhizal infection in root systems, *New Phytol.*, 91, 669, 1982.
18. **Hepper, C. M.,** Influence of age of roots on the pattern of vesicular-arbuscular mycorrhizal infection in leek and clover, *New Phytol.*, 101, 685, 1985.
19. **Amijee, F.,** Colonization of Root Systems by Vesicular-Arbuscular Mycorrhizal Fungi, Ph.D. Thesis, University of Leeds, 1986.
20. **Becker, W. N. and Gerdemann, J. W.,** Colorimetric quantification of vesicular-arbuscular mycorrhizal infection in onion, *New Phytol.*, 78, 289, 1977.
21. **Jones, F. R.,** A mycorrhizal fungus in the roots of legumes and some other plants, *J. Agric. Res.*, 29, 459, 1924.
22. **Hepper, C. M.,** A colorimetric method for estimating vesicular-arbuscular mycorrhizal infection in roots, *Soil Biol. Biochem.*, 15, 9, 1977.
23. **Hepper, C. M., Sen, R., Azcon-Aguilar, C., and Grace, C.,** Variation in certain isozymes amongst different geographical isolates of the vesicular-arbuscular mycorrhizal fungi *Glomus clarum, Glomus monosporum* and *Glomus mosseae, Soil Biol. Biochem.*, 20, 51, 1988.
24. **Hepper, C. M., Azcon-Aguilar, C., Rosendahl, S., and Sen, R.,** Fungal competition, in *Rothamsted Experimental Station Report for 1986,* Lawes Agricultural Trust, Harpenden, U.K., 1987, 141.
25. **Mosse, B.,** Plant growth responses to vesicular-arbuscular mycorrhiza. IV. In soil given additional phosphate, *New Phytol.*, 72, 127, 1973.
26. **Sanders, F. E.,** The effect of foliar-applied phosphate on the mycorrhizal infections of onion roots, in *Endomycorrhizas,* Sanders, F. E., Mosse, B., and Tinker, P. B., Eds., Academic Press, London, 1975, 261.
27. **Schubert, A. and Hayman, D. S.,** Plant growth responses to vesicular-arbuscular mycorrhiza. XVI. Effectiveness of different endophytes at different levels of soil phosphate, *New Phytol.*, 103, 79, 1986.
28. **Amijee, F., Tinker, P. B., and Stribley, D. P.,** The development of endomycorrhizal root systems. VII. A detailed study of effects of soil phosphorus on colonization, *New Phytol.*, 111, 435, 1989.
29. **Boatman, N., Paget, D., Hayman, D. S., and Mosse, B.,** Effects of systemic fungicides on vesicular-arbuscular mycorrhizal infection and plant phosphate uptake, *Trans. Br. Mycol. Soc.*, 70, 443, 1978.
30. **Jabaji-Hare, S. H. and Kendrick, W. B.,** Response of an endomycorrhizal fungus in *Allium porrum* L. to different concentrations of the systemic fungicides, metalaxyl (Ridomi®) and fosetyl-Al (Aliette®), *Soil Biol. Biochem.*, 19, 95, 1987.
31. **Furlan, V. and Fortin, J.-A.,** Effects of light intensity on the formation of vesicular-arbuscular endomycorrhizas on *Allium cepa* by *Gigaspora calospora, New Phytol.*, 79, 335, 1977.
32. **Hayman, D. S.,** Plant growth responses to vesicular-arbuscular mycorrhiza. VI. Effect of light and temperature, *New Phytol.*, 73, 71, 1974.
33. **Chilvers, M. T. and Daft, M. J. F.,** Effects of low temperature on development of the vesicular-arbuscular mycorrhizal association between *Glomus caledonium* and *Allium cepa, Trans. Br. Mycol. Soc.*, 79, 153, 1982.
34. **Tinker, P. B.,** Modelling mycorrhizal development, in *Proceedings of the 6th North American Conference on Mycorrhizae,* Molina, R., Ed., Forest Research Laboratory, Corvallis, OR, 1985, 140.
35. **Sanders, F. E. and Sheikh, N. A.,** The development of vesicular-arbuscular mycorrhizal infection in plant root systems, *Plant Soil,* 72, 233, 1983.

36. **Gilligan, C. A.,** Construction of temporal models. III. Disease progress of soil-borne pathogens, *Adv. Plant Pathol.*, 3, 67, 1985.
37. **Buwalda, J. G., Stribley, D. P., and Tinker, P. B.,** The development of endomycorrhizal root systems. V. The detailed pattern of development of infection and the control of infection level by the host in young leek plants, *New Phytol.*, 96, 411, 1984.
38. **Sonneveld, C.,** Changes in chemical properties of soil caused by steam sterilization, in *Soil Disinfestation,* Mulder, D., Ed., Elsevier, Amsterdam, 1979, 39.
39. **Cawse, P. A.,** Microbiology and biochemistry of irradiated soils, in *Soil Biochemistry,* Vol. 3, Paul, E. A. and McLaren, A. D., Eds., Marcel Dekker, New York, 1975, 213.
40. **Fitter, A. H.,** Functioning of vesicular-arbuscular mycorrhizas under field conditions, *New Phytol.*, 99, 257, 1985.
41. **Warnock, A. J., Fitter, A. H., and Usher, M. B.,** The influence of a springtail *Folsomia candida* (insecta, Collembola) on the mycorrhizal association of leek *Allium porrum* and the vesicular-arbuscular mycorrhizal endophyte *Glomus fasciculatus, New Phytol.*, 90, 285, 1982.
42. **Finlay, R. D.,** Interaction between soil micro-arthopods and endomycorrhizal associations of higher plants, in *Ecological Interactions in Soil: Plants, Microbes and Animals,* Fitter, A. H., Ed., British Ecological Society Special Publication 4, Blackwell, Oxford, 1985, 318.
43. **Pacovsky, R. S.,** Micronutrient uptake and distribution in mycorrhizal or phosphorus-fertilized soybeans, *Plant Soil,* 95, 379, 1986.
44. **Smith, D. C. and Douglas, A. E.,** *The Biology of Symbiosis,* Edward Arnold, London, 1987, chap. 1.
45. **Snellgrove, R. C. and Stribley, D. P.,** Effects of pre-inoculation with a vesicular-arbuscular mycorrhizal fungus on growth of onions transplanted to the field as multi-seeded peat blocks, *Plant Soil,* 92, 387, 1986.
46. **Mosse, B., Hayman, D. S., and Ide, G. J.,** Growth responses of plants in unsterilized soil to inoculation with vesicular-arbuscular mycorrhiza, *Nature (London),* 224, 1031, 1969.
47. **Mosse, B. and Hayman, D. S.,** Plant growth responses to vesicular-arbuscular mycorrhiza. II. In unsterilized field soils, *New Phytol,* 70, 29, 1971.
48. **Stribley, D. P., Tinker, P. B., and Snellgrove, R. C.,** Effect of vesicular-arbuscular mycorrhizal fungi on the relations of plant growth, internal phosphorus concentration and soil phosphate analyses, *J. Soil Sci.*, 31, 655, 1980.
49. **Yost, R. S. and Fox, R. L.,** Contribution of mycorrhizae to P nutrition of crops growing on an oxisol, *Agron. J.*, 71, 903, 1979.
50. **Mengel, K. and Kirkby, E. A.,** *Principles of Plant Nutrition,* International Potash Institute, Bern, Switzerland, 1978, 66.
51. **Owusu-Bennoah, E. and Wild, A.,** Autoradiography of the depletion zone of phosphate around onion roots in the presence of vesicular-arbuscular mycorrhizas, *New Phytol.*, 82, 133, 1979.
52. **Capaccio, L. C. M. and Callow, J. A.,** The enzymes of polyphosphate metabolism in vesicular-arbuscular mycorrhizas, *New Phytol.*, 91, 81, 1982.
53. **Sanders, F. E. and Tinker, P. B.,** Phosphate flow into mycorrhizal roots, *Pestic. Sci.*, 4, 385, 1973.
54. **Sanders, F. E. and Tinker, P. B.,** Mechanism of absorption of phosphate from soil by *Endogone* mycorrhizas., *Nature (London),* 233, 278, 1971.
55. **Tinker, P. B.,** Soil chemistry of phosphorus and mycorrhizal effects on plant growth, in *Endomycorrhizas,* Sanders, F. E., Mosse, B., and Tinker, P. B., Eds., Academic Press, London, 1975, 353.
56. **Jurinak, J. J., Dudley, L. M., Allen, M. F., and Knight, W. G.,** The role of calcium oxalate in the availability of phosphorus in soils of semiarid regions: a thermodynamic study, *Soil Sci.*, 142, 255, 1986.
57. **Bolan, N. S., Robson, A. D., Barrow, N. J., and Aglmore, L. A. G.,** Specific activity of phosphorus in mycorrhizal and non-mycorrhizal plants in relation to the availability of phosphorus to plants, *Soil Biol. Biochem.*, 16, 299, 1984.
58. **Hedley, M. J., White, R. E., and Nye, P. H.,** Plant-induced changes in the rhizosphere of rape (*Brassica napus* var. Emerald) seedlings. III. Changes in L value, soil phosphate fractions and phosphatase activity, *New Phytol.*, 91, 45, 1982.
59. **Greenwood, D. J., Gerwitz, A., Stone, D. A., and Barnes, A.,** Root development of vegetable crops, *Plant Soil,* 68, 75, 1982.
60. **Itoh, S. and Barber, S. A.,** Phosphorus uptake by six plant species as related to root hairs, *Agron. J.*, 75, 457, 1983.
61. **Barber, S. A.,** *Soil Nutrient Bioavailability,* John Wiley and Sons, New York, 1984, chap. 9.
62. **Snellgrove, R. C.,** Effects of Vesicular-Arbuscular Mycorrhizas on the Carbon and Phosphorus Physiology of *Allium* species, Ph.D. Thesis, Council for National Academic Awards, London, 1987.
63. **Smith, S. E., St. John, B. J., Smith, F. A., and Bromley, J.-L.,** Effects of mycorrhizal infection on plant growth, nitrogen and phosphorus nutrition in glasshouse-grown *Allium cepa* L., *New Phytol.*, 103, 359, 1986.
64. **Smith, S. E., St. John, B. J., Smith, F. A., and Nicholas, D. J. D.,** Activity of glutamine synthetase and glutamate dehydrogenase in *Trifolium subterraneum* L. and *Allium cepa* L: effects of mycorrhizal infection and phosphate nutrition, *New Phytol.*, 99, 211, 1985.

65. **Tinker, P. B. and Gildon, A.,** Mycorrhizal fungi and ion uptake, in *Metals and Micronutrients,* Robb, D. A. and Pierpoint, W. S., Eds., Academic Press, London, 1983, 21.
66. **Gildon, A. and Tinker, P. B.,** Interactions of vesicular-arbuscular mycorrhizal infections and heavy metals in plants. II. The effects of infection on uptake of copper, *New Phytol.,* 95, 263, 1983.
67. **Cox, G., Sanders, F. E., Tinker, P. B., and Wild, J. A.,** Ultrastructural evidence relating to host-endophyte transfer in a vesicular-arbuscular mycorrhiza, in *Endomycorrhizas,* Sanders, F. E., Mosse, B., and Tinker, P. B., Eds., Academic Press, London, 1975, 261.
68. **Lösel, D. M. and Cooper, K. M.,** Incorporation of ^{14}C-labelled substrates by uninfected and VA mycorrhizal roots of onion, *New Phytol.,* 83, 415, 1979.
69. **Cooper, K. M. and Lösel, D. M.,** Lipid physiology of vesicular-arbuscular mycorrhiza. I. Composition of lipids in roots of onion, clover and ryegrass infected with *Glomus mosseae, New Phytol.,* 80, 143, 1978.
70. **Stribley, D. P., Tinker, P. B., and Rayner, J. H.,** Relation of internal phosphorus concentration and plant weight in plants infected by vesicular-arbuscular mycorrhizas, *New Phytol.,* 86, 261, 1980.
71. **Mosse, B.,** Advances in the study of vesicular-arbuscular mycorrhiza, *Annu. Rev. Phytopathol.,* 11, 171, 1973.
72. **Whipps, J. M. and Lynch, J. M.,** Energy losses by the plant in rhizodeposition, in *Plant Products and the New Technology,* Annual Proceedings of the Phytochemical Society of Europe, Vol. 26, Fuller, K. W. and Gallon, J. R., Eds., Clarendon, Oxford, 1985, chap. 5.
73. **Snellgrove, R. C., Splittstoesser, W. E., Stribley, D. P., and Tinker, P. B.,** The distribution of carbon and the demand of the fungal symbiont in leek plants with vesicular-arbuscular mycorrhizas, *New Phytol.,* 92, 75, 1982.
74. **Snellgrove, R. C., Stribley, D. P., Tinker, P. B., and Lawlor, D. W.,** The effect of vesicular-arbuscular mycorrhizal infection on photosynthesis and carbon distribution in leek plants, in *Proceedings of the First European Mycorrhiza Congress,* Gianinazzi-Pearson, V. and Gianinazzi, S., Ed., INRA, Paris, 1986, 421.
75. **Roberts, A. M. and Walters, D. R.,** Stimulation of photosynthesis in uninfected leaves of rust-infected plants, *Ann. Bot.,* 56, 893, 1986.
76. **Stribley, D. P. and Snellgrove, R. C.,** Physiological changes accompanying mycorrhizal infection in leek, in *Proceedings of the 6th North American Conference on Mycorrhizae,* Molina, R., Eds., Forest Research Laboratory, Corvallis, OR, 1985, 395.
77. **Stribley, D. P. and Snellgrove, R.,** Vesicular-arbuscular mycorrhizas and the phosphorus physiology of plants, *J. Sci. Food Agric.,* 37, 11, 1985.
78. **Tinker, P. B., Stribley, D. P., and Snellgrove, R. C.,** The relationship between phosphorus concentration and growth in plants infected with vesicular-arbuscular mycorrhizal fungi, in *Proceedings of the Ninth International Plant Nutrition Colloquium, Warwick 1982,* Vol. 2, Commonwealth Agricultural Bureaux, Slough, U.K., 1982, 670.
79. **Nelsen, C. E.,** The water relations of vesicular-arbuscular mycorrhizal systems, in *Ecophysiology of VA Mycorrhizal Plants,* Safir, G. R., Ed., CRC Press, Boca Raton, FL, 1987, 71.
80. **Hardie, K.,** The effect of removal of extraradical hyphae on water uptake by vesicular-arbuscular mycorrhizal plants, *New Phytol.,* 101, 677, 1985.
81. **Nelsen, C. E. and Safir, G. R.,** Increased drought tolerance of mycorrhizal onion plants caused by improved phosphorus nutrition, *Planta,* 154, 407, 1982.
82. **Dunham, R. J. and Nye, P. H.,** The influence of soil water content on the uptake of ions by roots. III. Phosphate, potassium, calcium and magnesium uptake and concentration gradients in soil, *J. Appl. Ecol.,* 13, 967, 1976.
83. **Poss, J. A., Pond, E., Menge, J. A., and Jarrell, W. M.,** Effect of salinity on mycorrhizal onion and tomato in soil with and without additional phosphate, *Plant Soil,* 88, 307, 1985.
84. **Bagyaraj, D. J.,** Biological interactions with VA mycorrhizal fungi, in *VA Mycorrhiza,* Powell, C. L. and Bagyaraj, D. J., Eds., CRC Press, Boca Raton, FL, 1984, chap. 7.
85. **MacGuidwin, A. E., Bird, G. W., and Safir, G. R.,** Influence of *Glomus fasciculatum* on *Meloidogyne hapla* infecting *Allium cepa, J. Nematol.,* 17, 389, 1985.
86. **Plenchette, C., Fortin, J. A., and Furlan, V.,** Growth responses of several plant species to mycorrhizae in a soil of moderate P-fertility. I. Mycorrhizal dependency under field conditions, *Plant Soil,* 70, 199, 1983.
87. **Plenchette, C., Fortin, J. A., and Furlan, V.,** Growth responses of several plant species to mycorrhizae in a soil of moderate P-fertility. II. Soil fumigation induced stunting of plants corrected by reintroduction of the wild endomycorrhizal flora, *Plant Soil,* 70, 211, 1983.
88. **Hale, K. A. and Sanders, F. E.,** Effects of benomyl on vesicular-arbuscular mycorrhizal infection of red clover (Trifolium pratense L.) and consequences for phosphorus inflow, *J. Plant Nutr.,* 5, 1355, 1982.
89. **Nye, P. H. and Tinker, P. B.,** *Solute Movement in the Soil-Root System,* Blackwell Scientific Publications, Oxford, U.K., 1977, chap. 7.
90. **Barraclough, P. B.,** The growth and activity of winter wheat roots in the field: nutrient inflows of high-yielding crops, *J. Agric. Sci., Cambr.,* 106, 53, 1986.

91. **Baldwin, J. P., Nye, P. H., and Tinker, P. B.,** Uptake of solutes by multiple root systems from soil. III. A model for calculating the solute uptake by a randomly dispersed root system developing in a finite volume of soil, *Plant Soil,* 38, 621.
92. **Abbott, L. K. and Robson, A. D.,** The role of vesicular-arbuscular mycorrhizal fungi in agriculture and the selection of fungi for inoculation, *Aust. J. Agric. Res.,* 33, 389, 1982.
93. **Jeffries, P.,** Use of mycorrhizae in agriculture, *CRC Crit. Rev. Biotechnol.,* 5, 319, 1987.
94. **Owusu-Bennoah, E. and Mosse, B.,** Plant growth responses to vesicular-arbuscular mycorrhiza. XI. Field inoculation responses in barley, lucerne and onion, *New Phytol.,* 83, 671, 1979.
95. **Sasa, M., Zahka, G., and Jakobsen, I.,** The effect of pretransplant inoculation with VA mycorrhizal fungi on the subsequent growth of leeks in the field, *Plant Soil,* 97, 279, 1987.
96. **Wamocho, L. S., Skipper, H. D., and Loomis, E. L.,** Effect of *Glomus etunicatum* and native VAM fungi on direct-seeded and transplanted green onions, in *Proceedings of the 6th North American Conference on Mycorrhizae,* Molina, R., Ed., Forest Research Laboratory, Corvallis, OR, 1985, 250.
97. **Nix, J.,** *Farm Management Pocketbook,* Seventeenth Edition, Wye College (University of London), Ashford, U.K., 1987, 26.
98. **Martin, D.,** Comparison of systems for July and August marketing, presented at ''Onions with difficulty'', Arthur Rickwood Experimental Husbandry Farm, Mepal, U.K., 11 December, 1984, 5.
98a. **The Soil Association Symbol Scheme,** *Standards for Organic Agriculture,* The Soil Association Ltd, Bristol, U.K., 1987.
99. **Baltruschat, H.,** Evaluation of the suitability of expanded clay as a carrier material for VA mycorrhiza spores in field inoculation of maize, *Angew. Botanik,* 61, 163, 1987.
100. **Dehne, H.-W. and Backhaus, G. F.,** The use of vesicular-arbuscular mycorrhizal fungi in plant production. I. Inoculum production, *Zeitschrift Planzenkrankheiten Pflanzenschutz,* 93, 415, 1986.

Chapter 6

ROOT DISEASES

A. R. Entwistle

TABLE OF CONTENTS

I. INTRODUCTION

Root diseases are a major source of loss in *Allium* crops throughout the world. The causal fungal pathogens have several features in common with each other. For example, the foliar symptoms of plants with different root diseases are often those of water and nutrient stress. For example, there is loss of turgor, wilting, chlorotic foliar dieback, and stunting. Also, the methods of dissemination are similar: most root diseases can be carried in vegetative planting material, or by any action which moves soil containing the pathogen.

Another similarity is that all the diseases are difficult to control, partly because their presence in soil renders them inaccessible to fungicides and partly because they have resistant propagules which survive for long periods.

The reduction of fungal inoculum may provide opportunities for other methods of control which by themselves might be inadequate e.g. host resistance. The eradication of fungal pathogens by the use of soil fumigants is expensive and not always effective; soil solarization (Section III.F.2), however, is proving a valuable in hot countries. Increasingly, the control of soilborne diseases involves the use of several different approaches and may lead to the development of an integrated control program.

This chapter describes *Allium* white rot in detail because it provides an example of features common to other soilborne root diseases. Other root diseases are then described but in rather less detail (Table 1).

TABLE 1
Root Diseases of *Allium*

Pathogen	Disease	*Allium* host
Sclerotium cepivorum	White rot	all
Pyrenochaeta terrestris	Pink root	a c (p) s (sh)
Fusarium oxysporum		
f.sp. *cepae*	Basal rot	c
f.sp. *allii*	Basal rot	ch
Urocystis colchici	Smut	a c f p s
Sclerotium rolfsii	Southern blight	a c
Fusarium spp.		
Pythium spp.	Damping-off	c
Rhizoctonia solani		
Colletotrichum circinans	Smudge	a c p
Macrophomina phaseolina	Charcoal rot	c

Note: *Allium* hosts: a — *A. cepa* var. *ascalonicum;* c — *A. cepa;* ch — *A. chinense;* f — *A. fistulosum;* p — *A. porrum;* s — *A. sativum;* sh — *A. schoenoprasum*

II. ALLIUM WHITE ROT

A. DISTRIBUTION AND ECONOMIC IMPORTANCE

1. Worldwide

White rot is caused by the fungus *Sclerotium cepivorum* Berk. and was first reported in onions in the UK[1] and in garlic in Italy[2] (Table 2). The disease is now present in many areas of the world where *alliums* are cultivated and environmental conditions are favorable to the pathogen.[3,40-42] The economically damaging effects of white rot are mostly apparent in *alliums* cultivated for food although ornamental and other species are also affected.[43-45] White rot is a problem because of (1) the actual loss of crop and the additional costs associated with harvest, (2) the uncertainty associated with the possible increased damage when *alliums* are grown again in infested land, and (3) the possibility of new outbreaks in areas previously unaffected by the disease. Losses may be caused by death from the seedling stage onward and by cosmetic damage to otherwise marketable plants at harvest or in store (see chapter on Storage Diseases of Onions). There are also additional costs of harvesting and trimming in green onions and leeks and of culling dry bulb onions, onion sets, and garlic cloves before sale. Plant growth may be retarded by the deleterious effects of root infection on water and nutrient uptake[46] but go unnoticed in the absence of more recognizable symptoms.[47] Unfortunately, it is not possible to predict the effects of white rot from past experience. In extreme situations there is a total loss of crop, the land becomes unsuitable for *Allium* production and the pathogen spreads to new areas. In other situations damage is minimal.

White rot causes little or no economic damage in the states of Illinois, Kentucky, Louisiana, New York, and Virginia,[18,48] Czechoslovakia,[32] and parts of West Germany.[49] On a National or State basis, losses of 1 to 15% have been reported in bulb onions in New Jersey,[50] British Columbia and Ontario, Canada,[51,52] and the U.K.[53] These losses are small probably because the disease has affected only a small proportion of the land used for *Allium* production. Losses are much higher in other areas. White rot caused a 38% loss of crops ($\$2.8 \times 10^6$) in early 1980s in the combined areas of Walla Walla, Washington, and New Jersey.[18] Losses of 30 to 100% in individual fields were reported in shallots in Louisiana,[15] in garlic in California,[54] and bulb onions in British Columbia, Canada,[52] the U.K.,[55] and Egypt.[56,57] Severe losses have also been reported in individual crops of garlic in Italy and Mexico.[2,21]

TABLE 2
Outbreaks of *Allium* White Rot

Europe	
1841	UK[1]
1909—34	Italy,[2] France,[3] Spain,[4] Netherlands,[5] Germany[6]
1958,1968	Sweden,[7] Finland[8]
American continent	
1915—25	Argentina,[9] U.S.: Oregon,[3] Virginia,[3,10] Kentucky[11]
1934—42	Brazil;[12] U.S.: New Jersey,[13] California,[14] Louisiana;[15] Canada: British Columbia[16]
1951—61	U.S.: Walla Walla,[17] New York,[18] Tule Lake,[19] Canada: Manitoba[20]
1970—80	Mexico;[21] U.S.: Salinas Valley,[22] San Joaquin Valley,[19] Nevada;[22] Canada: Ontario[23]
1985,86	U.S.: Idaho;[22] Canada: Quebec[23]
Mediterranean and Africa	
1929—30	Egypt,[24] Cyprus,[25] South Africa[26]
Australia and New Zealand	
1922	New Zealand[27]
1928—41	Melbourne,[28] Victoria,[29] Queensland,[30] Tasmania[31]
1954	New Zealand: Pukohoe[27]

White rot is also present in: Czechoslovakia,[32] Poland,[33] Denmark,[7] Norway,[34] India,[35] Japan,[36] Columbia,[37] Rumania,[38] and Chile[39]

In Walla Walla, Washington, the disease was first detected in approximately 1951 and by 1961 most fields had localized infestations or were no longer used for *Allium* production.[17,58] As a result, production was moved to less fertile land which had not previously been used for onion production, hence was free from white rot.[18] However, the new areas were expected to become affected and production to become uneconomic within 5 years.[18]

In Britain, production of dry bulb onions in Lincolnshire and Cambridgeshire rose from 3745 ha in 1970 to 7250 ha in 1985—1986 due largely to improved methods of production[59,60] and by 1984—1986 fields in these two counties were being cropped with onions for the second and third times. Initially, white rot caused only minor damage but by 1984 approximately 10% of the onion production area was affected. These losses are relatively small but are causing concern for future production, as even small losses are sufficient to render production uneconomic.[54]

White rot is not always as severe or as widespread as expected. For example, on one farm in Louisiana, the disease had remained localized for over 20 years even though no efforts were made to contain it.[13] The same author drew attention to the lack of spread from an area of severe infection to an adjacent onion field separated only by a headland 4m wide. In the U.K., no white rot developed in experimental plots which had been deliberately located in an area where, just a few months previously, the disease had destroyed the crop, even though conditions were favorable to infection.[61] In Canada, white rot was not detected in some fields where sclerotia were present, possibly due to environmental or biological factors unfavorable to *S. cepivorum*.[62]

As expected, new outbreaks of the disease commonly occur in areas where the fungus is already prevalent locally. Thus, the only infection reported in Orange County, New York, since the original outbreak 24 years earlier, occurred in a field only 2 km away;[48] likewise in Egypt, infection spread within 2 years to areas adjoining the field where infection was first reported.[56] Outbreaks have, however, also occurred in areas which are geographically remote from other sources of infestation for example in Idaho,[63] Tasmania, Australia,[31] and Mexico[21] probably due to the use of infected planting material.

TABLE 3
Effect of White Rot on Onion Production in Egypt

Year	Production (1000 ha)	Production (1000 MT)	Export (1000 MT annum^{-1})	Ref.
All types of onion				
1956	47	NS[a]	197	64
1959—60	66	558	NS	65
1960—65	16	279	159	57,64
1966	86	638	NS	64
1975	21	603	NS	66
1979—81	20	594	NS	67,68
1983—85	25	846	NS	68,69
Winter Onions				
Pre-1965	20	NS	162	70
1975—79	10	159	40	70,71
1982	7	134	16	64,71
1985	9	191	NS	71

[a] NS — not stated.

2. Situation in Egypt

The record of white rot in Egypt demonstrates the potential of the pathogen to devastate onion production.

Allium white rot was first reported from a single field at Maghagha, 250 km south of Cairo, in 1929 and the neighboring district of Bani Mazar in 1930.[56] For the next 20 years, the pathogen was restricted to these two districts. Then, between 1952—1958, it progressively affected the area from El Saff to Qena, a distance of 570 km. The Fayyum oasis remained free of white rot, due to its isolation from the Nile Valley. Then, in 1978, infected bulbs were planted for seed crops and there has since been a marked decline in onion production in the Fayyum.[57] The Northern Delta region also remained free of white rot until 1979, again probably because of its isolation; the affected areas have now been removed from *Allium* production.[57]

Between 1965—1982, there has been a 65% reduction in the land area cropped with winter onions and a 90% reduction in exports (Table 3).[57,64-71] Onion production has ceased in whole villages in Minia and Beni Suef provinces in Middle Egypt.[72] The reduced production has resulted largely from the increase in white rot which followed the building of the Aswan High Dam.[70]

Historically, Egypt has been subject to the effects of the annual inundation by the River Nile. This has resulted in the development of direct and basin systems of irrigation, and the latter is thought to have restricted the effects of white rot. With the direct system of irrigation, water was removed directly from the Nile to a system of canals in the period following the inundation. Later in the year, as the level of the Nile fell, direct irrigation was supplemented by water from irrigation basins. The basins were filled earlier in the year as the Nile rose and emptied, usually about October. This left a thick deposit of silt soil.[73] Onions were planted in the silt from mid-October onward and received no water other than that retained by the silt. By the following March, the soil had dried out and onions were ready for harvest. After harvest, the soils were bare fallowed until the next inundation. The constraint on white rot resulted from the combined effects of (1) the deposition of silt,[70] (2) anaerobiosis during flooding,[70] and (3) the high soil temperatures in the basins after harvest and before the next flooding.

The effects of anaerobiosis and silt deposition were probably most evident in the irrigation basins where flooding was prolonged and where the silt deposition was maximal.[73] It appears that the silt was sufficient to cover previous infestations of white rot and reduce their effectiveness as inoculum.[64]

Anaerobiosis during flooding was reported to have had a deleterious effect on sclerotia.[70] However, there is some doubt about this as white rot infested soil placed in plant pots and flooded for 60 d, a period comparable to the flooding in irrigation basins, remained infective.[74]

The third and probably the most important factor which limited white rot was the occurrence of high soil temperatures in April—July when soil in the basins had dried out. Soil temperatures of 39°C at 5 cm depth have been recorded at Kharga from March onwards and at 10 to 20 cm depth, 1 to 2 months later; minimum soil temperatures were 30 to 33°C at 20 cm depth in June—August. In South Australia, such temperatures were attained during solarization and were sufficient to kill sclerotia.[75] The deleterious effects of high temperatures on soilborne pathogens are maximal in moist soil; the Egyptian soils were dry during the fallow period and the effects of high temperatures may have been somewhat less than in Australian conditions.

During the last century, there has been a progressive control of the Nile floods culminating in total control with the completion of the Aswam High Dam in 1968. This control appears to have removed a natural constraint on white rot and is considered to account for the increased losses from this disease.

B. PATHOGEN

1. Morphology

Hyphae are 9 to 18 μm in diameter and cells are mostly 30 to 100 μm in length, with apical cells 3 or 4 times longer.[76] Sclerotia are mostly spherical in shape and 200 to 700 μm in diameter[76] but may be smaller in culture[46] (Figure 1). Sclerotia are dumb-bell shaped or form in clusters in which sclerotia are joined together, especially when recovered directly from infected plants. The sclerotia have a dark, slightly roughened surface and are rubbery in texture. There is a two-layered cortex consisting of an outer, darkly pigmented layer 2 to 3 cells deep and a similar but nonpigmented inner layer. A central medulla comprises nonpigmented and highly refractive cells which are elongate and irregularly shaped in contrast to the cortical cells which are isometric.[77]

Large sclerotia approx. 1.0 to 1.5 cm in diameter have been reported from Egypt and the U.K.[78] These sclerotia have a compound structure. The large outer sclerotium has a cortex of typical structure but embedded in the central medullary tissue are sclerotia of normal size and structure.

Secondary sclerotia approx. 100 to 150 μm diameter may form within a parent sclerotium or adjacent to it.[79] Secondary sclerotia comprise 5 to 10% of the total[47] and are stimulated to germinate in soil in the presence of onion.[79] Possibly they are formed in response to unfavorable soil conditions, e.g., in the root systems of fungicide-treated plants.[80]

Spherical microconidia (spermatia) 1.6 to 3.4 μm in diameter[41,81] are produced on mycelium in soil[2,3,13,26,82] but there, as yet, is little evidence that they contribute to infection.

2. Germination of Sclerotia

S. cepivorum sclerotia germinate only in response to root exudates or extracts of plants in the *Allium* genus.[81] Thus, out of 63 types of plant from a wide range of families, sclerotial germination was stimulated only by *Allium*.[83]

Compounds which are mercaptans or mono- or disulfides having an *n*-propyl (alkyl) or an *n*-propenyl (allyl) radical stimulate sclerotia to germinate.[84] These compounds, which are volatile, are produced in soil by the microbial degradation of the nonvolatile water soluble precursors *n*-propyl cysteine and *n*-propenyl cysteine and their sulphoxides present in the tissues and root exudates of alliums.[85] Compounds which are stimulatory to *S. cepivorum*

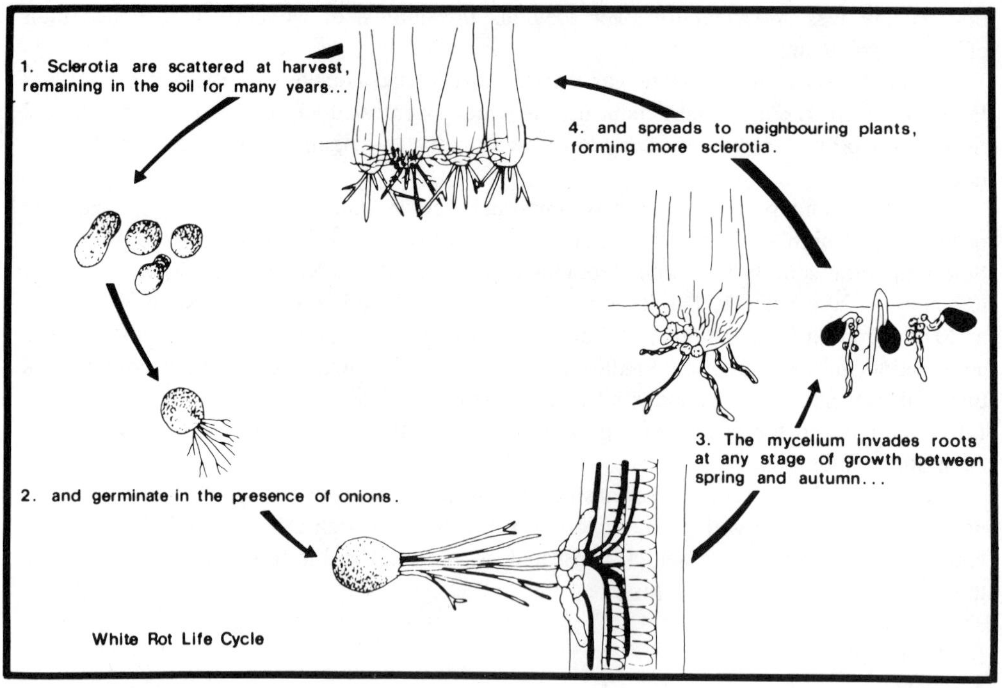

FIGURE 1. *Allium* white rot (*Sclerotium cepivorum*).

are major constituents of the flavor and odor compounds in *Allium* (see chapter on Flavor Biochemistry).

S. cepivorum isolates appear to vary in their responsiveness to stimulants. Some respond rapidly and the final germination percentage is high; other isolates fail to respond or respond less readily.[83] Sclerotia which have been cultured aseptically for long periods show greater variation in response to stimulants than sclerotia produced on onion in soil.[83] The responsiveness of sclerotia to germination stimulants is affected by the substrate on which they are produced. A faster rate and higher final germination percentage occurs in sclerotia produced on onion bulbs in soil, in comparison with sclerotia produced on maize meal or sugar beet seed. When first produced, sclerotia generally are unresponsive to stimulants, i.e., are dormant, but generally become responsive after 1 to 3 month burial in soil.[86] Freezing (−22°C) has been reported to overcome dormancy[87] but its significance in nature is not known.

The mode of action of this specific stimulation is not fully understood. Some constituents of *Allium* are antibiotic, e.g., *n* propyl sulfide, and these might suppress the activity of the soil microflora, thereby reducing soil fungistasis and overcoming exogenous dormancy. Sclerotia germinate, however, at concentrations of stimulant which are too low for antibiotic activity. Moreover, diallyl disulfide (DADS) has no antibiotic properties yet is highly stimulatory to *S. cepivorum* sclerotia.[88]

Different alliums vary in their capacity to stimulate sclerotia and this is related to the total quantity of sulfide precursors and the proportions of the stimulatory constituents. The lower capacity of leek (*A. ampeloprasum* var. *porrum* (L.) Gay) to stimulate sclerotial germination in comparison with onion, garlic or *A. fistulosum* L. is accounted for by the lower quantity of stimulatory compounds in leek compared with the other three species.[89]

3. Survival of Sclerotia

In a field experiment, sclerotia survived for 18 years.[90] The sclerotia had been cultured on artificial media, therefore sclerotia produced on onion bulbs in soil were also tested. There was some decay of sclerotia shortly after they had formed on the onion but sclerotia which survived this initial stage were apparently capable of long periods of survival.[90] The survival of immature sclerotia, i.e., those in which rind formation was incomplete, did not differ greatly from sclerotia with fully formed rinds. Survival was somewhat reduced, however, at two of the four sites tested. The cause of this lower survival is not known but it is not thought to be due to the presence of mycoparasites.[90]

The survival of sclerotia collected from diseased onions was only 30% after 3 months burial in soil, with a further 10% loss nine months later.[91] Sclerotia cultured on maize meal showed 90% survival in a similar period. A reduced survival of sclerotia has also been reported in hot soils. For example at Irymple, S. Australia, sclerotial viability in nonsolarized soil was reduced to 0 to 5% at 0 to 11cm depth in soil and to 37 to 62% at 14 to 26 cm.[75] This was probably due to the effects of high soil temperatures with mean maxima of 33 to 42°C being recorded. At a second site, soil temperatures were 8 to 12°C lower and there was rather less effect on sclerotial viability. These temperatures were similar to those in Egypt, where high soil temperatures were also thought to be a factor in the natural control of white rot. Flooding also reduced sclerotial survival: up to 80% of naturally produced sclerotia failed to survive in peat soils between December 1979 and March 1981 in Canada, with the highest rates of loss occurring in the winter periods.[52,92] The effect was even more marked when the sclerotia were dried before burial with >70% of sclerotia lost after 4 months.[92] Additional experiments in 1985—1986 demonstrated that flooding for a period of 3 weeks in mid-March resulted in an 80% reduction in the numbers of sclerotia in soil in comparison with a 30% reduction in nonflooded soil. There was little further loss of sclerotia 3 months later in either soil. Anaerobiosis during flooding was considered to account partly for the reduced survival but clearly other, as yet unknown, factors also affected sclerotia in the nonflooded conditions.[52]

In these rather special circumstances, i.e., hot or flooded soil, there is the opportunity for reducing soil sclerotial populations especially if the reduced survival continues with additional cycles of the respective conditions.

C. SYMPTOMS

Plants can become infected at any stage of growth providing environmental conditions are favorable. However, the type of symptom and its severity are dependent on the growth stage of the host at infection and the duration of soil conditions favoring the pathogen. There is little evidence that infection is affected directly by differences in tissue maturity or in plant size. However, large plants have large root systems and these may give rise to a higher rate of stimulation of sclerotial germination and hence to infection. Mechanical damage to roots, insect, or nematode attack may also increase root exudation, thus enhancing the stimulation of sclerotium germination.[81] In South Africa, infection of onion bulbs planted in severely infected soil was entirely dependent on injury, e.g., by insects.[26]

Infected plants may die before emergence (Figure 2A) or soon after[93,94] and the symptoms are similar to those caused by damping-off fungi, e.g., *Pythium* spp. In New Jersey, and in the U.K., seedlings produced in autumn (August — October) for overwintering become infected and often fail to survive until spring.[95,96] Eye-shaped lesions were reported on the sheathing leaf bases between the stembase and the soil surface on overwintering seedlings (Figure 2B).[97] Similar symptoms occur in leeks which are grown from transplanted seedlings and planted more deeply than plants which are direct sown. Infection causes young plants to wilt (Figure 2C) and the symptoms are similar to those of onion fly.[48] As the plants become larger, the wilt may be confined to the older leaves. If plants become water deficient, e.g., under hot or dry conditions, foliar symptoms develop even in large plants. Characteristically, the tips of the older leaves turn yellow and lose turgor, and ultimately the plant dies. When the hot conditions persist, the dead plant desiccates and the remains can blow away leaving little evidence of infection.[48] In contrast, when the water supply is abundant there may be little evidence of foliar symptoms in older plants even when root infection is severe.[98]

Infection is accompanied by the development of mycelium on the stembase. The mycelium is grayish or greenish white and ''Cottony'' or ''felty'' in appearance[3] and is most evident when plants are removed from the ground (Figure 2D).

Black sclerotia are produced within mycelium on the stembase and are the only known means of survival (Figure 2E). Sclerotia may also develop some distance along the aerial parts of the leaves in onion, e.g., in moist conditions and have also been reported on the inflorescence of leeks;[99] occasionally sclerotia form within roots. Generally the whole of the stembase is affected but sometimes symptoms are confined to one side.[13]

The mycelial stage is not always readily apparent. The mycelium lyses when the foodbase is exhausted[100] and in soil which is excessively hot or excessively cold and wet. Then, the only evidence of white rot may be a blackening of the stem base tissue or the presence of collapsed roots. In the absence of sclerotia, diagnosis can be uncertain, since it is the author's experience that the mycelium may fail to regenerate when tissue is transferred to conditions favorable to the pathogen. The presence of root collapse is not necessarily indicative of white rot because similar symptoms also result from senescence or from other soilborne pathogens.

The symptoms of white rot in other *Allium* spp. are similar.[3] In Japanese bunching onion (*A. fistulosum*) and leek the infected plants remain more upright than in *A. cepa*, therefore foliar symptoms are less evident. in these circumstances the presence of white rot could be overlooked unless plants are removed from soil.

As an alternative to direct sowing, onions may be grown in modules for transplanting. There is evidence of a higher incidence of white rot in plants grown in modules or peat blocks than in direct sown plants.[101] The most probable explanation is that plants grown in this manner have a well-developed root system at a time when soil conditions are favorable to the pathogen. In the U.K., direct-sown plants take 2 months longer to reach similar stage of growth and conditions are then frequently less favorable for infection. Also, with block raised plants root damage at the time of transplanting may enhance sclerotial germination and hence infection.[102]

Crops grown at high population densities, e.g., salad and pickling onions, provide opportunities for the growth of hyphae from infected to healthy plants. As a result, infection commonly affects groups of adjacent plants; spread of infection is also favored by the close contact afforded by plants grown in multiseeded modules.

After harvest, infected onion bulbs may mummify[10] or rot.[103] Storage losses have been reported in the U.S. in onions, garlic, and shallots[3,104] and in onions in Australia,[103] India,[105] and Czechoslovakia,[106] Otherwise, there is little information on this aspect of the disease. There appear to be no reports on the potential for storage losses from white rot arising from subclinical infection when bulbs or cloves, which are apparently healthy, are harvested from infected crops.

White rot typically occurs in patches (''hot spots'') in the field;[13,21,54,61] these may be a few square metres in area (Figure 2F) or they may be much larger and coalesce with adjacent patches.[53] In the U.K., the presence of some patches was attributed to the previous dumping of infected waste but the cause of others was unknown.[55] There is little information on the effects of cultivation and cropping on the size or shape of these patches. If they are associated with localized high populations of sclerotia, there is the possibility that the mechanical movement of soil during cultivation or leveling will cause the patches to expand.

D. SOURCES OF OUTBREAKS AND DISSEMINATION

1. Planting Material

The only known source and means of dissemination of *S. cepivorum* is via infected plant material or sclerotia.

The significance of dormant vegetative material as a source of white rot was recognized early this century. Garlic cloves imported from South Europe[3] and onion bulbs from Central Europe[9] are thought to have been responsible for the first outbreaks of white rot in Oregon and Virginia. Recently, garlic cloves were suspected as the source of white rot in Bajio, Mexico.[21] In 1968, white rot was introduced into Finland on onion sets[8] and, in 1978, into the Fayyum area in Egypt on bulbs from Upper Egypt.[57]

Shallot transplants were both the source of and the method of dissemination of white rot in Louisiana in the 1940s; shallot sets decay after infection and are not a source of white rot.[13] Similarly, the movement of infected onion transplants was the cause of the spread of white rot along the San Joaquin Valley, California in the 1970s.[19] There is a similar risk with other transplanted *Allium* crops, e.g., in South Africa[26] or Egypt[64] and in home gardens.

Contamination of seed by sclerotia or mycelium is unlikely to be a source of infection because seed production generally occurs in areas too hot for white rot. In an effort to explain the source of ''spontaneous'' outbreaks of white rot in Evesham, U.K., more than 100 samples of seed were examined for sclerotia but without success and this possibility was discounted.[61] Nevertheless, there is one report of leek seed contaminated by *S. cepivorum*.[61]

2. Weeds

White rot was reported in *A. canadense* L., a weed in the border areas of garlic and shallot fields in Louisiana.[13] It was not recorded whether the weed had been a source of infection for the cultivated plants or whether both the cultivated and weed alliums had common sources of infection. *A. vineale* L. is an occasional weed in the U.K. but it is doubtful if it is an important source of white rot.[61] It appears that white rot rarely affects wild alliums.

3. Water and Soil

Dispersal of sclerotia in irrigation water has been demonstrated in the U.S.[17,107] and in wind-blown plant remains and infested soil in the U.S., Germany, and Canada.[48,49,52] Sclerotia survive ingestion by sheep and goats when infected plants are used as fodder.[56] Sacks, bins, pallets, farm machinery, or footwear contaminated by infected plant waste or soil are other recognized methods of dispersal.[52,108]

4. Unresolved Outbreaks

The cause of some outbreaks is unknown. Leeks and onions, grown for the first time on recently ploughed grassland or old orchards, suffered 90% losses from white rot.[61] There was no evidence that the pathogen had been introduced in soil or on the planting material. The only explanation appears to be the long term survival from infected material present before the grassland or orchards were planted.

E. EPIDEMIOLOGY

1. Environmental Factors

Growth *in vitro* was optimum at 20 to 24°C and only slightly less at 15°C.[39,109] The minimum and maximum temperatures for growth were approximately 5 and 29°C, respectively.[109]

Onion seedlings grown at a constant temperature were killed within 3 weeks at 14 and 18°C. High rates of infection also occurred at 8 to 10°C in onion but not garlic.[109] At 20 to 22°C, infection was variable.[109] Symptom development was affected very little over the range of soil moistures at which infection occurred.[110]

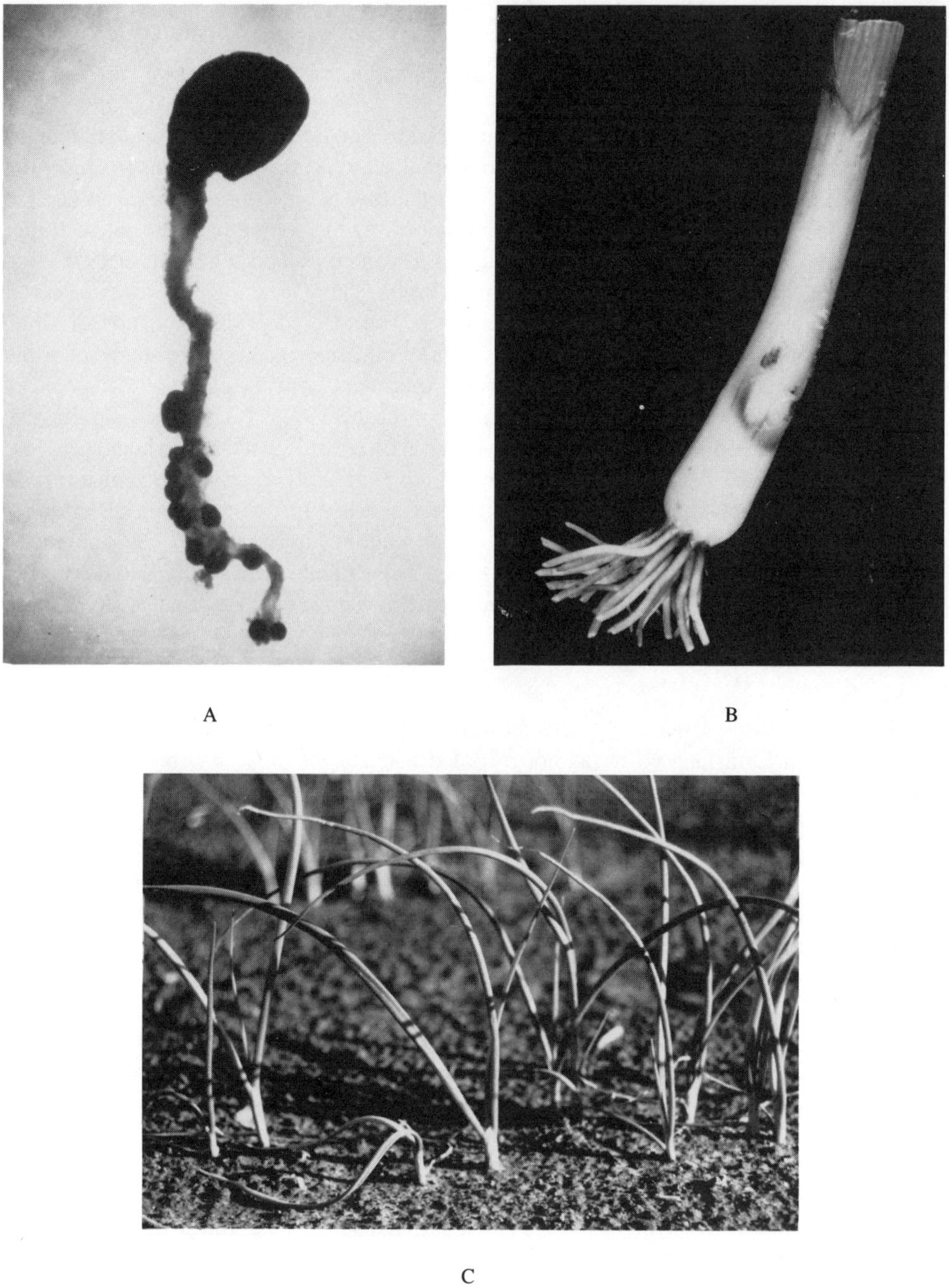

FIGURE 2. Symptoms of *Allium* white rot. (a) pre-emergence seedling infection; note sclerotia, (b) eye-shaped lesions on pseudostem below soil surface, (c) seedling wilt resulting from pseudostem lesion in 2b.: autumn-sown bulb onions Oct-Nov, U.K., (d) mycelium on roots: spring-sown salad onions, U.K., (e) sclerotia on maturing dry bulb onion: Egypt, (f) white rot present as a small patch (''hot spot''): spring-sown dry bulb onions, U.K.

Host stimulated sclerotial germination was maximal at 15 to 18°C and soil matric potentials of −85 to −300 mb. Considerable germination also occurred at 15 to 21°C at soil matric potentials of −100 mb to −1 bar, and at 12°C at −300 mb. Very few sclerotia germinated at 9°C at any soil moisture content.[110]

D

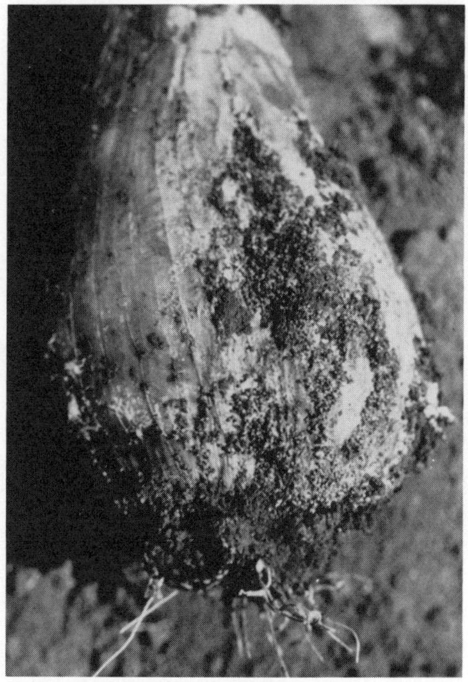

E

F

White rot incidence has been reported to be higher in dry conditions.[111,112] One explanation might be that infection had already occurred but foliar symptoms became evident only when plants became stressed through lack of water. In the U.K. salad onions, the risk of white rot is reported to increase when plant growth is checked. Improvements to soil structure, and to fertilizer and irrigation practices have been advocated as ways of reducing such checks to growth, thus minimizing the effects of white rot.[111,112]

2. Plant Spacing and Inoculum Position

Hyphal growth, thus secondary infection, occurs between roots of the same or adjacent plants[82,113] and is an important aspect of white rot development.[94,102] Secondary infection was fastest when roots or bulbs were in contact and became progressively slower as the spacing increased.[82,113] The amount of plant to plant spread will, therefore, depend on the time and depth at which infection occurs and the proximity of adjacent plants. Sclerotia deep in the soil are activated later in the season than shallow sclerotia because of the greater time needed for root growth. By this stage, however, the root systems will be larger thus providing increased opportunities for secondary infection. In garlic, many roots grow occurs horizontally for at least 30 cm at approximately 6 cm depth in soil and infection arising, say, from a low sclerotial population of 1 sclerotia kg^{-1} soil can give rise to 50% infection due to secondary infection.[113]

In Egypt, infection in onions was much reduced when sclerotia were present at >15 cm deep in soil.[114]

3. Soil Sclerotial Populations and Disease Incidence

Information on the relationships between the size of soil sclerotial populations has been used to forecast the incidence of white rot in onions[115-117] and to assess the performance of fungicidal control measures.[118,119]

The effects of various naturally occurring populations of sclerotia on the incidence of white rot in bulb onions were first investigated by Tichelaar.[115] There was a clear relationship between the size of soil sclerotial populations at the beginning of the season and the incidence of white rot at harvest (Table 4). This information formed the basis of advice to farmers for planning onion production.[116] A similar relationship was observed in bunching onions (*A. cepa* × *A. fistulosum*) in New Jersey in 1977—1979; in the following year, however, infection at 50 and 100 sclerotia kg^{-1} soil was 50% lower than in the previous 2 years.[117]

Soil sampling has also been used to identify areas of locally high soil sclerotial populations in Mexican garlic fields.[21] Areas with >20 sclerotia kg^{-1} soil had a high correlation with areas where white rot symptoms developed earliest. The presence of white rot in patches in California was also reported to be due to the presence of locally high populations of sclerotia.[54] The situation is less clear in the U.K. Sclerotia were recovered from 35 out of 46 soil samples from white rot patches in comparison with 11 out of 52 samples collected outside the patches. Soil sclerotial populations were generally higher inside the patches (0 to 377 sclerotia kg^{-1} soil) than outside (0 to 114 sclerotia kg^{-1} soil).[55] However, there was no consistent relationship between sclerotial populations and white rot incidence within these patches. In Brazil, 86% infection was recorded in garlic grown where there were 10 to 49 sclerotia kg^{-1} soil, but 3.2×10^3 sclerotia kg^{-1} soil were recorded in some fields.[119] Similarly large populations of sclerotia have been used in laboratory, greenhouse, and field experiments.[40,96,122]

There are several reports of low sclerotial populations resulting in high disease incidence. For this reason, growers in the Salinas Valley, California are advised against garlic production in fields with >1 sclerotia kg^{-1} soil.[118] Such low sclerotial populations are close to the practical limit of detection with existing methods for sampling soil and extracting sclerotia. An apparent absence of sclerotia in soil samples may, therefore, be a reflection of the inadequacy of the method of sclerotial analysis. Possibly this provides an explanation for the occurrence of white rot in soils where sclerotia are apparently absent. Thus, infection was 9% in garlic in the Salinas Valley,[120] 40% in New York State,[123] and up to 100% in module-grown bulb onions in Lincolnshire, U.K.[55] when sclerotial recovery from soil was zero. In these situations, the host plants provided a more sensitive indicator for the presence of *S. cepivorum* than did soil sclerotial analysis.

More information is needed on the effects of sclerotial number on the incidence and

TABLE 4
Relationship Between Soil Sclerotial Populations of *S. cepivorum* and the Incidence of *Allium* White Rot in Naturally Infested Fields[55]

Netherlands[115] (bulb onions)		U.S.: New Jersey[117] (bunching onions) 1977—79 1979—80			U.S.: California[120] (garlic) 1977		Egypt[121] (garlic) 1980—81	
SSP[a]	%AWR[b]	SPP	%AWR	%AWR	SPP	%AWR	SPP (×10[3])	%AWR
0	NR[c]	0	14	6	0	9	—	—
		1	14	8	1.5	40	1	5—8
2—10	25	5	15	12	2.3	70	5	25—40
		10	16	18	4.8	78		
12—20	31	20	19	29				
22—30	40							
32—40	46							
42—60	54	50	26	52			6	38—45
62—80	60							
>80	69	100	36	65			14—16	90—100

[a] SSP — soil sclerotial population (sclerotia kg^{-1} soil).
[b] AWR — *Allium* white rot.
[c] NR — not recorded.

patterns of infection in field before soil sclerotial analysis can be generally used for forecasting. Experiments in controlled conditions, using known numbers of sclerotia, may provide some understanding of the processes involved.[40,121] However, the variability in sclerotial numbers which occurs between soil samples taken close together indicates that studies of the field situation are essential.[117] An improved method of soil sclerotial analysis, which reduced the time involved, would clearly be an aid to such research. The method developed for elutriating nematode cysts from soil might also be applicable to sclerotia thus speeding analysis.[124]

F. CONTROL

1. Restriction of Dissemination from Infected Areas

a. Planting Material

Evidence that infected shallot transplants, onion bulbs, sets, transplants, and garlic cloves have been sources of white rot in the field has been discussed earlier. Clearly, the production of vegetative planting material should be confined to areas known to be free from white rot. Also, the growing crop and vegetative planting material should be inspected for evidence of infection, particularly before export to areas free of the disease.[8,21,26,125]

There is the possibility of eradicating infection in dormant material, e.g., by soaking garlic cloves in formaldehyde;[21] when sclerotia are present, a temperature of >48°C in the dip is needed for effective control.[126]

b. Quarantine

This has not generally proved successful at containing white rot in commercial situations. In Norway, restrictions, both on the import of diseased onions and the sale of home-produced diseased sets and transplants, failed to prevent the further dissemination of white rot.[127] Quarantine failed to contain the first outbreak of white rot at Lake Labish, Oregon and its use was discontinued 7 years later.[63] In Tasmania, Australia, an attempt to provide additional protection, by sowing white rot affected areas with grass, failed because the restrictions proved unacceptable in an area of high land values.[31] Success with quarantine measures is dependent on the rigor with which they are enforced; but strict enforcement may be impractical with intensive farming. In 1985, a quarantine order was implemented to maintain the white rot free status of three counties in Washington State;[128] white rot had had serious effect on onion production elsewhere in the state.[17] The quarantine order regulated the import of *Allium* propagation material, the sanitation of farm vehicles and cultivation equipment, and the movement of livestock from white rot affected areas.[128] In 1986, white rot was reported in one field in Ohio and comprehensive quarantine procedures were to be rigorously enforced; the affected area has been sown with grass and surrounded by a fence and ditch.[129]

Ideally, farm machinery and footwear should be washed free of soil, preferably at the source of contamination[21,48,128] There may be little that can be done to prevent the mechanical and windblown movement of contaminated soil during harvest.[48,49] There is also a risk of spreading contaminated soil when fields are leveled or graded. Similarly, the return of infected waste to the field of origin, but to a different area from the original infection, can result in new infection foci.[53] Precautions should also be taken to prevent infected material being blown from trucks at harvest or when stores are emptied.

c. *Infected Waste*

Temperatures of >35°C have been recorded during the aerobic composting of vegetable waste.[130,131] These temperatures are similar to those which kill sclerotia of *S. cepivorum*.[132] There is also the possibility that sublethal temperatures might weaken sclerotia and reduce their effectiveness as inoculum.[75,133]

Alternatively, sclerotia present in waste vegetable matter or soil might be stimulated to germinate in conditions where sclerotial regeneration is prevented. Thus, sclerotial germination was stimulated when onions were left in the field and covered with polythene sheeting and is being actively investigated as a means of control.[134] A similar technique could prove useful for the treatment of waste from stores, especially if onion juice or diallyl disulfide was added to optimize the effects on sclerotial germination, hence eradication. Probably, the material would have to be left for 1 to 3 months to minimize the inhibitory effects of dormancy on sclerotial germination.[135] Treatment of waste with compounds which generate methyl isothiocynate (MIC), e.g., metham sodium or dazomet is another possibility. Infected material could also be burnt.[21]

2. Eradication of Sclerotia from Fields

a. *Roguing*

Roguing, the removal of infected plants during the growing season, has potential for the eradication of soilborne pathogens which form sclerotia.[136] In Mexico, the maximum benefits were obtained when plants were rogued at th early stage of infection because sclerotia were trapped by mycelium, and thus removed.[21] Optimum results were obtained by also removing neighboring plants which may not have had foliar symptoms but nevertheless were infected. Roguing should have the combined effects of (1) reducing soil sclerotial populations by the stimulatory action of roots on sclerotium germination and (2) removing sclerotia which have formed on the host, before they become incorporated into soil. This practice merits consideration, even though it is labor intensive, providing the numbers of infected plants is small and they can be identified by foliar symptoms. It would appear to have merit in home gardens and smallholdings.

b. *Solarization*

Several soilborne plant pathogens have been controlled or partly controlled by mulching fields with polythene sheeting during the summer months prior to cropping, thus raising soil temperatures.[137] This process is termed solarization. The effects are most noticeable in hot, sunny regions and when the soil is moist. In South Australia, solarization killed 96% sclerotia at 25 cm depth when the mean maximum temperature was approx. 38°C;[75] there was total kill at 5 to 15 cm where the temperatures were higher. Similar effects on sclerotial viability were observed in North Australia at a soil depth of 5 cm; at 10 and 15 cm the mean maximum soil temperatures were 29 to 34°C and had less effect on sclerotia. *S. cepivorum* sclerotia were killed by a diurnal regime of 6h at 45°C and 18 h at 25°C maintained for 20 d. Some of the effects of solarization occurred at temperatures which were not directly lethal. Antagonistic fungi and bacteria were more abundant under these conditions and probably parasitized sclerotia which had been weakened but not killed.[75,133] Solarization also reduced white rot incidence by 50 to 67% in Sohag, Egypt.[138]

c. *Stimulants of Sclerotial Germination*

The specific dependence on *Allium* exudates for sclerotial germination provides opportunities to eradicate sclerotia from soil.[85] Methods which have been tested are (1) trap crops, (2) plant extracts, (3) onion oil, and (4) diallyl disulfide.

Onions were grown as a trap crop, in naturally infested soils, then killed with herbicide when seedlings were at the two-leaf stage. At this stage the seedlings were infected but new sclerotia had not formed.[139] This method failed to reduce soil sclerotial populations probably because the seedlings were too small to cause sufficient stimulation of sclerotial germination.

When plants were protected by iprodione, sclerotia were stimulated to germinate within the root system but infection was restricted and new sclerotia did not form.[140] Therefore, the repeated cropping with fungicide-treated plants should theoretically lead to a reduction in sclerotial numbers.[80,140] This method may be limited if germinating sclerotia form secondary sclerotia[80,141] or enhanced degradation of iprodione removes the protection.[142,144] The filtrates of cultivated alliums were highly stimulatory to sclerotial germination.[81] Therefore it may be possible to use *Allium* waste as a cheap source of stimulatory material.

Onion oil, injected into a naturally infested field reduced sclerotial populations by 50 to 97%.[29] In three experiments there was a 75% reduction in the incidence of white rot. White rot was not reduced in a fourth experiment probably because the residual population of sclerotia was still too high. Apparently the higher adsorptive capacity of the soil used in the fourth experiment reduced the effectiveness of onion oil.[29] Onion oil caused a 59% reduction in sclerotial numbers and a 73% reduction in white rot incidence in Canada.[145] However, there was no effect in Tasmania, Australia; the reasons for this are not understood.[31] This technique would need to be more consistent in its effects before it could be adopted in commerce.

d. *Soil Fumigants*

Several compounds have the potential for killing sclerotia and thus for controlling white rot and reducing further dissemination (Table 5). On other occasions, however, there may be little or no effect. For instance, metham sodium was alternately very effective,[149] or was ineffective.[147] A similar variability in effect was noted with dazomet.[148,150]

The effects of soil fumigants are likely to be governed by the duration and concentration of active ingredient to which sclerotia are exposed. For example, the continuous injection of metham-sodium into sprinkler irrigation, thus prolonging exposure of sclerotia to active ingredient, was far more effective at killing sclerotia, than a single injection of the chemical.[149]

Soil absorptive capacity and soil structure[151] affect the release of MIC and its subsequent diffusion. Distribution of dazomet granules was improved by the use of a purpose-built rotovator ("spading machine") for incorporating chemical.[148] The use of polythene sheets to cover soil is proving increasingly necessary in the UK to optimize retention of fumigants which are rotovated or injected into soil.[148]

The high cost of soil fumigants may restrict their use to high value salad or bunching

TABLE 5
Effects of Soil Fumigants on *Allium* White Rot

Chemical	Rate[a]	Application[b]	Sclerotia	White rot	Ref.
			% control[c]		
Metham sodium	2400	i p	NS[d]	18	146
	1200	i p	29—47	9	146
	456	s	0	0	147
	300	i p	21—75	25—27	148
	234	s	NS	10	31
	234	s	83	9	149
Dazomet	570	r p	100	100	148
	500	r p	NS	76	31
	380	r p	75—94	61—91	148
Methyl bromide	350	i p	100	100	147
	100	i p	95—100	89	146
	75	i p	96—100	69	146
Dichloropropene	225	i p	56—86	83	146
Chloropicrin	350	i p	0	0	147
Formaldehyde	760	s	NS	0	119

[a] Rate: Metham sodium, dichloropropene, chloropicrin, formaldehyde — liters commercial product ha^{-1}; dazomet, methyl bromide — kg commercial product ha^{-1}

[b] Application: i — injection; p — polythene sheeting; s — with sprinkler irrigation; r — rotovated.

[c] % control is data presented as percentages of values in untreated plots.

[d] NS — not stated.

onions. Costs could be reduced if it proved possible to treat patches of white rot rather than whole fields. However, it would be necessary to determine that the patches were due to localize populations of sclerotia rather than, for example, to the effects of soil or other factors suppressing the activity of sclerotia present in other areas of the field.

3. Cultural Practices

a. Rotation

Crop rotations of 4 to 10 years[6,8,25,152] have been advocated to restrict further dissemination of white rot. There appears to be little experimental evidence to support this policy; indeed, an 8-year rotation was ineffective in the U.K.[153] Possibly, similar amounts of disease could be expected in, say, a third onion crop which had been grown on the same site in consecutive years (''monoculture'') compared with an equivalent crop grown with a 4-year rotation; clearly any increase in white rot would be first evident with continuous rotation. This suggests two alternative strategies of crop management. First, the relocation of *Allium* production to areas that are geographically isolated or distant and free from white rot. For this reason, garlic production is being transferred from the Salinas Valley to the San Joaquin Valley, California.[63] There may be opportunities to move to white rot free land, even in districts where the disease is generally prevalent. For example, although white rot is widespread in the Vale of Evesham, U.K., land which had been under pasture for many years, proved to be free of white rot when brought into cultivation.[112]

Alternatively, *Alliums* might be grown every year on land that was free of white rot. There would be a risk of increased damage from nematodes (*Dityllenchus dipsaci* L.) and the larvae of onion fly (*Delia* spp.),[154] and this would need careful monitoring. In Canada, it has been possible to monoculture onions on organic soils, without serious problems from pests or diseases.[23] Monoculturing alliums might be a method for maximizing the benefits of eradicant treatments against white rot should these prove expensive or inconvenient, e.g., soil fumigants,[149] sclerotial germination stimulants.[52]

b. Avoidance

The development of white rot is restricted to soil temperatures of 10 to 20°C.[109,110] As a result, it has been possible to grow alliums outside this temperature range and thereby avoid the worst effects of the disease. In Louisiana white rot does not affect shallots before December 1.[13] Generally, shallot sets were planted in September—October and harvested in December—January. However, by planting infected areas in late August to early September, when soil temperatures are high, the crops could be harvested in October—November before infection develops.

In the U.K., crops of overwintered salad onions are sown from July—September for harvest the following February—May. White rot generally does not develop in U.K. winters. Accordingly, if infection is present in autumn, the crop could be harvested before April the following year when conditions again become favorable to white rot.[61]

In Egypt, onions are grown from seedlings transplanted in October—December. The crops mature in January—early March when *S. cepivorum* is most active. Onions grown from sets mature 2 months earlier than the transplanted crop, thereby avoiding the worst effects of white rot.[64,155] Several cultivars and strains of onion have been grown as sets with the objectives of breeding early maturing cultivars which avoid white rot (Table 6).

In South Africa, the seedlings of late maturing cultivars were produced when temperatures were increasingly unfavorable.[26] In these situations, infected transplants generally recovered from white rot. The production of seedlings of early maturing cultivars occurred when temperatures were still low enough for the development of white rot emphasizing the need to keep such areas free from infestation.[26] Also, in New Jersey spring planted onions can be grown safely because temperatures are too high for the disease.[95]

4. Protection by Chemical Fungicides

a. Historical

The various methods used to apply fungicide (Table 7) have been chosen for simplicity of application, economy and effectiveness, and the absence of phytotoxic effects on plant growth.

Certain conclusions can be made from a review of the literature on the use of fungicides for the control of white rot:

1. There are a number of compounds which in some situations are capable of providing a high standard of control; the most important examples are calomel, dicloran, PCNB and benzimidazoles and dicarboximide fungicides.
2. Sometimes these fungicides are effective for part of the growing season only. Thus, calomel and thiophnate methyl controlled white rot in the autumn period of overwintered salad onions, in the U.K., but not the following spring when rising temperatures again favor the disease.[159] The use of compounds with a more persistent effect, e.g., iprodione or vinclozolin or alternatively the application of stembase treatments in early spring extended the period of protection.[96]
3. Fungicides may be effective on one site but not on others. This is somewhat difficult to evaluate, since the original research may have been carried out on experimental plots whereas in commercial situations application rate and timing may differ. Dicloran, thiophanate methyl and iprodione are examples.

b. Dicloran

In the 1960—1970s, dicloran provided a high standard of control as a dip treatment for garlic,[162] as a broadcast treatment for direct-seeded dry bulb onions,[7] or a dip treatment for

TABLE 6
Avoidance of White Rot in Egypt by the Use of Early Maturing Cultivars[155]

Cultivar	Maturity date (1982—83)	Percentage white rot
Beth Alpha	Dec 20	4
Extra Early Yellow Bermuda	Dec 20	2
Shandaweel	Jan 20	20—24
Giza 6[a]	Feb 3	25—34
Other cultivars and strains (six)	Feb 3	30—57

[a] Cultivar generally used in commerce.

TABLE 7
Application of Fungicides for Control of *Allium* White Rot

Method	Fungicide
Broadcast on Soil	Dicloran,[17,156] iprodione, PCNB,[119,156] thiram, vinclozolin[156]
In-furrow spray	
To soil	Dicloran, PCNB,[17] iprodione, vinclozolin[21]
To garlic cloves	Dicloran, iprodione, vinclozolin[147]
Below seed	Procymidone[31,157]
Seed treatment	Benomyl, calomel,[93,158] dicloran,[158] iprodione,[96] thiabendazole,[158] thiophanate methyl[159]
Dip treatments	Benomyl,[160]
Onion and shallots sets, garlic cloves	Benomyl,[160] dicloran,[161,162] iprodione,[32,119,163]
	PCNB, vinclozolin[8]
Seedling transplants	Calomel, dicloran[17] benomyl, thiophanate methyl[164]
Fluid gel	Iprodione[165]
Modules	
Incorporated in compost	Iprodione, procymidone
Preplant drench	PCNB, thiabendazole, vinclozolin[166]
Stem base spray	Iprodione,[96] thiophanate methyl[159] vinclozolin[163]

dry bulb onions grown from transplanted seedlings[164] or for onions grown from sets,[161] as an in-furrow spray to soil for dry bulb onions[17] and as a seed treatment in salad onions.[158] In other situations the standard of control was rather variable.[167] Dicloran failed completely to control white rot in transplanted dry bulb onions in 1967 at Walla Walla, Washington.[17,168,169]

c. *Benomyl and Thiophanate Methyl*

Both fungicides have given effective control as a seed treatment in salad onions,[159,170] as a preplant furrow treatment in bunching onions and as a dip treatment for onion sets[101] or garlic cloves.[160] The use of thiophanate methyl at Wellesbourne, U.K., was subsequently discontinued in favor of iprodione, which was more effective. When tested again some years later thiophanate methyl gave no protection against white rot at Wellesbourne.[142]

d. *Iprodione and Vinclozolin*

Iprodione (as experimental compound 26019RP) showed promise for the control of white rot when tested as a seed treatment in 1974.[91] The effectiveness was further improved by the use of stembase treatments, and this led to the successful commercial adoption of combined seed and stem base treatments in the U.K.[96] Since then iprodione has been used in a variety of *Allium* crops.

Vinclozolin was also effective in onions in the U.K.,[171] in garlic in France,[172] the U.S.[147] and Brazil,[119] and in direct seeded dry bulb onions in the U.S.[173]

In 1981—1983, iprodione failed to control white rot at Wellesbourne, U.K., at increased doses of 18 kg a.i ha^{-1} (0.6 g a.i. m^{-1} row).[142,171] However, some control was observed in experimental plots which had not been treated previously with iprodione. The sensitivity of *S. cepivorum*, isolated from the sclerotia and mycelium of infected treated plants was unchanged even though dicarboximide-tolerance occurs readily *in vitro*.[174]

Control was restored by pre-treatment of experimental plots with methyl bromide, but the effect was temporary.[142] A partial and temporary improvement in control was also obtained when iprodione was applied in combination with thiram or fentin acetate.[175] These two compounds had previously been shown to retard enhanced degradation of carbendazim in soil and thus to restore the control of *Sclerotinia* in lettuce.[176]

Variability in the standard of fungicidal control might arise from deviations from the optimum method of application, rates, and timing of the treatment. The low standard of control observed with thiophanate methyl in Canada[156] may reflect the lower rate of application (approximately 0.09 g a.i. m^{-1} row^{-1}) in comparison with experiments in the U.S. or U.K. (approximately 0.15 g a.i. m^{-1} row^{-1}).[91,177] In the U.K., however, other factors must also have been involved because control failed in later years on the same site, even at iprodione concentrations four times the former rate.

The standard of control is also affected by the numbers of sclerotia in soil. Iprodione or dicloran were less effective at controlling garlic white rot at 10 sclerotia kg^{-1} soil or above.[118] At low sclerotial populations, iprodione and vinclozolin give similar standards of control whereas at high sclerotial populations, vinclozolin was the more effective chemical.[173] A similar effect of sclerotial populations on control was noted in Brazil but at much higher sclerotial populations[119] (Table 8). The characteristically patchy occurrence of white rot in the field probably reflects difference in soil sclerotial populations. It follows that these differences in sclerotial patterns might be expected to give rise to variations in the effectiveness of control treatments within a field.

The physical and chemical properties of soil also affect the availability and persistence of organic chemicals applied to soil. Silt and peat soils are highly absorptive and increased rates of chemical may be needed. This may be another factor accounting for the reduced effectiveness of thiophanate methyl in Canada in peat soil in comparison with mineral soil.[156,159,177] Dicarboximide fungicides are less stable at high pH and persistence may be reduced if either alkaline water is used to prepare sprays[178] or soils have a naturally high pH or have been treated with large amounts of lime.

The repeated application of an organic fungicide to soil can result in an increase in the degradation rate — termed enhanced or accelerated degradation — thereby reducing persistence. Dicloran had a higher degradation rate in soil from commercial fields, which had been treated several times with dicloran, in comparison with soil which had not been previously treated with the chemical. Additional applications of dicloran increased the rate of degradation still more.[169] The effect was apparently due to the presence of dicloran-degrading bacteria and was reversed by a variety of soil sterilization treatments *in vitro*. Enhanced degradation of dicloran was associated with the loss of control of white rot by dicloran in 1967,[169] but there is no data which directly links the enhanced degradation with reduced disease control. An increase in dicloran tolerance in the pathogen may also have been a factor in the loss of control of white rot.[168]

An increased degradation rate of iprodione was also associated with loss of control of white rot by iprodione and vinclozolin although, again, there were no data directly linking the two processes.[143]

TABLE 8
Effect of Soil Sclerotial Populations on the Control of White Rot in Garlic by Iprodione[119]

Iprodione applied to		**Percentage garlic plants with white rot**	
Cloves (g a.i. kg^{-1} cloves)	Soil (g a.i. m^{-2})	Sclerotia kg^{-1}	soil ($\times 10^3$)
		0.1—0.2	1.6—3.2
1000	4	2	30
1000	0	2	42
0	0	94	92

Recently, it was noted that there was a marked reduction in the persistence of iprodione or vinclozolin in soil from an area where iprodione had been applied as a series of four sprays at 0.5 kg ha^{-1} $application^{-1}$ for the control of foliar *Botrytis*.[144]

Simultaneously with the loss of control at Wellesbourne, two other commercial fields were identified where there was a similar failure; it is not known, however, whether control had ever been achieved with iprodione on the latter two sites. On these sites, the rate of degradation of iprodione was higher in soil from the cultivated areas compared with soil from uncultivated border areas.[143]

No other examples of loss of control by iprodione were known at that time. Then, in 1986, there were reports of failures of control with iprodione in the Salinas Valley, U.S.,[22] Kent, U.K.[111] Germany,[49] and Tasmania;[31] and with vinclozolin near Evesham, U.K.[112] Again, there were no data which directly linked enhanced degradation of iprodione or vinclozolin to the loss of white rot control, but it seems increasingly likely that this is so.

The mechanism by which enhanced degradation is initiated in the field is unknown. Seed, furrow and stembase treatments affect only a small zone of soil where the change presumably occurs, yet the change is readily disseminated, for example the experimental field at Wellesbourne, U.K. and the induction of enhanced degradation by a 10% inoculum of degrading soil.[144]

Two of the recent examples of loss of control in the U.K. have occurred only after prolonged use of iprodione or vinclozolin applied as seed and stembase treatments against white rot and as sprays (iprodione) against foliar *Botrytis*.[111,112] This possibility indicates that the change occurred slowly and was detected only when losses had become sufficiently obvious on a field scale. Alternatively, there may have been no change until a threshold quantity of fungicide had been applied, when the change was sudden and total.

Currently there are no practical measures for reversing the effect by enhanced degradation. Soil sterilization *in vitro* inhibits degradation[169,176] but in the field the effect on white rot was temporary[142] possibly because of reinvasion or contamination from adjacent areas. The addition of thiram to iprodione offers promise and further improvements may be possible.[175]

e. Procymidone

This dicarboximide fungicide is active against white rot in salad onions, dry bulb onions grown either from seed[31,53,157] or from sets,[179] and in leeks.[53] At Wellesbourne, U.K., there was 100% control in conditions where iprodione was ineffective.[53]

Seed treatments have given a 90% reduction in white rot in dry bulb onions, but reductions of 24 to 65% were more usual. In contrast, procymidone was ineffective in Victoria, Australia (Table 9). Soil treatments or combined seed and soil treatments usually gave the best control (Table 9). The lack of control in Victoria was considered to be an effect of soil type;[157] a

TABLE 9
Effects of Procymidone on White Rot in Dry Bulb Onions Grown from Seed in Six Experiments in Naturally Infested Soil 1984—86

Fungicide applied to	Rate	Victoria,[157] Australia	Tasmania,[31] Australia			Netherlands[179]	
Seed	50	68	6	25	30	—	—
	25	—	—	—	—	57	18
Soil	4	32	—	—	—	—	—
	2	54	—	—	—	11	10
Seed + soil	50 + 4	43	—	14	31	—	—
	50 + 2	38	—	—	—	—	—
	25 + 2	—	—	—	—	7	7
Untreated		73	61	82	86	83	38

[a] Rate: seed treatment — g a.i. kg^{-1} seed; soil treatment — kg a.i. ha^{-1}.

reduced level of effectiveness in Tasmania, was thought to be a result of high soil sclerotial populations.[31] The Australian data are of special interest because the fungicide was applied, in combination with fertilizer, as a band placed below the seed thus simplifying application and avoiding phytotoxicity. This technique may have use in the U.K. where starter solutions of fertilizer are being investigated as a means of improving establishment in onions.[180]

Procymidone is a dicarboximide fungicide and, therefore, there is the possibility of enhanced degradation. The use of seed and dip treatments might conceivably minimize this possibility by reducing the amounts of chemical used.

f. Other Chemicals

Calomel has been used successfully, chiefly in the U.K.,[17,61,93] but is likely to become less important because of the undesirable environmental properties of mercury. Calomel has value as a standard treatment for comparison with experimental materials in screening experiments.[96,159,165]

PCNB,[17] prochloraz,[181] thiabendazole,[182] propiconazole, and penconazole[183] have the capacity to reduce white rot. Triadimefon was ineffective at low doses,[156] and at high doses it was partially effective but phytotoxic.[184] Thiram was ineffective as a seed treatment at 10 to 150 g a.i. kg^{-1} seed[81,96] or broadcast at 45 kg ha^{-1}.[156]

G. FUTURE POSSIBILITIES FOR CONTROL

1. Microbial Control

a. Background

The experimental use of antagonistic microorganisms to control white rot has shown promise in several field trials.[185-187] However, the method has not proved sufficiently effective or reliable for use in commerce.

The white rot disease cycle presents several targets for microbial control. There are (1) sclerotia remaining in soil from previous outbreaks, (2) stimulation of sclerotial germination, (3) growth of hyphae from sclerotia to root, (4) root infection, (5) growth of hyphae from infected host tissue (i.e., nutrient-rich food base) to adjacent roots and plants, and (6) production of new sclerotia.

Control might be achieved by eradicating sclerotia in soil before planting or by applying treatments which protect the growing plant. Also, the dispersal of *S. cepivorum* would be restricted if sclerotia were eradicated from infected plant residues for example those remaining in the field after harvest or after grading or storage.

b. Eradication of Sclerotia before Planting

At this stage, sclerotia are likely to be widely dispersed in soil particularly if there has been a long period since the previous outbreak of white rot. In these circumstances, there are obvious practical difficulties in bringing biological control organisms into contact with sclerotia. Therefore, *Coniothyrium minitans* Campbell might fail to control white rot even though it is an effective parasite of sclerotia under favorable conditions.[188] A more appropriate choice might be *Sporidesmium sclerotivorum* Uecker, Ayers, and Adams.[189] *S. sclerotivorum* has the capacity to produce large numbers of macroconidia and to grow through soil from parasitized to healthy sclerotia. This can result in a progressive reduction in sclerotial populations.[190] However, isolates of *S. sclerotivorum* differ in their activity against *S. cepivorum* sclerotia and not all are effective.[189] For use in commerce, isolates would be needed which could grow rapidly *in vitro* and produce large numbers of macroconidia.[189]

An alternative method of eradicating *S. cepivorum* would be to apply antagonists to the growing crop. New sclerotia form on stembase tissue and thus are more accessible to treatment than sclerotia which have been dispersed in soil.[113] Antagonists might attack sclerotia directly or become incorporated within the rind during development thus reducing survival.[191] *C. minitans* and *Trichoderma harzianum* Rifai are active against hyphae as well as sclerotia[12] and the growth of hyphae from infected to adjacent healthy plants might also be halted by treatment of the growing crop.

Eradication of *S. cepivorum* from the crop before harvest would have the advantage of minimizing the dispersal of *S. cepivorum* to unaffected areas. The treatment of white rot patches, identifiable by foliar symptoms, is one obvious possibility.[53] An alternative would be the treatment of infected plant residues after the healthy plants have been harvested for example by spraying infected material with sclerotial parasites or by incorporating biological control organisms into dumps of plant waste.

c. Protection of Growing Plants

Often, microorganisms have been used to protect plants from infection. A common procedure has been to identify antagonistic organisms on the basis of their reaction in dual confrontation on agar medium.[12,185,192] Selected isolates are then applied to the seed,[185,188] to the roots of seedlings,[12] or to the furrow or planting hole.[188] Alternatively, organisms are broadcast over the soil surface then rotovated.[185] Additional possibilities would be the incorporation of organisms into modules or into to gel used for fluid drilling.[193] Sometimes biological control organisms have been added with a food base[186,194] otherwise the food base has been excluded.[12,185]

The results of such tests are dependent on the identification of antagonism, the choice of a suitable method of application and the compatibility and survival of the antagonist in soil. Organisms isolated from local soils — residents — would be expected to be better adapted to these soils than nonresident organisms.[12] In Brazil, fungi were isolated from soil from areas with a history of severe white rot.[12] The isolates were then tested for antagonism to *S. cepivorum* on nutrient-rich and nutrient-deficient medium. Twenty-four isolates proved antagonistic out of fifty which were tested. *T. harzianum* produced diffusible antibiotics *in vitro* which restricted mycelial growth. Later, *T. harzianum* hyphae coiled round *S. cepivorum* hyphae disrupting their contents.[12] *Paecilomyces lilacinum* (Thom.) Samson and a *Penicillium* sp. also restricted *S. cepivorum* growth *in vitro* but without hyphal contact. The application of 2.5×10^7 spores of *T. harzianum* to the roots and planting hole reduced white rot incidence in garlic from 100 to 80%; *P. lilacinum*, *Penicillium* sp. and *C. minitans* 2A had little effect on white rot.[12]

Organisms which were antagonistic to *S. cepivorum in vitro* were compared with non-antagonistic organisms for their effect on white rot in a pot test.[192] *Penicillium nigricans* Bain. (Thom.) was both very antagonistic *in vitro* and very effective at controlling white rot.[192,194] Eight other isolates of other species also proved antagonistic *in vitro*, three reduced white rot to a moderate extent the remainder were ineffective. Out of five nonantagonists, one reduced white rot and the others had no effect.[192] Another isolate of *P. nigricans* isolated from sclerotia from soil in Canada also proved antagonistic *in vitro* but was ineffective when tested against white rot in the field. The success of *P. nigricans* in the former example may have been due to the use of a food base whereas in the second, spores of the antagonist were applied as a seed treatment, i.e., there was no food base.[185]

It can be seen that isolates identified on the basis of their antagonism to *S. cepivorum in vitro* do not often fulfill their promise in the field even when the isolates were recovered from and tested in local soils. It is not possible, however, to determine whether lack of success is because of inadequacies in the *in vitro* screening method for identifying suitable organisms or whether factors such as the nutrient or physiological status of the antagonist or its compatibility to the environment are involved.

In the field, organisms which initially are effective, may later fail to provide control for reasons which are not understood. For example, *B. subtilis* controlled white rot in two successive years but was ineffective in the following two years.[187]

In Egypt, *T. harzianum* was raised in the dark on barley seed and gliotoxin fermentation medium and used to treat a total of 230 ha in 1980 — 1983.[195] An average of approximately 80% control of white rot was achieved. *T. harzianum* was less effective when it was added in cool conditions, or when there was low soil organic content and crops were irrigated more than once.[195]

An improvement in the selection of biological control organisms was achieved with *B. subtilis* by measuring *in vitro* antagonism to a range of *S. cepivorum* isolates and then comparing the effects in seedling infection tests. Thus, isolates of *B. subtilis*, able to control white rot caused by a range of *S. cepivorum* isolates, were identified from their *in vitro* characteristics.[196]

2. Host Resistance

a. Background

A low incidence of white rot, has been interpreted as evidence of host resistance in commercial cultivars, e.g., cvs. Up-to-Date,[34] Ailsa Craig,[197] in species, e.g., *A. cernuum* Roth., *A. ampeloprasum* L.s.l.[198] *A. obliquum* L., and *A. tuberosum* Rottl.ex.Spreng.[34] and in Plant Introductions (PIs) particularly PI 264650.[62,199]

At other times, however, this material has shown little evidence of resistance, e.g., cv. Ailsa Craig,[200] *A. cernuum*, and *A. tuberosum*.[40]

At the present time, there are no cultivars in commerce with either sufficient resistance to *S. cepivorum* or with resistance which is consistently expressed. Nevertheless, there are indications that a reduced incidence of white rot is regularly associated with certain seed lots of cv. Ailsa Craig and with PI 264650 and this may indicate resistance.

A low white rot incidence in a screening test may be due to (1) small sclerotial populations or a lack of response to stimulatory host exudates, (2) a restriction of secondary infection, e.g., caused by wide plant spacing, or (3) a difference in the composition and production of stimulatory host root exudates.

b. Causes of Variability in Response

Populations of <1 sclerotium kg^{-1} soil are sufficient to cause disease in some fields.[120] Therefore, the experimental use of high sclerotial populations in screening tests for resistance requires explanation. In controlled conditions favorable to *S. cepivorum*, 40% or more seedlings remained healthy in the presence of 50 to 500 sclerotia in the root zone of plants grown in small pots[45,87] or 10^4 sclerotia kg^{-1} soil in seed trays.[198] Possibly, dormancy was a factor and some sclerotia did not germinate in response to root exudates.[87] In contrast, a high disease incidence was obtained from the use of large amounts of sclerotial inoculum in similar tests in controlled conditions.[40] There is the disadvantage that small amounts of host resistance might be masked by a high "infection pressure" and a less severe test might be preferable.[34]

The growth of hyphae between roots[113] or bulbs (secondary infection)[82] is markedly restricted by increases in plant spacing and this is reflected in a lower disease incidence. Such reductions can be mistaken for evidence of host resistance.[94] The precise spacing of plants would minimize variation in infection due to variations in plant to plant spread.[197] Nevertheless, some variability in plant populations is largely unavoidable due to differences in seed vigor, sowing depth, and soil structure, etc. In these situations, regression analysis and analysis of covariance can be used to compensate for variations in disease incidence.[94] For example, five cultivars and four breeding lines, tested in the field, varied in infection from 5 to 56% and these differences were statistically significant. However, when the variations in plant populations had been compensated for, the differences proved to be nonsignificant.[94] In a second trial, comprising a total of 25 PIs, breeding lines, and cultivars, differences in white rot incidence remained statistically significant even after adjustment for variations in plant population. There were, however, changes in the ranking of some test lines. Most of the changes were of one or two places only, but three of the lines changed position by four places.[94]

Sclerotium germination, hence root infection, is affected by stimulatory precursors in *Allium*.[201] Therefore, differences in disease incidence may reflect differences in stimulatory capacity rather than tissue resistance.[89,202] *A. aflatunense* B. Fedtsch., *A. cristophii* Trautv., and *A. karataviense* Regel do not stimulate sclerotial germination and this was considered to account for a lack of infection when these species were grown in severely infested soils.[203] Leek has consistently been reported to be less susceptible to white rot than onion, garlic or *A. fistulosum* L.[44,45] but there is no detectable tissue resistance *in vitro*.[202] The rate of sclerotial germination is lower with leek than with the other three species due to a smaller content of stimulatory precursors and a higher proportion of the inactive methyl-radical in the precursor compounds.[89] Therefore, the lower disease incidence in leek is most probably due to differences in the stimulation of sclerotial germination, rather than host resistance.[202]

An added complication is that the amount of sclerotial germination may be related to the size and vigor of the root systems. This makes it difficult to compare cultivars from different geographic regions in a single set of conditions because their patterns of growth will differ.[109] Egyptian cultivars grown in the U.K. grew faster in the early part of the year compared with European cultivars. The faster growth was accompanied by higher rates of sclerotial germination, thus, disease development but there was no evidence of increased tissue susceptibility.[204] It follows that a lower incidence of white rot in inbred lines, which often demonstrate weaker growth in comparison with their parent, cannot be interpreted as evidence of host resistance without additional supportive data.[205]

In spite of these drawbacks, evidence of resistance does appear to have been obtained with cv. Ailsa Craig and PI 264650 origin cv. Zittauer Gelb.[206] With cv. Ailsa Craig, however, resistance was limited to plants from certain seed lots only.[199] The low incidence of white rot observed with PI 264650 occurred both at high plant populations and in the progeny of hybrids with susceptible cultivars.[199,206]

Future research on host resistance needs to distinguish between differences in infection resulting from variations in sclerotial germination and from tissue resistance. Thus a screening test would need to account for possible variation in (1) the pathogenicity of *S. cepivorum* isolates,[196] (2) sclerotial dormancy,[81] (3) secondary infection,[94] and (4) the effects of microflora in different soils.[207]

III. PINK ROOT

A. DISTRIBUTION AND ECONOMIC IMPORTANCE

Pink root was first reported in Texas in 1917 but had probably been present for at least 20 years previously.[208] The disease is present in areas with high soil temperatures, and mainly affects onions and garlic.[209] New outbreaks continue to be reported,e.g., Mendoza, Argentina,[210] Arequipa and Lima, Peru,[211]; and the Sudan.[212] Pink root also affects a wide range of vegetable and cereal crops[213,214] and, therefore, is not confined to areas of onion production.[215]

The disease occurs in onions grown for the first time[215] and becomes increasingly damaging in successive onion crops, particularly with short rotations.[216-218] In extreme situations, many plants are lost through seedling infection, the remainder become stunted[217,218] and total and marketable yields are reduced.[219] Pink root is mainly a field disease and the damaging effects of pink root are mainly on growing plants but losses also occur during transport and in store.[208,209]

Pink root often occurs in association with *Fusarium* basal rot (Section IV.B). In these circumstances it may be difficult to determine the relative importance of each disease.[220]

B. PATHOGEN

Pink root is caused by the fungus *Pyrenochaeta terrestris*.[221] The fungus was first assigned to the genus *Phoma* (as *P. terrestris*)[216] because setae were not present on the pycnidia of isolates from California. Later, another isolate from California proved to be atypical in comparison with isolates from other parts of the U.S. and the presence of setae was found to be normal.[221] Accordingly, the fungus was reassigned to the genus *Pyrenochaeta*.[221]

P. terrestris is diagnosed *in vitro* by the presence of setae on pycnidia and the production of a pink coloration on sterile wheat straw.[222] Pycnidial production is faster when cultures are irradiated with UV light.[223] The pink color is most intense at pH 7.9 and turns to yellow at pH 7.0 or less.[224]

Mature pycnidia have rarely been seen on wild or cultivated hosts but form occasionally when onions are grown in sand cultures inoculated with *P. terrestris*.[221] Pycnidial primordia, however, form abundantly in epidermal and cortical root cells of onion and a range of other hosts.[213,221,225] In culture, pycnidia are subglobose, ostiolate, papillate, dark brown to black, 170 to 400 μm in diameter Setae are 60 to 180 μm in length, vary considerably in number and are generally grouped round the ostiole but occasionally form elsewhere on the pycnidium.[221] Hyphae are hyaline, septate, guttulate 1.0 to 4.5 μm in diam.[216,221,226] Conidia are unicellular, ovoid, biguttulate 4.0 to 7.0 $\times$ 1.5 $\times$ 2.3 μm.

Inoculum is thought to be in the form of infected organic residues, pycnidia (primordia?), pycnospores, and microsclerotia.[215]

C. SYMPTOMS

P. terrestris causes a characteristic pink coloration in the roots of *Allium* and a wide range of cultivated and wild hosts.[213-216,225] *Allium* roots also turn pink when infected by *F. roseum*[227] (Section IV.D.3) but the two fungi are not generally found together. *Fusarium* basal rot (Section IV.B) commonly occurs with *P. terrestris* but does not cause roots to go pink.[216] Initially, pink roots are firm but later become soft and collapse and the color becomes purplish.[216] The wrapper scales of bulb onions are also colonized and become purplish.[213] Occasionally, *P. terrestris*-infected roots are yellow rather than pink[224] making diagnosis difficult if roots are senescent.[222]

Pink root disease causes a check at any stage of growth, providing soil temperatures favor infection. Thus, seedling growth is retarded and stunting become progressively worse during the season.[219,228] Leaf number and size are reduced, bulbing starts early and final bulb size is reduced.[219] Foliar symptoms are similar to those of a drought stress, that is, chlorotic leaf dieback and wilting.[216]

D. SOURCES OF OUTBREAKS AND DISSEMINATION

The first outbreaks of pink root in Texas were attributed to infected onion sets imported from Bermuda.[208] There is also the possibility that *P. terrestris* could already have been present as a normal component of the soil microflora.[215]

Examples of planting material with potential for the dissemination of pink root are onion seedlings[215,222] and sets,[208,215] shallot sets,[229] and garlic cloves.[208,209] The presence of *P. terrestris* in infected vegetative planting material, or in adhering soil, would be expected to increase damage from pink root in areas where populations of the pathogen in soil are low. The introduction of *P. terrestris* in this manner could be of particular importance in irrigated desert soils where the paucity of an antagonistic microflora might favour a rapid increase in *P. terrestris*.

Dissemination also results from the movement of spores, pycnidia, infested soil and infected plant residues by agricultural machinery, wind and water.[215,217] *P. terrestris* is not seed transmitted.[215]

E. EPIDEMIOLOGY

1. Environmental Factors

P. terrestris is a common component of the flora in some soils, thus a reservoir of inoculum and has been recovered from uncultivated desert, mountain, and forest soils.[215] In onion fields, a seedling bioassay was used to demonstrate the presence of the pathogen at depths of 45 cm.[230] *P. terrestris* was, however, most abundant in the upper 15 cm of soil probably reflecting the effects of cultivation on the distribution of infected onion roots. *P. terrestris* has been recovered from the root systems of a range of graminaceous[214] and other plants.[215]

P. terrestris, like *Fusarium oxysporum* f.sp. *cepae* (Section IV.B), is most active at 24 to 28°C in soil.[216,231] Therefore, the damaging effects of pink root are determined by the seasonal pattern of soil temperatures and the growth period of the host. When high soil temperatures only occur late in the season, damage from pink root in minimized by growing a cultivar which matures early thus escaping infection.[232] Late maturing cultivars continue to grow at soil temperatures favoring pink root. Root infection then checks plant growth, shortens the growing season and reduces yield.[219,228,232]

Pink root is also more damaging in plants grown from low vigor seed and this may reduce protection by host resistance.[232,234] Some cultivars, however, produce new roots at a rate sufficient to keep pace with infection and thus, produce a marketable crop.[235]

Populations of *P. terrestris* in soil may be low when onions are grown for the first time, or following soil fumigation.[217] Populations rise rapidly in the presence of onion and there is a corresponding rise in pink root incidence.[217,235] Pink root incidence declines with long rotation[208,218] presumably because of a decrease in populations of activity of *P. terrestris*.

2. Relationship between Pink Root and Fusarium Basal Rot

Pink root and basal rot (*Fusarium oxysporum* f.sp. *cepae*; Section IV.B) are often present together as a disease complex.[209,231] As a result it can be difficult to distinguish which of the two diseases is the main cause of damage.[220]

The association of these two diseases has led to the suggestion that their presence together is linked.[213,231] Later, it was demonstrated that the presence of either pathogen is not dependent on the presence of the other.[213,216,236] Furthermore, a pink coloration of *Allium* roots is due to the presence of *P. terrestris* and is unaffected by the additional presence of *Fusarium* spp.[216,231] An exception is *F. roseum* infection in leek[227] (Section IV.D.3).

Only the roots and dried outer scales of onion are invaded by *P. terrestris* whereas with *F. oxysporum* f.sp. *cepae* both the roots and the fleshy leaf scales become infected.[213,231] Infection by *P. terrestris*, however, provides an opportunity for invasion by saprophytic forms of *F. oxysporum* and other *Fusarium* spp.[216,231,236]

F. CONTROL

1. Host Resistance

Leek, chives,[235] *A. fistulosum*,[233,235] and some garlic clones[208,209] are resistant to pink roots whereas shallot[228,234] is susceptible.

Onion resistance has been reported in Bermuda types, e.g., cvs. Yellow Bermuda, Red Bermuda,[208] Sweet Spanish types, e.g., cvs. Prizetaker, Sweet Spanish, Valencia[235] and the local Sudanese cv. Red.[212] A large number of other cultivars are very susceptible, e.g., cvs. Granex, Egyptian, Ben Shemen, (a selection of cv. Riverside),[219,233] Nassei,[212] Texas Early Grano 951, and San Joaquin.[237] Breeding for resistance has been either by the selection of resistance within onion cultivars or by the transfer or resistance from *A. fistulosum* to onion.[237]

Cultivars reported to be pink root resistant (PRR) may not prove resistant in all situations. Cultivars and selections of cultivars of American origin and reported to be PRR were compared in Israel with cultivars in regular commercial use. When tested in microplots, pink root had little effect on the time taken to maturity of either PRR or regular cultivars, with the exception of cv. Spano which matured early.[232] However, the PRR cvs. Dessex, Granex and Laredo had larger bulbs and healthier roots than other cultivars including the PRR cv. Grano.[232] When tested in naturally infested field conditions, the differences in yield, between the three best PRR cultivars and one local cultivar, were small. Resistance appeared to involve two factors (1) the capacity of roots to resist infection by the pathogen and (2) the capacity of the plant to continue to produce new roots after infection has occurred. In addition, cultivars which mature early may escape infection by avoiding temperatures favorable to the pathogen.[232]

Research is in progress to transfer resistance from *A. fistulosum* to *A. cepa*. Thus the progeny of *A. fistulosum* × male sterile *A. cepa* were markedly resistant to pink root.[238] Some of the "F2" generation produced plants which bulbed — possibly the results of backcrosses between the F1 and *A. cepa* — which were indistinguishable in vegetative morphology from onion.[238]

There is the possibility of resistance in garlic. Seven clones of garlic from five countries were tested in Wisconsin soil tanks. Symptoms occurred 35 d after planting in two clones but foliage remained green in four clones and of these two had 10% of less pink root whereas a susceptible onion cultivar was destroyed.[209]

Preliminary tests indicate the host resistance to pink root may be partly due to tolerance to pyrenocines produced by the pathogen.[239]

2. Solarization

This technique has proved of special value against pink root in Israel because the high soil temperatures are conducive to solarization. For example, in S. Israel, solarization reduced pink root incidence to <10% at harvest and symptoms were mild.[228] In nonsolarized plots, however, severe pink root affected most of the plants within 8 weeks of sowing.[228] Similar results were obtained in cooler conditions in the north of the country, except that bulb quality was improved. In S. Israel, pink root and other soilborne diseases often caused marked reductions in plant populations which are compensated for by increasing the seed drilling rate.[219,228] Therefore, the control of pink root and other diseases in seedlings can result in excessive plant populations and bulb size and quality is reduced. The use of solarization to control pink root should provide opportunities for improving the accuracy of plant populations.[219]

3. Soil Fumigation

The use of dichloropropene (Telone, Dow Chemical Company) at 450 l ha^{-1} to control pink root increased both the yield of the profitable large-sized onions and total yields in the Treasure Valley area of Idaho and East Oregon.[217] Fumigants were applied in performed beds in autumn and then seeded the following spring.[217] More recently, the use of a dichloropropene-chloropicrin (40/60%) mixture at 84 l ha^{-1} has given good control.[218] Changes in populations of *P. terrestris* may be measured by a soil dilution technique as well as by host plant bioassay.[246] Chloropicrin, methyl bromide, and MIC, applied in controlled laboratory conditions, markedly reduced *P. terrestris* populations for more than 4 months.[217] In field conditions, populations were reduced early in the season but by harvest had returned to normal.[217]

4. Rotation

A rotation of 5 to 6 years in the Treasure Valley, U.S. sufficiently restricts pink root to allow profitable onion production.[218] Long rotations may be impractical, however, if there is insufficient alternative fertile land for onions or alternative profitable crops are not available.[218]

IV. FUSARIUM BASAL ROT

A. NOMENCLATURE AND ECOLOGY

Fusarium basal rot of onion is caused by *F. oxysporum* Schlect. emend. Snyder and Hansen f.sp. *cepae* (Hanzawa). The complexity of *Fusarium* taxonomy has resulted in a variety of synonyms as follows:

F. cepae Hanz.,[241,242] *F. cepae* (Hanz.) emend. Link and Bailey (not *F. cepae* (Hanz.) emend. Walker and Tims),[236] *F. oxysporum* Schlect. forma 7 Wollenweber,[242] *F. vasinfectum* Atkinson v. *zonatum* (Sherbakoff) f.1 (Link and Bail.) Wollenw.,[242] *F. vasinfectum* Atk. v. *zonatum* (Sherb.) f.2 (Link and Bail.) Wollenw.,[242] *F. zonatum* (Sherb.) Wollenw.[243] *F. zonatum* (Sherb.) Wollenw. f.1,[231,236,244] *F. zonatum* (Sherb.) Wollenw. f.2 Link and Bail.[236,244]

The taxon forma specialis was introduced to distinguish forms of the *Elegans* section of *Fusarium*.[242,245] These forms are recognized by their selective pathogenicity, usually to a single host species.[242] For example, f.sp. *cepae* is identified by testing isolates of *F. oxysporum* for their pathogenicity to onion slices[246] or by inoculating plants grown in either methyl bromide-treated soil,[228] sterilized (pasteurized) or nonsterilized soil.[247] It should be recognized, however, that there is the possibility that a forma specialis may also give rise to disease symptoms in hosts other than the standard host. Another possibility is that infection could occur but without recognizable disease symptoms, i.e., the plant species is a "symptomless host".[248] The chlamydospores of *Fusarium* are resistant to treatments used to surface sterilize plant tissue. Therefore, it may be difficult to distinguish, with certainty, between a symptomless host and a symptomless carrier, i.e., where proliferation occurs within the root zone but there is no infection.[249]

Thus, determining the identity and the involvement of *Fusarium oxysporum* isolated from diseased *Allium* tissue is not straightforward. The following examples illustrate some possibilities:

1. *F. oxysporum* isolates from onion basal rot may prove to be nonpathogenic.[236,246] The original isolate of *F. oxysporum* might have been pathogenic but later became masked or displaced by saprophytic *Fusarium* spp. Alternatively, the onions may have been wounded, e.g., by maggots, and pathogenic *Fusaria* may never have been involved.

2. Forma specialis *cepae* causes disease symptoms in Welsh onion[250] and garlic[209] as well as in onion (Section IV.B.)
3. *Oxalis corniculata* is a symptomless host for f.sp. *cepae*[251] (Section IV.A.5).
4. Basal rot symptoms can develop in onion inoculated with f.sp. *allii*[250] (Section IV.B). Also, f.sp. *conglutinans* (Wollenw.) Snyder and Hansen causes seedlings to wilt in W. Bengal, India.[252] Slight symptoms may develop from inoculation with other non *cepae* formae speciales.[253]
5. The possible occurrence of symptomless carriers for f.sp. *cepae* or f.sp. *allii* appears not to have been resolved.
6. Isolates of *F. oxysporum*, not identified as f.sp. *cepae*, from several rotation crops, proved pathogenic to onion seedlings.[254] Similarly, *F. oxysporum* isolates from onion proved pathogenic to ten rotation crops[255] (Section IV.D.1). In other reports, *F. oxysporum* is not pathogenic[236] (Section IV.D.1).

B. FUSARIUM BASAL ROT OF ONION AND GARLIC

1. Distribution and Economic Importance

F. oxysporum f.sp. *cepae* (nomenclature: Section IV.A) is present in most parts of the world where onions are grown.[246] There can be up to 90% loss of seedlings,[256] yields of susceptible cultivars can be halved,[246] and there can be a 30% loss in store.[257] Seed yield is also reduced.[258] The disease causes serious losses in garlic.[209]

2. Pathogen

F. oxysporum f.sp. *cepae* forms three types of spore (Table 10). Microconidia form between the scales of infected bulbs[259] but production is sparse in culture;[236,244] microconidia are usually unicellular and sphaeroid.[236,244]

Macroconidia are produced *in vitro* on sporodochia and pionnotes; they are fusiform or ellipsoidal and mostly 3 to 4 septate, range 0 to 8 septate.[244] Chlamydospores are produced both from hyphae and macroconidia. Hyphal chlamydospores are terminal or intercalary and form within 10 to 15 d *in vitro*; conidial chlamydospores form from any cell of the macroconidium and are produced within 1 month.[244] In nature, chlamydospores are 7.5 to 10.0 μm in diameter and form in soil and roots.[251,259] Hyphae are 2 to 4.6 μm in diameter and cells are 7 to 35 μm in length; branches occur at <200 μm intervals.[244]

The majority (97%) of isolates from New York State, formed sporodochia and sclerotia when first isolated from soil or infected tissue.[260,261] Sporodochial forms mutated in infected tissue or culture and the resulting morphological cultural forms were described as "ropy", pionnotal, or mycelial and did not produce sclerotia or sporodochia.[260] These morphological variants were less virulent than sporodochial forms.[260]

Forma specialis *cepae* is not totally restricted to *A. cepa*, for example, causing basal rot in garlic in field situations.[209] In steam sterilized compost, small amounts of disease develop in *A. chinense* and *A. fistulosum*[250] (Section IV.C.2).

3. Symptoms

There is delayed seedling emergence,[256] seedling damping off[262] stunted growth and basal rotting in growing plants[228,246,259] accompanied by premature maturation[255] and basal rotting in store.[257,263] A mealy decay may also be present on the scales.[246,256] Infection of seedlings and maturing plants is accompanied by chlorotic leaf dieback and dessication with the leaves characteristically remaining upright.[246,256]

Elongation, stem distortion and leaf curling are caused by f.sp. *cepae* in the San Francisco Valley, Brazil[266] (see chapter on Leaf Diseases of Onions: *Glomerella cingulata*).

Usually, pink coloration in *Fusarium* infected roots occurs only when *Pyrenochaeta terrestris* (Section III) is also present.[216,267] However, pinkness in the roots of stunted plants

TABLE 10
Spore Characteristics of *Fusaria* Causing *Fusarium* Basal Rot

Spore type and septation	Frequency (%) and size (μm) of spores			
	F. zonatum (Sherb.) Wr.		*F. cepae*. Hanz. emend. Link and Bailey[236]	*F. zonatum* (Sherb.) 2 Wr. Forma 1 (Phillipines)[244]
	Forma 1[236]	Forma 2[236]		
Macroconidia	3-septate			
Frequency	62	72	66	NR[a]
Length (l)	37.6	38.5	35.7	32.3
	(27.2—46.5)	(24—48)	(25—46)	(16—48.8)
Width (w)	3.8	3.7	3.8	4.0
	(2.8—4.6)	(2.4—4.6)	(3—4.5)	(2.4—5.6)
	4-septate			
Frequency	13	14	14	NR
l	41.5	39.9	40.9	38.2
	(34.4—50)	(31.5—53)	(32—51)	(31.2—48)
w	3.2	3.8	3.7	4.2
	(3.2—4.6)	(3.2—4.6)	(3.5)	(2.8—5.6)
Microconidia-	0-septate			
Frequency	Most	Most	Most	NR
l	7.8	7.9	7.5	7.6
	(5—11.6)	(5—11.6)	(6—9)	(4—12)
w	2.7	2.8	2.7	2.7
	(1.6—4.2)	(1.7—4.6)	(2—3.3)	(1.2—4.8)
	1-septate			
l	13.2	14	12.7	15.5
	(8—18.2)	(8.3—18.3)	(11.5—17)	(7.2—2.8)
w	2.7	3	2.8	3.3
	(1.6—3.3)	(1.7—4.6)	(2—3)	(2—4.8)
Chlamydospores[b] -	0-septate			
Frequency	Most	Most	Most	Most
l	6.5	NR	6.4	6—8[a]
	(5.6—8)		NR	(4—14)
w	6.7	NR	6.4	5.2[a]
	(4.8—8)		NR	(3—8.4)

[a] NR — not recorded.
[b] Includes values for spherical hyphal chlamydospores and for barrel-shaped and spherical-shaped conidial chlamydospores.

has been reported in the apparent absence of *P. terrestris*, thus twice as much *F. oxysporum* was isolated from pink roots as from healthy roots.[255]

4. Infection

Propagules close to the root germinate and the germ tubes penetrate healthy root cells or invade wounds in the root.[259,268] Colonization of the root vascular tissues is via undifferentiated vascular tissue near to the root tip or directly via wounds.[259] Further growth of the pathogen occurs within the root cortex and vascular system, which are continuous with those of the stem tissue, followed by the stembase tissue. Infection also occurs directly from soil into the stembase (stemplate) via the natural wounds, which develop when roots emerge from within the stembase tissue, when the pathogen population is high.[259]

Infection remains localized within the stembase for several weeks or months before spreading to the outer fleshy leaf scales and causing the characteristic basal rot.[269]

5. Epidemiology

a. Environmental Factors

The optimum temperature for the growth of f.sp. *cepae* on solid media is 24 to 27°C.[241,251] Growth of the pathogen is reduced at lower temperatures and ceases at approximately 9°C; at 36°C the fungus is killed.[251] Egyptian isolates grow best at 30°C.[270]

Temperature is the main factor governing the activity of f.sp. *cepae* in the field and the types of symptom and disease incidence.[241] Generally, the growing crop is not affected until soil temperatures are high,[264] approximately 25°C. However, root pruning occurs at lower temperatures and restricts plant growth.[271] High soil temperatures occurring early in the season result in seedling damping off.[251,256] Often, high temperatures occur later in the season. As a result, foliar symptoms and basal rotting become evident midseason onward in New York, Colorado and Georgia, in Valencia, Spain, and Hungary.[241,254] Differences in plant spacing do not affect the incidence of basal rot.[263]

Symptoms continue to develop rapidly in store at 20 to 30°C.[270] The lower temperature results in a wet rot, whereas at the higher temperature there is a dry rot.[241] Bulbs which appear healthy at harvest may nevertheless develop basal rot in store. This gives the misleading impression that the pathogen has spread through contact with infected bulbs; it is more likely that there were slight symptoms at harvest which were overlooked.[236,247,257,263,270] At low storage temperatures (8°C), there is minimal basal rotting but bulbs sprout prematurely causing additional loss of crop.[241]

The incidence of *Fusarium* basal rot increases with the frequency of cropping with onions.[246,254,256,272] Therefore, rotations of 3 or more years have been recommended.[244,246,264] The increase in disease incidence is probably due to an increase in the numbers of chlamydospores, formed from hyphae or macroconidia, in the root zone and in infected host tissue.[244,259] A lack of increased basal rot incidence with successive onion crops may be an indication of soil suppressivity.[273] The incidence of basal rot is not always related to f.sp. *cepae* populations in soil although the reasons for this are not understood.[261] For example, some organic soils in New York State contain high f.sp. *cepae* populations but disease incidence is less than other soils with similar f.sp. *cepae* populations.[261] Another example occurred in Colorado. Soil sampled midseason from a field, where previously there had been 17% basal rot, had a *F. oxysporum* population of 3×10^3 colony forming units (cfu) g^{-1} soil of which 73% isolates were f.sp. *cepae*;[246] 3 weeks later, however, f.sp. *cepae* populations had declined to 514 cfu g^{-1} and were then comparable with the population in soil from a field with no history of basal rot 377 cfu g^{-1} f.sp. *cepae*.[246]

b. Effects of Host Wounds

The incidence of *Fusarium* basal rot is increased when tissue is wounded,[241] e.g. when seedlings are trimmed before transplanting, or when roots are damaged during cultivation.[246] Wounds also result from the emergence of roots from within the stem base tissue and these provide an entry for f.sp. *cepae*.[259] The increased incidence of basal rot, when plants are under stress from high temperatures and drought, may be a response to tissue damage.[246]

Insect larvae are commonly associated with *Fusarium* basal rot of onion. This has led to the suggestion that the larvae damage the onion tissue, thus, facilitating infection by f.sp. *cepae*.[246,263] However, there can be severe basal rotting even in the absence of larvae.[256] and it is likely that the feeding is on tissue already damaged by f.sp. *cepae* rather than on healthy tissue.[246]

The relationship between *Fusarium* basal rot and pink root was discussed in Section III.

6. Sources of Outbreaks and Dissemination

Infected onion transplants,[246] sets,[244] and possibly garlic cloves[209] are both a potential source and a method of dissemination of f.sp. *cepae*. Vegetative planting material may be infected with f.sp. *cepae* but without evidence of symptoms; alternatively adhering soil may contain *Fusarium* spores.[244] The pathogen has also been isolated from onion seed;[258] there have been other reports of seedborne *F. oxysporum* but it is not clear if f.sp. *cepae* was involved.[274,275]

Chlamydospores and dormant hyphae in decayed tissue and chlamydospores present in soil survive between seasons and are a source of inoculum for the crop which follows.[244,246,251]

Oxalis corniculata is a symptomless host for f.sp. *cepae* and might contribute to survival and dispersal of the pathogen.[251] Also, f.sp. *cepae* was isolated from *O. pes caprae* and *O. zeekoevleyensis* in South African onion fields and these species may also be a reservoir of f.sp. *cepae*.[276]

In general, methods of dissemination common to other soilborne plant pathogens,e.g., irrigation water, in soil blown by wind or adhering to farm machinery, are also applicable to f.sp. *cepae*.[246]

7. Control

a. Planting Material and Fungicides

Vegetative planting material may be infected by f.sp. *cepae* but without recognizable symptoms or there may be infested soil adhering to healthy plants. Ideally, therefore, planting material should be produced in areas where f.sp. *cepae* is absent. This may not be practicable because f.sp. *cepae* is widespread.

Alternatively, transplants could be produced in fields which have been fumigated (Section IV.B.7).

Seed treatment with fungicides has been reported to reduce basal rot incidence[257] and may have value where the disease is not severe. *Pseudomonas cepacia*, a root colonizing bacterium of onions, provided protection of onion seedlings when the bacteria was tested as a seed treatment.[277] The treatment, however, has proved unreliable.[278]

b. Eradication of the Pathogen from the Field

Methyl bromide, chloropicrin or metham sodium reduce populations of *Fusarium* pathogenic to onion seedlings.[254] Soil fumigation could therefore restrict the dissemination of basal rot and other soilborne pathogens from areas used regularly for transplant production.[254,279] The effects may not be long lived, however, because low populations of f.sp. *cepae* increase rapidly in the presence of onions.[251]

c. Host Resistance

Cultivars with a low level of resistance to basal rot have reduced losses in New York State since 1960, compared with cultivars used previously.[263,280]

Breeding for resistance from commercial cultivars has given promising results.[231] Healthy plants were selected from severely infected crops then selfed and the cycle of selection and selfing repeated. After 4 cycles, 7 out of 21 selections developed 0 to 8% basal rot compared with the maximum of 32% rot. More recently, a single cycle of selfing resulted in a 150% improvement in resistance and this was reported to be due to the accumulation of minor genes.[281]

Tests for resistance have been based on the reaction of seedlings[265,282] or older plants[247,265,282] in sterilized or nonsterile soil.[247] Alternatively, plants have been tested in field conditions with natural inoculum and symptoms assessed at intervals during growth and in store.[231,247,263,270] In some tests there has been good correlation between seedling infection in a glasshouse test and basal rot incidence in the field.[270] In others, however, there was much less correlation between tests with seedlings and tests with older plants. For example, out of 47 Indian cultivars, eleven proved to be resistant both as seedlings and mature plants.[265] An additional 24 cultivars, however, were resistant only as seedlings. There was a similar lack of correlation in South African cultivars and breeding lines.[281]

The following cultivars and breeding lines have been reported as very resistant: Granex, L690,[247] Elba Globe, Iowa 44,[263] Yellow Sweet Spanish (Zuckerman 1959 strain), B2264 (inbred line from Brigham Yellow Globe), Texas Early Grano 951 (inbred line),[283] Hybrid Early Premium,[284] Caledon Globe (B13-6 selection).[282] Resistant and susceptible cultivars did not differ in the time at which infection occurred and the incidence of root infection.[280] Differences in resistance were an expression of the capacity of the pathogen to grow from the stembase to the fleshy leaf scales.[259,268]

The sugar content of fleshy onion leaf tissue is considerably higher than in the stembase.[285] The pathogen produces endo-pectin-*trans*-eliminase (endo-PTE), an enzyme which macerates onion tissue.[269] The induction of endo-PTE is much less with fleshy leaf tissue than with stembase tissue due,[286] it is thought, to the repressive effects of the higher sugar content in the apoplast of fleshy leaf tissue.[285,287] As the plant matures, apoplast sugar content of the fleshy scale declines and accumulates in the symplast, thereby removing the repression of endo-PTE and allowing fungal penetration of the scale tissue.[287] There was no difference in the sugar content, or its location in resistant and susceptible selections of cv. Caledon Globe.[285] The slow rate of induction of endo-PTE by fleshy leaf tissue, however, is thought to provide an opportunity for other resistance factors to operate.[285,286] Possibly, a difference in the structure of epidermal cells of roots also contributes to resistance.[268]

C. FUSARIUM BASAL ROT OF *A. CHINENSE*

1. Distribution and Economic Importance

Severe bulb rotting occurred in *A. chinense* G. Don (*A. bakeri* Regel) in Japan in 1969—1970 and the new form of *F. oxysporum* named *F. oxysporum Schlect.* f.sp. *allii* Matuo, Tooyama and Isaki.[250]

Fusarium basal rot is now one of the most damaging diseases of *A. chinense* in Tottori and Fukui Prefectures, Japan. Bulb rot caused by a different *Fusarium* sp. had previously been reported in Tochigi Prefecture, Japan but the symptoms were mild and superficial and f.sp. *allii* was not involved.[250]

2. Pathogen

Forma specialis *allii* forms colonies on agar medium which are characteristic of *F. oxysporum* Schlect. emend. Snyder and Hansen. Microconidia and macroconidia are abundant in culture. Microconidia are mostly 1-septate and 14.3 to 16.0 × 2.8 to 3.2 μm (range 12 to 20 × 2.5 to 4.0 μm); unicellular microspores are 5 to 14 × 2 to 4 μm. Macroconidia are falcate and mostly 3-septate; macroconidia formed on sporodochia are 37.0 × 3.7 μm (range 32 to 44 × 3.0 to 4.5 μm) and those formed on nonsporodochial hyphae are 30.0 × 3.7 to 4.1 μm (range 20 to 38 × 3 to 5.0 μm). Chlamydospores are terminal or intercalary and spheroid-elliptic and hyaline to pale brown.

Forma specialis *allii* is not totally specific to *A. chinense*, at least not in steam sterilized compost. Inoculation resulted in disease symptoms in onion, Welsh onion, asparagus, gladiolus, and tulip but to a much lesser extent than in *A. chinense*.[250] Forma speciales *cepae*, *tulipae*, *asparagi*, and *gladioli* also gave rise to disease in *A. chinense* but the number of successful inoculations was small and symptoms were again mild.[250]

3. Symptoms and Infection

Fusarium basal rot occurs at two periods in the life of the crop. Sets are planted in July-September and plants die within 30 to 40 d. The fleshy leaf tissue next to the stembase tissue rots, becomes water soaked and the roots decay. A yellow-brown rot spreads to the rest of the set which, in dry conditions, becomes mummified or, in wet conditions, forms a wet rot. This stage of the disease probably results from the use of infected planting material.[250]

The second appearance of the disease is in May—June shortly before harvest. There is a foliar wilt, vascular discoloration, and decay of the bulb and roots.

The newly introduced cv. Tamarakkyo is less susceptible to f.sp. *allii* than cv. Rakuda.[250] Apart from the usual precautions taken with pathogenic fusaria and soilborne pathogens in general, no other control measures have been reported.

D. DISEASES CAUSED BY OTHER *FUSARIUM* SPP.

1. Onions

F. oxysporum (not reported as f.sp. *cepae*) has been reported as nonpathogenic to onion.[236] Nevertheless, *F. oxysporum* has also been recorded as a cause of seedling damping-off;[228,252,254,262] reduced yields at harvest;[228,288] and storage losses.[288,289] For example, in Zambia, the presence of *F. oxysporum* in onion sets resulted in over 80% infection of the young plants (''seedlings'').[288] Infected seedlings established poorly when transplanted and there was a high incidence of basal rot in the surviving plants. Storage losses were as high as 94%.[288] In Finland, *F. oxysporum* was also transmitted via infected onion sets but storage losses were not serious probably because of low temperatures.[289]

Dipping infected sets in benomyl suspension before planting reduced losses considerably.[288,289]

F. oxysporum Schlect., *F. solani* (Mart.) App. and Wr., *F. equiseti* (Corda) Sacc., *F. moniliforme* Sheld. were associated with a range of disease symptoms in onion.[255] *F. oxysporum* had previously been described as nonpathogenic to onion, and *F. moniliforme* as occasionally pathogenic.[236]

F. oxysporum was most abundant where there had been two or three previous crops of onion and in plants which were stunted or had basal rotting and pink roots. *F. solani* was isolated mainly from plants which had senesced prematurely and had pink roots. In India[290] and Czechoslovakia[32] *F. solani* causes a root rot. In Israel, *F. equiseti* was isolated from stunted plants with pink roots.[255]

These *Fusarium* spp. were not specific to onions. Thus, *F. oxysporum* and *F. equiseti* isolated from a range of vegetable and other crops resulted in 14 to 39% wilt in onion seedlings. *F. solani* and *F. moniliforme* isolated from a similar range of host species caused 2 to 10% seedling wilt in onion. Also, *Fusarium* isolates from onions caused seedling wilt in a wide variety of plant species; the effect was particularly noticeable with *F. oxysporum* and *F. equiseti* which caused 17 to 38% wilting in tomato and egg plant.[255]

F. solani-related rotting in onion becomes more prevalent when bulbs are stored due to increased tissue susceptibility.[291,292]

F. roseum is a source of minor loss in France.[227] *F. semitectum*[293] and *F. moniliforme* var. *subglutinans* Wr. and Reink.[294] cause seedling damping-off in India and Italy respectively.

2. Garlic

F. solani (Mart.) App. and Wr. emend. Synd. and Hans. causes wilting, root collapse and a storage dry rot in Egypt.[295] Infected cloves used for planting material may fail to develop leaves, then decay and become covered in hyphae.[295]

F. roseum causes minor damage in France.[227]

3. Leek

F. roseum causes approx. 30% loss in summer-grown crops in France; infected roots turn pink.[227]

F. culmorum causes considerable damage in the Netherlands; dipping the seedlings in benomyl before transplanting gave a two- to threefold increase in marketable yields.[296]

4. Shallot

F. oxysporum has been reported in New Zealand.[297] *F. roseum* has been reported in France, where damage is slight and is associated with a pink coloration in the roots.[227]

V. ONION SMUT

A. DISTRIBUTION AND ECONOMIC IMPORTANCE

The disease was first reported in the U.S. in the 1850s and in Europe 20 years later and is probably present in most temperate *Allium* growing areas.[243,299] Smut causes economical losses in direct-seeded onions but also occurs on *A. cepa* var. *ascalonicum* (shallot),[243] *A. porrum*, and[243,298] *A. sativum*.[213,298,299] The pathogen also infects *A. vineale*, *Camassia*,[299] *Colchicum*,[298,299] and *Polygonatum*.[299]

B. PATHOGEN

Smut is caused by the soilborne fungus *Urocystis colchici* (Schlect.) Rabenh. (= *U. cepulae* Frost;[300] *U. magica* Pass.[301]).

In field conditions, *U. colchici* is an obligate parasite of alliums and some other genera.[298,299] The fungus can be cultured on malt extract agar or in liquid media containing aspartic acid or L-asparagine as a nutrient source.[302,303] The fungus persists for many years in soil as teliospores[304] and has been reported after 25 years' absence of alliums.[305]

Teliospores germinate in soil and infect roots and cotyledons but true leaves are immune.[243,299,306] Smut hyphae develop within cotyledon and leaf cells, forming dark internal pustules containing teliospores. The teliospores are released when the host tissue ruptures or when infected leaves die and rot and have little or no period of dormancy.[305] The mycelium appears to be unable to survive in soil.[305]

C. SYMPTOMS

Dark streaks form within cotyledons and true leaves, later the host tissue ruptures releasing masses of dark teliospores. Seedlings are killed at emergence and the survivors remain stunted.[299] The bulbs of infected plants fail to reach marketable size[306] and may be disfigured by dark pustules.[299] Rotting occurs if smut infection is followed by secondary invading organisms.[299]

D. EPIDEMIOLOGY AND DISSEMINATION

The onion is susceptible to smut only in the period between seed germination and the appearance of the first true leaf, a period of about 2 to 3 weeks.[306] Therefore, factors which delay seedling growth indirectly favor increased infection.

Teliospore germination *in vitro* is maximum at 20 to 24°C with approximately 50 and 80% germination occurring at 4 and 6 d respectively at the higher temperature.[302] Germination is markedly less at 16 and 28°C and is minimal at 12 and 32°C.[302] Mycelial growth in liquid culture is at maximum at 17 to 24°C with malt extract as a nutrient source; growth rate on sucrose-asparagine medium is similar except at 24°C when it is higher.[303] Maximum teliospore germination occurs at pH 5.5 to 6.5 with slightly lower rates of germination at pH 5.0 and up to pH 8.0.[302] Maximum mycelial growth is at pH 5.8 to 6.4 but growth occurs in the range pH 3.0 to 8.0.[303]

The incidence of infection is governed by factors affecting both the growth of onion seedlings and the pathogen. Seedling growth at lower temperatures prolongs the period of exposure of susceptible host tissue, hence increasing infection in comparison with higher temperatures.[307] Infection rates are low at 12 and 28°C: at 12°C the period of susceptibility is prolonged but pathogen activity is low whereas at 28°C the period of susceptibility is very short and the pathogen, although moderately active, has little effect.[307]

Increased depths of seeding in dry soils result in higher rates of infection due either to the prolonged exposure of host tissue to infection, a greater amount of susceptible host tissue or to a reduction in host reserves hence greater susceptibility.[306,307]

E. CONTROL

The effects of smut are minimized by environmental conditions and management practices which favor rapid seedling emergence. Such factors include optimum soil moisture, temperature and soil structure, and shallow seeding. Alternatively, the effects of the disease can be avoided by the use of transplants grown in smut-free soil or in modules.

Seed treatment with thiram provides control with seed sown down to 2.5 cm but at greater depths, additional furrow treatment may be needed.[306]

Crop rotation is of no value because of the long survival capacity of the pathogen.

VI. SOUTHERN BLIGHT

A. DISTRIBUTION AND ECONOMIC IMPORTANCE

Southern blight is capable of causing considerable losses in garlic, onion and shallot, both in the field and store,[297,308-312] in other situations losses are minor.[255]

B. PATHOGEN

Southern blight is caused by the soil-borne fungus *Sclerotium* (*Corticium*) *rolfsii* Sacc.

Sclerotia are initially white and soft, later becoming firm and pale to dark brown and are 0.5 to 2 mm in diameter.

C. SYMPTOMS

The stem base and roots become infected and rot. The early foliar symptoms of chlorotic die back are similar to those of *Fusarium*.[313] There is a characteristic growth of pale mycelium on the soil surface around the stem base, on which sclerotia form.

The disease may, like *Allium* white rot, occur in patches in the field due to the clustering of sclerotia.[313,314]

The pathogen affects a very wide taxonomic range of plants in addition to *Allium*.[308,315] It is active only in the upper 12 cm of soil[314] and is dependent for activity on the presence of dead or senescent plant tissue.[313]

D. EPIDEMIOLOGY AND DISSEMINATION

S. rolfsii activity is optimum at 27 to 30°C[315] and the disease occurs only in warm areas particularly where there is high seasonal rainfall or a monsoon period[308] or where crops are irrigated.[316] Therefore, the disease is markedly seasonal, for example, occurring in April — May in Louisiana[309] and damaging transplant production in south Georgia when soil temperatures may reach 30 to 39°C at 1 to 2 cm depth.[254]

E. CONTROL

Control is based on the combined effects of (1) reducing soil sclerotial populations by crop rotation, fumigants, or solarization, (2) sanitation by minimizing crop and weed residues or ploughing the residues below the depth where *S. rolfsii* is active, and (3) the use of protectant fungicides such as PCNB.

When onions are grown in Israel, during the cool season *S. rolfsii* is inactive and infection does not occur.[316] In this situation, *S. rolfsii* sclerotial populations decline. *Allium* root exudates are inhibitory to some soil microorganisms and, as a result, sclerotia are thought to become susceptible to invasion by antagonistic fungi.[316] The incorporation of cereals or sweet potato into the rotation also resulted in a reduction of sclerotial populations.[313]

Metham sodium applied in irrigation, chloropicrin, or methyl bromide injected into soil reduced sclerotial populations.[313] Also, solarization reduced soil sclerotial populations of *S. rolfsii* and is a promising method of control.[75,137]

PCNB applied as a dip treatment or as a combined furrow and band treatment to soil reduced southern blight incidence in shallots.[310]

There is no evidence of genetic resistance in *Allium* to *S. rolfsii*.

VII. DAMPING-OFF

A. DISTRIBUTION AND ECONOMIC IMPORTANCE

F. oxysporum and *F. oxysporum* f.sp. *cepae* cause delayed seedling emergence and seedling damping-off[256] in addition to root and basal rots.[209,254,256,262,272,277] *F. moniliforme*, *F. roseum*, and *F. solani* also cause damping-off in onion in areas of intensive vegetable production.[254,274] *Colletotrichum circinans* causes damping-off in warm humid conditions.[243]

Pythium species, various *Fusarium* spp., and *Rhizoctonia solani* (Kuhn) can cause damping-off problems in areas of intensive production of direct seeded alliums and other vegetable crops or on areas reserved for the production of transplants.[61,254,264]

B. PATHOGENS

Species of *Pythium* pathogenic to onion seedlings are *P. afertile* Kanouse and Humphrey,[317] *P. butleri*,[318] *P. coloratum* Vaartaja,[317,319] *P. debaryanum* Hesse, *P. echinulatum* Matthews,[317] *P. graminicola* Subramaniam,[320] *P. irregulare* Buisman (= *P. polymorphon* Sideris),[317,319,320] *P. mamillatum* Meurs,[320] *P. spinosum* Sawada,[320] *P. sylvaticum* Campbell and Hendrix,[319] *P. torulosum* Coker and Patterson,[317] and *P. ultimum* (Trow.).[317,319,321]

P. irregulare was also a cause of retarded growth in onions.[321,322] In addition, *P. acanthicum* Drechsler and *P. rostratum* Butler were isolated from onions but did not prove pathogenic.[319,321] There is considerable variation in the frequency of occurrence of the various *Pythium* spp. in different situations, in the pathogenicity to onion of different isolates of the same species and in the species responsible for damping-off in particular situations.[317,319,321] In Canada, for instance, *P. irregulare* was the main cause of damping-off in onions[321] whereas in New Zealand[317] and in New York State[319] *P. coloratum* was also damaging. *Pythium* species isolated from *Allium* may be pathogenic to other commonly grown vegetable species and isolates from a range of vegetable species are pathogenic to *Allium*.[317,319]

There are fewer reports of *Rhizoctonia* causing damping-off in alliums but, in Israel, this fungus may cause major seedling loss.[228]

C. CONTROL

Pythium and *Rhizoctonia* propagules in soil can be substantially reduced by eradicant treatments. In Australia, for example, solarization reduced *P. irregulare* populations down to 6 to 11 cm.[75] In south Georgia, methyl bromide or metham sodium resulted in similar reductions in the populations of *Pythium* (mainly *P. irregulare* and *P. ultimum*), *Rh. solani*, and *Fusarium* (mainly *F. roseum*, *F. oxysporum*, and *F. solani)* in areas used for the production of onion transplants.[254]

Protectant fungicides applied as soil or as seed treatments may also be effective. Thus, *P. butleri* was controlled by drenches of captafol or thiram and *Rh. solani* by drenches of carbendazim or PCNB.[318]

VIII. SMUDGE

A. DISTRIBUTION AND ECONOMIC IMPORTANCE

Smudge does not affect roots but is included here because the pathogen is soilborne. Onion smudge was first reported in 1851 in the U.K., then elsewhere in Europe by the early

1860s.[325] The first report in the U.S. was in 1874 and smudge is now widespread there. In Cordoba, Argentina the disease was first reported in 1970.

The disease has little effect on growth or yield of onions but reduces cosmetic quality because of spotting of the outer wrapper scales. Also, in store, bulbs shrink and sprout prematurely.

B. PATHOGEN

The disease is caused by the fungus *Colletotrichum circinans* (Berk.) Vogl. (= *C. dermatium* (Pers. ex Fr.) Grove f. *circinans* [Berk.] Arx.)[298] The fungus affects onions, shallots, leek,[325] and *A. fistulosum*[297] but not garlic.

Hyphae are 2 to 8 μm in diameter, septate, initially hyaline, and guttulate. Later, the hyphal walls thicken and become dark green, and septation is more frequent. Thickened hyphae grow together to form a dark, greenish mat (stroma) less than 1 mm in diameter. Stromata form between the cuticle and epidermal cell wall. Short hyaline conidiophores develop on the outer surface of the stromata, the host cuticle ruptures, thus forming an acervulus.

Numerous needle-shaped dark hyphae (setae) 80 to 315 μm in length and 0 to 3 septate develop on the acervulus.

Conidia, 17 to 21 (14 to 30) × 3 to 4 (3 to 6) μm are budded singly from the conidiophores and in the mass are cream colored. Conidia form 1 to 3 germ tubes at germination and septae commonly develop.

In Japan, a *Colletotrichum* sp. from onion differed in conidial size from *c. circinans* and formed a *Physalospora* ascospore stage.[327]

C. SYMPTOMS

Smudge is mostly confined to the dried wrapper scales of white-skinned cultivars. Stromata are recognizable as dots, initially dark green, later black, scattered over the surface of the bulb or grouped in concentric rings.

In moist conditions, ascervuli develop and the cream colored spore masses and setae are easily recognized. In brown and colored cultivars, the symptoms are confined to the nonpigmented areas near the neck of the bulb. The pathogen may spread to the underlying fleshy scales resulting in yellow crater-like depressions.

Smudge symptoms have been confused with *Allium* white rot, with *Phoma* spp. (which form pycnidia), and with smut (which forms dark streaks rather than spots, and which produces powdery spores). Neither of these latter diseases and fungi have setae, and *C. circinans* forms no fruiting structure other than the acervulus.

D. SOURCE OF OUTBREAKS AND DISSEMINATION

When first described, smudge rarely occurred in first-grown onion crops. Subsequently, smudge incidence increased rapidly, probably because onions were commonly grown in the same fields in successive years. In this situation, infection would be initiated from stromata surviving in onion scales from a previous crop. Therefore, opportunities for dissemination also occur with the movement of infested soil. Infected seedling transplants and onion sets are also important methods of dissemination. Spread of the pathogen in store is minimal.

E. EPIDEMIOLOGY

Hyphal growth in vitro is optimum at 25°C and abundant at 15 to 30°C. Growth is minimal at 2 and 31°C. Conidial germination is optimum at 20°C and occurs readily at 13 to 25°C. In the field, smudge develops rapidly at a soil temperature of 20°C, providing conditions are moist. Secondary infection, is dependent on water splash.

Stromata are resistant to dessication and freezing whereas conidia do not survive in these conditions. Therefore, survival over winter and between seasons is by stromata whereas secondary infection results form the formation of acervuli in humid conditions and the subsequent dispersal of conidia by water splash. Spores which have been dispersed by water splash but which have not resulted in visible symptoms at harvest, may give rise to lesions in store giving the misleading impression that the pathogen has spread during storage.

F. CONTROL

Damage is mainly to pale cultivars, e.g., salad and silverskin onions. Therefore the use of brown or colored cultivars minimizes infection. Otherwise the only method of control is to ensure that bulbs are dried quickly after harvest.

IX. SUMMARY

Root diseases of *Allium*, caused by soilborne fungal pathogens, are often a major source of crop loss and in extreme situations they limit commercial production. Some diseases however, apparently do little damage, e.g., *Macrophomina phaseolina*.[328,329]

Damage is governed by (1) the method and the extent that the pathogen is introduced — often via vegetative planting material, (2) the environmental conditions, e.g., soil temperature, and (3) the rotational frequency of *Allium* and other susceptible host crops.

Improved methods of control are needed. The effectiveness of fungicides is limited by the difficulties of ensuring contact with the pathogen in soil. Enhanced degradation is potentially another limitation. On the other hand, resistance of the pathogen to fungicides has not been a problem. Soil fumigants are not always effective at eradicating fungal propagules. Soil solarization, however, shows considerable promise but not in cool climates. Host genetic resistance when available, provides only partial protection.

There is now an increased interest in the development of integrated control measures, i.e., the use of several different control methods, which alone would give insufficient protection.

REFERENCES

1. **Berkeley, M. J.,** Notices of British fungi (*Sclerotium cepivorum* Berk.), *Ann. Mag. Nat. Hist.*, 6, 355, 1841.
2. **Voglino, P.,** Sul parassitismo e lo sviluppo dello *Sclerotium cepivorum* Berk. nell' *Allium sativum* L., *Stazioni Sperimentali Agrariae Italiane,* 36, 89, 1903.
3. **Walker, J. C.,** White rot of *Allium* in Europe and America, *Phytopathology,* 14, 315, 1924.
4. **Caballero, A.,** El Bioxat, o enfermedad de los ajos, en Banolas, *Bol. de la R. Soc. Esp. de Hist. Nat.*, 22, 210, 1922.
5. **van Poeteren, N.,** Verslag over de Werkzaamheden van den Plantenziektenkundigen Dienst in het Jaar 1926, *Verslagen en Mededeelingen van den Plantenziektenkundigen Dienst te Wageningen,* 51, 20, 1928.
6. **Bremer, H.,** The flour disease of onions, *Nachrichtenbl. Deutsch. Pflanzenschutzdienst.*, 4, 37, 1934.
7. **Haegermark, U.,** *Sclerotium cepivorum* Berk., A fungus disease of onion, *Vaxtskyddsnotiser Stockholm*, 22, 14, 1958.
8. **Osara, K.,** Cultivation of *Allium* species and onion white rot in Finland, in *Proc. 3rd Int. Workshop on Allium White Rot,* Entwistle, A. R., Ed., T. W. Printing Associates Ltd, Leamington Spa, U.K., 1987, Section 1, 50.
9. **Hauman-Merck, L.,** *Zbl. Bakt.*, 43, 447, 1915.
10. **Wingard, S. A.,** White rot of onions in Virginia, *Plant Dis. Rep.*, 14, 104, 1930.
11. **Valleau, W. D.,** White rot (*Sclerotium cepivorum*) of onion reported from Kentucky, *Plant Dis. Rep.*, 9, 46, 1925.
12. **De Oliveira, V. L., Bellei, M. M., and Borges, A. C.,** Control of white rot of garlic by antagonistic fungi under controlled environmental conditions, *Can. J. Microbiol.*, 30, 884, 1984.

13. **Tims, E. C.,** White rot of shallot, *Phytopathology,* 38, 378, 1948.
14. **Gardner, M. W.,** *Sclerotium cepivorum* found on garlic in California, *Plant Dis. Rep.*, 23, 36, 1939.
15. **Tims, E. C.,** White rot of *Allium* in Louisiana, *Plant Dis. Rep.*, 26, 219, 1942.
16. **Ormrod, D. J. and Conroy, J. F.,** First record of white rot of onion in coastal British Columbia, *Can. Plant Dis. Survey,* 50, 110, 1970.
17. **Locke, S. B.,** Experimental control of onion white rot by means of soil chemicals, *Plant Dis. Rep.*, 52, 272, 1968.
18. **Adams, P. B.,** *Allium* white rot in the United States, in *Proc. 2nd Int. Workshop on Allium White Rot,* Adams, P. B. and Entwistle, A. R., Eds., 1983, 1.
19. **Somerville, A. and Hall, D. H.,** Economic importance and distribution of white rot of onions and garlic in California, in *Proc. 2nd Int. Workshop on Allium White Rot,* Adams, P. B. and Entwistle A. R., Eds., 1983, 6.
20. **Creelman, D. W.,** Summary of the prevalence of plant diseases in Canada in 1962, *Can. Plant Dis. Survey,* 43, 90, 1963.
21. **Laborde, J. A.,** Coexistence of garlic white rot with commercial production in Central Mexico, in *Proc. 3rd. Int. Workshop on Allium White Rot,* Entwistle, A. R., Ed., T. W. Printing Associates Ltd, Leamington Spa, U.K., 1987, Section 1, 24.
22. **Crowe, F. J.,** White rot control in the western U.S., with reference to the influence of inoculum density of *Sclerotium cepivorum, Proc. 3rd. Int. Workshop on Allium White Rot,* Entwistle, A. R., Ed., T. W. Printing Associates Ltd, Leamington Spa, U.K., 1987, Section 3, 1.
23. **Valk, M.,** Canadian onion production and the status of *Allium* white rot infestation, in *Proc. 3rd. Int. Workshop on Allium White Rot,* Entwistle, A. R., Ed., T. W. Printing Associates Ltd, Leamington Spa, U.K., 1987, Section 1, 71.
24. **Nattrass, R. M., The occurrence of the 'white rot' of the onion (*Sclerotium cepivorum* Berk.) in Egypt, *Bulletin Ministry of Agriculture, Egypt. Technical and Scientific Service (Plant Protection Section),* 107,** 9, 1931.
25. **Nattrass, R. M.,** The white rot disease of onions in Cyprus, *Cyprus Agric. J.*, 28, 98, 1933.
26. **Du Plessis, S. J.,** White mould in onions, *Farm. S. Afr.*, 7, 112, 1932.
27. **Beeson, M. M.,** Onion white rot now known to be present in Pukekohe, *N.Z. Commercial Grower,* 15, 2, 1960.
28. **Somerville, A. and Merriman, P. R.,** Distribution and economic importance of white rot of onions in Australia, in *Proc. 2nd Int. Workshop on Allium White Rot,* Adams, P. B. and Entwistle A. R., Eds., 1983, 11.
29. **Merriman, P. R., Isaacs, S., MacGregor, R. R., and Towers, G. B.,** Control of white rot in dry bulb onions with artificial onion oil, *Ann. Appl. Biol.*, 96, 163, 1980.
30. **Hughes, I. K. and Wong, W. C.,** Biological control of *Allium* white rot (*S. cepivorum*) in Queensland, Australia, in *Proc. 3rd. Int. Workshop on Allium White Rot,* Entwistle, A. R., Ed., T. W. Printing Associates Ltd, Leamington Spa, U.K., 1987, Section 4, 51.
31. **Wong, J. A. -L. and Maynard, J. R.,** Application of procymidone to seed and fertilizer for the control of *Allium* white rot in bulb onions, in *Proc. 3rd. Int. Workshop on Allium White Rot,* Entwistle, A. R., Ed., T. W. Printing Associates Ltd, Leamington Spa, U.K., 1987, Section 3, 86.
32. **Rod, J.,** *Allium* crops and *Allium* white rot in Czechoslovakia, in *Proc. 3rd. Int. Workshop on Allium White Rot,* Entwistle, A. R., Ed., T. W. Printing Associates Ltd, Leamington Spa, U.K., 1987, Section 1, 64, 1986.
33. **Rondomanski, W. and Doruchowski, R. W.,** Introductory searches on onion resistance to white rot, *Biul. Warzywniczy Instytut Warzywnictwa-Skierniewice,* 22, 99, 1978.
34. **Semb, L., Vik, J., and Hirvonen-Semb, A.,** Resistance of onion (*Allium cepa* L.) and other *Allium* spp. to the white rot fungus (*Sclerotium cepivorum* Berk.), *Biul. Warzywniczy Instytut Warzywnictwa-Skierniewice,* 22, 105, 1978.
35. **Dhyani, S. C. and Chauhan, S. K.,** Effect of soil moisture on *Fusarium* wilt and *Sclerotium* rot of onion *Allium cepa, Indian Sci. Cong. Assoc. Proc.*, 62, 56, 1976.
36. **Yamamoto, W.,** Species of the sclerotiniaceae in Japan, *Trans. Mycol. Soc. Jpn.*, 2, 2, 1959.
37. **Leguizamon, C. J. and Barriga, O. R.,** Garlic (*Allium sativum* L.) diseases in Cundinamarca and Boyaca, *Noticias Fitopatolgicas,* 5, 4, 1976.
38. **Savulescu, T.,** Phytosanitary situation in the Romanian Peoples Republic in 1957-58, *Met. Inst. Cerc. Agron. Roman.*, 105, 1960.
39. **Ali, A. E. R., El Yamany, T., and Kamel, M.,** White rot of onion in Arab Republic of Egypt. I. Physiological studies on an isolate of the causal fungus *Sclerotium cepivorum* Berk., *Agric. Res. Rev.*, 49, 299, 1971.
40. **Adams, P. B. and Papavizas, G. C.,** Effect of inoculum density of *Sclerotium cepivorum* and some soil environmental factors on disease severity, *Phytopathology,* 61, 1253, 1971.

41. **Asthana, R. P.,** Studies on sclerotium-forming fungi. I. *Sclerotium cepivorum* Berk. and *Sclerotium tuliparum* Klebahn. I. Cultural studies, *Proc. Indian Acad. Sci.,* 26, 93, 1947.
42. **Commonwealth Mycological Institute,** Distribution maps of Plant Diseases, Ed. 5., *Commonwealth Mycological Institute,* Kew, Surrey, UK, 1985, Map 5.
43. **Coley-Smith, J. R.,** Studies of the biology of *Sclerotium cepivorum* Berk. III. Host range; persistence and viability of sclerotia, *Ann. Appl. Biol.,* 47, 511, 1959.
44. **Coley-Smith, J. R. and Esler, G.,** Infection of cultivars of onion, leek, garlic and *Allium fistulosum* by *Sclerotium cepivorum, Plant Pathol.,* 32, 373, 1983.
45. **van der Meer, Q. P., van Bennekom, J. L., and van der Giessen, A. C.,** Screening for resistance to white rot caused by *Sclerotium cepivorum* Berk. in onions (*Allium cepa* L.) and leeks (*Allium porrum* L.), *Euphytica,* 32, 697, 1983.
46. **Patil, B. G. and Rane, M. S.,** White rot of garlic, *College of Agriculture Nagpur Magazine,* 45, 86, 1973.
47. **Merriman, P. R., Samson, I. M., and Schippers, B.,** Stimulation of germination of sclerotia of *Sclerotium cepivorum* at different depths in soil by artificial onion oil, *Netherlands J. Plant Pathol.,* 87, 45, 1981.
48. **Lorbeer, J. W. and Clarke, T. T.,** *Allium* white rot in New York State, USA: The Present Situation, in *Proc. 3rd. Int. Workshop on Allium White Rot* Entwistle, A. R., Ed., T. W. Printing Associates Ltd, Leamington Spa, U.K., 1987, Section 1, 41.
49. **Mattusch, P.,** The phytopathological situation in onions in the Federal Republic of Germany, in *Proc. 3rd. Int. Workshop on Allium White Rot,* Entwistle, A. R., Ed., T. W. Printing Associates Ltd, Leamington Spa, U.K., 1987, Section 1, 44.
50. **Johnston, S. A.,** *Allium* white rot in USA (New Jersey), in *Proc. 2nd Int. Workshop on Allium White Rot,* Adams, P. B. and Entwistle A. R., Ed., 1983, 5.
51. **Oudemans, P. and Edgington, L. V.,** Distribution of white rot of onions in Ontario, in *Proc. 2nd. Int. Workshop on Allium White Rot,* Adams, P. B., and Entwistle, A. R., Eds., 1983, 10.
52. **Valdes, E. and Edgington, L. V.,** Effect of flooding on *Sclerotium cepivorum* in organic soil, in *Proc. 3rd. Int. Workshop on Allium White Rot,* Entwistle, A. R., Ed., T. W. Printing Associates Ltd, Leamington Spa, U.K., 1987, Section 6, 49.
53. **Entwistle, A. R. and Munasinghe, H. L.,** Epidemiology and control of white rot disease of onions, in *Plant Disease Epidemiology,* Scott, P. R. and Bainbridge, A. Eds., Blackwell, Oxford, 1978.
54. **Kurtz, E. A.,** *Allium* white rot: geographic distribution and economic importance, in *Proc. 2nd Int. Workshop on Allium White Rot,* Adams, P. B. and Entwistle A. R., Eds., 1983, 8.
55. **Entwistle, A. R.,** Relationships between soil sclerotial populations of *Sclerotium cepivorum* and the incidence of *Allium* white rot, in *Proc. 3rd. Int. Workshop on Allium White Rot,* Entwistle, A. R., Ed., T. W. Printing Associates Ltd, Leamington Spa, U.K., 1987, Section 6, 21.
56. **Mikhail, S., Stewart, D. M., El Haggagy, M. Kh., and Wilkinson, R. E.,** The role of grazing animals in the dissemination of the onion white rot pathogen in Egypt, *Plant Port. Bull. Food Agric. Org. U. N. Rome,* 22, 37, 1974.
57. **Georgy, N. I.,** *Allium* white rot: Geographic distribution and economic importance, in *Proc. 2nd Int. Workshop on Allium White Rot,* Adams, P. B. and Entwistle A. R., Eds., 1983, 16.
58. **Rogers, H. T.,** Onion wipe out threatens Walla Walla, *Am. Veg. Grower,* March 20, 78, 1972.
59. **MAFF,** Key horticultural statistics for the United Kingdom, calendar years 1970-1979, HMSO, London, 1980.
60. **Entwistle, A. R.,** *Allium* white rot in the UK, 1984-86, in *Proc. 3rd. Int. Workshop on Allium White Rot,* Entwistle, A. R., Ed., T. W. Printing Associates Ltd, Leamington Spa, U.K., 1987, Section 1, 11.
61. **Croxall, H. E., Sidwell, R. W., and Jenkins, J. E. E.,** White rot (*Sclerotium cepivorum*) of onions in Worcestershire with special reference to control by seed treatment with calomel, *Ann. Appl. Biol.,* 40, 166, 1953.
62. **Utkhede, R. S. and Rahe, J. E.,** Screening world onion germplasm collection for resistance to white rot, *Can. J. Plant Sci.,* 58, 819, 1978.
63. **Collins, R. L.,** The status of white rot in onions and garlic in the States of California, Idaho and Washington, USA, in *Proc. 3rd. Int. Workshop on Allium White Rot* , Entwistle, A. R., Ed., T. W. Printing Associates Ltd, Leamington Spa, U.K., 1987, Section 1, 9.
64. **van der Meer, Q. P.,** Onions in modern Egypt, *Onion World,* 2, (May), 9, 1986.
65. **Jones, H. A., and Mann, L. K.,** *Onions and Their Allies,* Leonard Hill [Books] Ltd., London, 1963, chaps. 1, 2.
66. **Food and Agriculture Organization**, *FAO Production Yearbook,* Vol. 31, Food and Agriculture Organization of the United Nations, Rome 1977, Table 49.
67. **Food and Agriculture Organization**, *FAO Production Yearbook,* Vol. 34, Food and Agriculture Organization of the United Nations, Rome, 1980, Table 49.
68. **Food and Agriculture Organization**, *FAO Production Yearbook,* Vol. 37, Food and Agriculture Organization of the United Nations, Rome, 1983, Table 57.

69. **Food and Agriculture Organization**, *FAO Production Yearbook,* Vol. 39, Food and Agriculture Organization of the United Nationa, Rome, 1985, Table 57.
70. **Abd El Moity, T. H.,** Situation of white rot disease of onions in Egypt, in *Proc. 1st. Int. Workshop on Allium White Rot,* Wageningen, Netherlands, Aug 28-29, 1979, 14.
71. **Fawzy, R.,** Onion cultivation in Egypt, in *Proc. 3rd. Int. Workshop on Allium White Rot,* Entwistle, A. R., Ed., T. W. Printing Associates Ltd, Leamington Spa, U.K., 1987, Section 1, 13.
72. **Sirry, A. R., Higazy, M. F. H., and Georgy, N. I.,** Studies on white rot of onion. I. Effect of fertilizers on disease severity, plant growth and its mineral content, *Agric. Res. Rev.,* 52, 47, 1974.
73. **Crary, D. D.,** Irrigation and land use in Zeiniya Bahari, Upper Egypt, *Geo. Rev.,* 39, 568, 1949.
74. **Ali, A. A.,** The effect of certain environmental factors and varietal test trials on the incidence of white rot disease of onions caused by *Sclerotium cepivorum* Berk. in the Arab Republic of Egypt, *Agric. Res. Rev.,* 54, 43, 1976.
75. **Porter, I. J. and Merriman, P. R.,** Effects of solarization of soil on nematode and fungal pathogens at two sites in Victoria, *Soil Biol. Biochem.,* 15, 39, 1983.
76. **Mordue, J. E. M.,** *Sclerotium cepivorum, CMI Descriptions of Pathogenic Fungi and Bacteria,* Commonwealth Mycological Institute, Kew, Surrey, U.K., 1976, 512.
77. **Townsend, B. B. and Willetts, H. J.,** The development of sclerotia of certain fungi, *Trans. Br. Mycol. Soc.,* 37, 213, 1954.
78. **Georgy, N. I. and Coley-Smith, J. R.,** Variation in morphology of *Sclerotium cepivorum* sclerotia, *Trans. Br. Mycol. Soc.,* 79, 534, 1982.
79. **Entwistle, A. R., and Munasinghe, H. L.,** Formation of secondary sclerotia by *Sclerotium cepivorum, Trans. Br. Mycol. Soc.,* 77, 432, 1981.
80. **Coley-Smith, J. R., Esler, G. A., and New, C. M.,** Possibilities for biological and integrated control of white rot disease of *Allium,* in *Proc. 1981 British Crop Prot. Conf. Pests and Diseases,* 1981, 459.
81. **Coley-Smith, J. R.,** Studies of the biology of *Sclerotium cepivorum* Berk. IV. Germination of sclerotia, *Ann. Appl. Biol.,* 48, 8, 1960.
82. **Scott, M. R.,** Studies of the biology of *Sclerotium cepivorum* Berk. II. The spread of white rot from plant to plant, *Ann. Appl. Biol.,* 44, 584, 1956.
83. **Coley-Smith, J. R. and Holt, R. W.,** The effect of species of *Allium* on germination in soil of sclerotia of *Sclerotium cepivorum* Berk., *Ann. Appl. Biol.,* 58, 273, 1966.
84. **Coley-Smith, J. R. and King, J. E.,** The production by species of *Allium* of alkyl sulphides and their effect on germination of sclerotia of *Sclerotium cepivorum* Berk., *Ann. Appl. Biol.,* 64, 289, 1969.
85. **King, J. E. and Coley-Smith, J. R.,** Production of volatile alkyl sulphides by microbial degradation of synthetic alliin and alliin-like compounds, in relation to germination of sclerotia of *Sclerotium cepivorum* Berk., *Ann. Appl. Biol.,* 64, 303, 1969.
86. **Coley-Smith, J. R., Taylor, I. M., and Parfitt, D.,** Dormancy in the sclerotia of *Sclerotium cepivorum,* in *Proc. 3rd. Int. Workshop on Allium White Rot,* Entwistle, A. R., Ed., T. W. Printing Associates Ltd, Leamington Spa, U.K., 1987, Section 6, 17.
87. **Zinkernagel, V.,** Inoculation techniques with isolates of *Sclerotium cepivorum* attacking onion varieties, selections and crossings, in *Proc. 3rd. Int. Workshop on Allium White Rot,* Entwistle, A. R., Ed., T. W. Printing Associates Ltd. Leamington Spa, U.K., 1987, Section 5, 18.
88. **King, J. E. and Coley-Smith, J. R.,** Effects of volatile products of *Allium* species and their extracts on germination of sclerotia of *Sclerotium cepivorum* Berk., *Ann. Appl. Biol.,* 61, 407, 1968.
89. **Coley-Smith, J. R.,** A comparison of flavour and odour compounds of onion, leek, garlic and *Allium fistulosum* in relation to germination of sclerotia of *Sclerotium cepivorum, Plant Pathol.,* 35, 370, 1986.
90. **Coley-Smith, J. R. and Sansford, C. E.,** Survival of sclerotia of *Sclerotium cepivorum,* in *Proc. 3rd. Int. Workshop on Allium White Rot,* Entwistle, A. R., Ed., T. W. Printing Associates Ltd, Leamington Spa, U.K., 1986, Section 6, 11.
91. **Entwistle, A. R. and Munasinghe, H. L.,** White rot disease of salad onions, *Ann. Rep. Natl. Veg. Stn. 1974,* 110, 1975.
92. **Leggett, M. E. and Rahe, J. E.,** Factors affecting the survival of sclerotia of *Sclerotium cepivorum* in the Fraser Valley of British Columbia, *Ann. Appl. Biol.,* 106, 255, 1985.
93. **Fletcher, J. T.,** The control of onion white rot using a pure calomel seed dressing and a new sticker, *Plant Pathol.,* 13, 182, 1964.
94. **Littley, E. R. and Rahe, J. E.,** Effect of host plant density on white rot of onion caused by *Sclerotium cepivorum, Can. J. Plant Pathol.,* 9, 146, 1987.
95. **Adams, P. B. and Springer, J. K.,** Time of infection of fall-planted onions by *Sclerotium cepivorum, Plant Dis. Rep.,* 61, 722, 1977.
96. **Entwistle, A. R. and Munasinghe, H. L.,** The effect of seed and stem base spray treatment with iprodione on white rot disease (*Sclerotium cepivorum*) in autumn-sown salad onions, *Ann. Appl. Biol.,* 97, 269, 1981.
97. **Entwistle, A. R.,** White rot disease of onions, *Ann. Rep. Natl. Veg. Stn.,* 1984, 66, 1985.

98. **Entwistle, A. R. and Blakeman, R.,** Use of aerial photography for the detection of *Allium* white rot, in *Proc. 3rd. Int. Workshop on Allium White Rot,* Entwistle, A. R., Ed., T. W. Printing Associates Ltd, Leamington Spa, U.K., 1987, Section 2, 6.
99. **Moore, W. C.,** Report on fungus, bacterial and other diseases of crops in England and Wales for years 1933-42, HMSO, London, 1944.
100. **Scott, M. R.,** Studies of the biology of *Sclerotium cepivorum* Berk. I. Growth of the mycelium in soil, *Ann. Appl. Biol.,* 44, 576, 1956.
101. **Entwistle, A. R.,** Differences in the incidence of *Allium* white rot in direct-drilled and module-grown bulb onions, in *Proc. 3rd. Int. Workshop on Allium White Rot,* Entwistle, A. R., Ed., T. W. Printing Associates Ltd, Leamington Spa, U.K., 1987, Section 2, 1.
102. **Coley-Smith, J. R., Reese, R. A., and Georgy, N. I.,** Differential stimulation of germination of sclerotia of *Sclerotium cepivorum* by cultivars of onion and its effect on white rot disease, *Plant Pathol.,* 36, 246, 1987.
103. **Hughes, I. K.,** Control of white rot of onions in storage, *Queensl. J. Agric. Anim. Sci.,* 27, 393, 1970.
104. **Tims, E. C.,** Further observations of white rot of *Allium* in Louisiana, *Plant Dis. Rep.,* 27, 380, 1943.
105. **Sohi, H. S.,** Storage rot of onion bulbs and its control, *Indian Phytopathol.,* 30, 149, 1978.
106. **Rod, J.,** Antagonistic effects of some fungi on fungal pathogens causing storage rots of onion (*Allium cepa* L.), *Ceska Mykologie,* 38, 235, 1984.
107. **Adams, P. B.,** A rapid method for quantitative isolation of sclerotia of *Sclerotinia minor* and *Sclerotium cepivorum* from soil, *Plant Dis. Rep.,* 63, 349, 1979.
108. **Harrison, D. E.,** White rot of onions in Victoria, *J. Dept. Agric.,* 52, 510, 1954.
109. **Walker, J. C.,** The influence of soil temperature and soil moisture upon white rot of *Allium, Phytopathology,* 16, 697, 1926.
110. **Crowe, F. J. and Hall, D. H.,** Soil temperature and moisture effects on sclerotium germination and infection of onion seedlings by *Sclerotium cepivorum. Phytopathology,* 70, 74, 1980.
111. **Hendry, S. and Henshall, A.,** *Allium* White Rot at Southfleet, UK, 1976-86, in *Proc. 3rd. Int. Workshop on Allium White Rot,* Entwistle, A. R., Ed., T. W. Printing Associates Ltd, Leamington Spa, U.K., 1987, Section 1, 15.
112. **Revill, G. W.,** *Allium* white rot at Woodfield Farm, Birlingham, Worcestershire, UK, in *Proc. 3rd. Int. Workshop on Allium White Rot,* Entwistle, A. R., Ed., T. W. Printing Associates Ltd, Leamington Spa, U.K., 1987, Section 1, 58.
113. **Crowe, F. J. and Hall, D. H.,** Vertical distribution of sclerotia of *Sclerotium cepivorum* and host root systems relative to white rot of onion and garlic, *Phytopathology,* 70, 70, 1980.
114. **Ali, A. E. R., El Yamany, T., and Bekhit, M. R.,** White rot of onion. II. Effect of some chemical and physical factors on sclerotia germination and onion plant infection by *Sclerotium cepivorum* Berk. the causal organism of white rot of onion in the Arab Republic of Egypt, *Agric. Res. Rev.,* 49, 313, 1971.
115. **Tichelaar, G. M.,** Witrot in zaai-uien, *Groenten en Fruit,* 16, 2107, 1965.
116. **Anon,** *Sclerotium cepivorum, Adviesbasis Aaltjesonderzoek.* Bedrijfslaboratorium voor Grond- en Gewasonderzoek, Oosterbeek, Netherlands, 1975, 15.
117. **Adams, P. B.,** Forecasting onion white rot disease, *Phytopathology,* 71, 1178, 1981.
118. **Hall, D. H. and Somerville, P. A.,** Effect of inoculum potential on control of white rot by fungicides, in *Proc. 2nd Int. Workshop on Allium White Rot,* Adams, P. B. and Entwistle, A. R., Eds., 1983, 113.
119. **Resende, M. L. V., Zambolim, L., and Da Cruz, J.,** Efficacy of fungicides in the control of garlic white rot, according to the level of *Sclerotium cepivorum* Berk. sclerotia in the soil, in *Proc. 3rd. Int. Workshop on Allium White Rot,* Entwistle, A. R., Ed., T. W. Printing Associates Ltd, Leamington Spa, U.K., 1987, Section 3, 36.
120. **Crowe, F. J., Hall, D. H., Greathead, A. S., and Baghott, K. G.,** Inoculum density of *Sclerotium cepivorum* and the incidence of white rot of onion and garlic, *Phytopathology,* 70, 64, 1980.
121. **Amein, A. M., El Shabrawy, A. M., and Abd El Razik, A. A.,** Density of *Sclerotium cepivorum* Berk. sclerotia in soil in relation to severity of white rot of garlic, *Assiut J. Agric. Sci.,* 11, 1982.
122. **Abd El Moity, T. H.,** Effect of single and mixture of *Trichoderma harzianum* isolates on controlling three different soil-borne pathogens, in *Egypt. J. Microbiol. Special Issue,* 1985, chap. 11, 111.
123. **Papavizas, G. C.,** Isolation and enumeration of propagules of *Sclerotium cepivorum* from soil, *Phytopathology,* 62, 545, 1972.
124. **Winfield, A. L., Enfield, M. A., and Foreman, J. H.,** A column elutriator for extracting cyst nematodes and other small invertebrates from soil samples, *Ann. Appl. Biol.,* 111, 223, 1987.
125. **Botting, G. W.,** White rot menaces our onion crops, *J. Agric. S. Aust.,* 66, 420, 1963.
126. **Somerville, P. A. and Hall, D. H.,** The effect of hot water, formaldehyde treatment for control of stem and bulb nematode, on the viability of sclerotia of *Sclerotium cepivorum,* in *Proc. 2nd. Int. Workshop on Allium White Rot,* Adams, P. B. and Entwistle, A. R., Eds., 1983, 127.
127. **Semb, L.,** Lokhvitrate-settelok, *Gartneryrket,* 56, 458, 1966.

128. **Anon,** Onion briefs: Washington white rot quarantine order signed, *Onion World* 1, (Oct), 13, 1985.
129. **Anon,** White rot suspect in Idaho, *Onion World,* 2, (June/July), 20, 1986.
130. **Sewell, G. W. F., Wilson, J. F., and Martin, D. G.,** Machine picking in relation to progressive *Verticillium* wilt of the hop. II. The effect of composting on the infectivity of machine-picked hop waste, *Rep. East Malling Res. Stn. 1961.* 102, 1962.
131. **Yuen, G. Y. and Raabe, R. D.,** Effects of small-scale aerobic composting on survival of some fungal plant pathogens, *Plant Dis.,* 68, 134, 1984.
132. **Adams, P. B.,** Effects of soil temperature, moisture, and depth on survival and activity of *Sclerotinia minor, Sclerotium cepivorum* and *Sporidesmium sclerotivorum, Plant Dis.,* 71, 170, 1987.
133. **Lifshitz, R., Tabachinik, M., Katan, J., and Chet, I.,** The effect of sublethal heating on *Sclerotium rolfsii, Can. J. Microbiol.,* 29, 1607, 1983.
134. **Anon**, White rot tests hopeful, *Onion World,* 3, (June/July) 7, 1986.
135. **Coley-Smith, J. R., Taylor, I. M., and Parfitt, D.,** The use of diallyl disulphide for reducing populations of sclerotia of *Sclerotium cepivorum* in soil, in *Proc. 3rd. Int. Workshop on Allium White Rot,* Entwistle, A. R., Ed., T. W. Printing Associates Ltd, Leamington Spa, U.K., 1987, Section 4, 35.
136. **Patterson, C. L. and Grogan, R. G.,** Differences in epidemiology and control of lettuce drop caused by *Sclerotinia minor* and *S. sclerotiorum, Plant Dis.,* 69, 766, 1985.
137. **Katan, J.,** Solar heating (solarization) of soil for control of soilborne pests, *Ann. Rev. Phytopathol.,* 19, 211, 1981.
138. **El-Yamani, T., Abdel-Rahim, M. F., Michael, K. Y., Grinstein, A., and Katan, J.,** Soil solarization — a potential method for controlling white rot and other soilborne diseases in onions, in *Proc. 2nd Int. Workshop on Allium White Rot,* Adams, A. B., and Entwistle, A. R., Eds., 1983, 126.
139. **Merriman, P. R. and Isaacs, S.,** Evaluation of onions as a trap crop for *Sclerotium cepivorum, Soil Biol. Biochem.,* 10, 339, 1978.
140. **Entwistle, A. R. and Munasinghe, H. L.,** The effect of iprodione on sclerotium germination, root infection and mycelial spread of *Sclerotium cepivorum* in salad onions, *Ann. Appl. Biol.,* 95, 329, 1980.
141. **Somerville, P. A. and Hall, D. H.,** Factors affecting sclerotial germination of *Sclerotium cepivorum,* secondary sclerotia formation, and germination stimulants to reduce inoculum density, *Plant Dis.,* 71, 229, 1987.
142. **Entwistle, A. R.,** Loss of control of *Allium* white rot by fungicides and its implications, *Aspects Appl. Biol.,* 12, 201, 1986.
143. **Walker, A., Entwistle, A. R., and Dearnaley, N. J.,** Evidence for enhanced degradation of iprodione in soils treated previously with this fungicide, *Br. Crop Prot. Council Monogr.,* No. 27, 117, 1984.
144. **Walker, A., Brown, P. A., and Entwistle, A. R.,** Enhanced degradation of iprodione and vinclozolin in soil, *Pestic. Sci.,* 17, 183, 1986.
145. **Utkhede, R. S. and Rahe, J. E.,** Treatment of muck soil with onion oil to control onion white rot, *Can. J. Plant Pathol.,* 4, 79, 1982.
146. **Keer, J. I.,** The assessment of soil partial sterilants for the control of *Allium* white rot in Lincolnshire, UK, in *Proc. 3rd. Int. Workshop on Allium White Rot,* Entwistle, A. R., Ed., T. W. Printing Associates Ltd, Leamington Spa, U.K., 1987, Section 3, 20.
147. **Greathead, A. S., Hall, C. H., and Kurtz, E.,** Control of white rot of onion and garlic in the Salinas valley and Tulelake areas of California, in *Proc. 2nd Int. Workshop on Allium White Rot,* Adams, P. B. and Entwistle A. R., Eds., 1983, 35.
148. **Davies, J. M. L. and Coley-Smith, J. R.,** Chemical control of white rot in module raised spring bulb onions — comparison of soil sterilants, in *Proc. 3rd. Int. Workshop on Allium White Rot,* Entwistle, A. R., Ed., T. W. Printing Associates Ltd, Leamington Spa, U.K., 1987, Section 3, 2.
149. **Adams, P. B. and Johnston, S. A.,** Factors affecting efficacy of metham applied through sprinkler irrigation for control of *Allium* white rot, *Plant Dis.,* 67, 978, 1983.
150. **Entwistle, A. R., White, J. G., and Coleman, K. E.,** Effects of dazomet on white rot on bulb onions, *Ann. Rep. Natl. Veg. Res. Stn.,* 1985, 64, 1986.
151. **Ben-Yephet, Y. and Frank, Z. R.,** Effect of soil structure on penetration by metham-sodium and of temperature on concentrations required to kill soilborne pathogens, *Phytopathology,* 75, 403, 1985.
152. **Valk, M.,** White rot problems in Canada, *Onion World,* 1, (Nov), 21, 1985.
153. **Ogilvie, L. and Hickman, C. J.,** Progress report on vegetable diseases, *Ann. Rep. Long Ashton Res. Stn. 1937,* 96, 1937.
154. **Finch, S., Foster, R. E., Cadoux, M. E., and Eckenrode, C. J.,** Reducing maggot damage, *Onion World,* 2, (May), 18, 1986.
155. **El Gammal, M. M. and Ibrahim, M. Y.,** Avoiding white rot disease through growing onions from sets, in *Proc. 2nd Int. Workshop on Allium White Rot,* Adams, P. B. and Entwistle A. R., Eds., 1983, 115.
156. **Utkhede, R. S. and Rahe, J. E.,** Evaluation of chemical fungicides for control of onion white rot, *Pestic. Sci.,* 10, 414, 1979.

157. **Porter, I. J.,** Evaluation of procymidone applied either as a seed treatment or banded with fertilizer, on white rot of dry bulb onions in victoria, in *Proc. 3rd. Int. Workshop on Allium White Rot,* Entwistle, A. R., Ed., T. W. Printing Associates Ltd, Leamington Spa, U.K., 1987, Section 3, 30.
158. **Lafon, R. and Bugaret, Y.,** Les traitements des semences d'oignon contre la pourriture blanche, *Phytiatrie-Phytopharmacie,* 2, 67, 1968.
159. **Entwistle, A. R. and Munasinghe, H. L.,** The effect of seed and stem-base spray treatments with thiophanate-methyl on the control of white rot disease (*Sclerotium cepivorum*) of autumn-sown salad onions, *Plant Pathol.,* 29, 153, 1980.
160. **Lafon, R. and Bugaret, Y.,** Efficacite du benomyl contre la pourriture blanche de l'ail et de l'oignon, *Phytiatrie-Phytopharmacie,* 19, 3, 1970.
161. **Ryan, E. W. and Kavanagh, T.,** White rot of onion (*Sclerotium cepivorum*). II. Control by fungicidal dusting of onion sets, *Irish J. Agric. Res.,* 15, 325, 1976.
162. **Lafon, R. and Bugaret, Y.,** Etude sur l'efficacite du dicloran contre la pourriture blanche de l'ail *(Sclerotium cepivorum Berk.), Phytiatrie-Phytopharmacie,* 3, 165, 1966.
163. **Ryan, E. W.,** *Allium* white rot in Ireland, in *Proc. 3rd. Int. Workshop on Allium White Rot* Entwistle, A. R., Ed., T. W. Printing Associates Ltd., Leamington Spa, U.K., 1987, Section 1, 66.
164. **Maloy, O. C. and Machtmes, R.,** control of onion white rot by furrow and root-dip application of fungicides, *Plant Dis. Rep.,* 58, 6, 1974.
165. **Entwistle, A. R. and Munasinghe, H. L.,** The control of white rot (*Sclerotium cepivorum*) in fluid-drilled salad onions, *J. Hortic. Sci.,* 56, 251, 1981.
166. **Davis, J. M. L. and Wafford, J. D.,** Chemical control of whiterot in module raised spring bulb onion, in *Proc. 3rd. Int. Workshop on Allium White Rot,* Entwistle, A. R., Ed., T. W. Printing Associates Ltd, Leamington Spa, U.K., 1987, Section 3, 6.
167. **Fletcher, J. T., Knight, B. C., Bate, E., and Cragg, I. A.,** The control of white rot (*Sclerotium cepivorum* Berk.) in salad onions with dicloran, *Plant Pathol.,* 20, 88, 1971.
168. **Locke, S. B.,** Botran tolerance of *Sclerotium cepivorum* isolants from fields with different botran-treatment histories, *Phytopathology,* 59, 13, 1969.
169. **Groves, K. and Chough, K. S.,** Fate of the fungicide 2,6-dichloro-4-nitroaniline (DCNA) in plants and soils, *J. Agric. Food Chem.,* 18, 1127, 1970.
170. **Ryan, E. W. and Kavanagh, T.,** White rot of onion (*Sclerotium cepivorum*) I. Control by fungicidal pelleting of onion seed, *Ir. J. Agric. Res.,* 15, 317, 1976.
171. **Entwistle, A. R.,** Chemical control of *Allium* white rot at NVRS, UK, 1981-83, in *Proc. 2nd. Int. Workshop on Allium White Rot,* Adams, P. B. and Entwistle, A. R., Eds., 1983, 27.
172. **Duchon-Doris, J.,** La lutte contre *Sclerotium cepivorum, Rev. Hortic.,* 200, 33, 1979.
173. **Hall, D. H., Somerville, P., and Greathead, A. S.,** Onion white rot investigations, in *Proc. 3rd. Int. Workshop on Allium White Rot,* Entwistle, A. R., Ed., T. W. Printing Associates Ltd, Leamington Spa, U.K., 1987, Section 3, 16.
174. **Entwistle, A. R.,** The effect of dicarboximide fungicides on *Sclerotium cepivorum*, in *Proc. 10th. Int. Congress of Plant Protection,* Brighton, U.K., 1983, 629.
175. **Entwistle, A. R., Munasinghe, H. L., and Coleman, K. E.,** Loss of control of onion white rot with iprodione, *Ann. Rep. Natl. Veg. Res. Stn. 1986/1987,* 60, 1987.
176. **Yarden, O., Katan, J., Aharonson, N., and Ben-Yephet, Y.,** Delayed and enhanced degradation of benomyl and carbendazim in disinfested and fungicide-treated soils, *Phytopathology,* 75, 763, 1985.
177. **Adams, P. B.,** Evaluation of fungicides in the field for control of white rot of onion, in *Methods for Evaluating Plant Fungicides, Nematicides, and Bactericides,* American Phytopathological Society, Minnesota, 1978, 61.
178. **Hock, W. K.,** Test water pH, *Am. Veg. Grower,* 33, 51, 1985.
179. **Blikman, L.,** *Allium* white rot in Holland, in *Proc. 3rd. Int. Workshop on Allium White Rot,* Entwistle, A. R., Ed., T. W. Printing Associates Ltd, Leamington Spa, U.K., 1987, Section 1, 1.
180. **Costigan, P. R., Rowse, H. R., Heaviside, M., and Grey, H.,** Starter fertilizer for maincrop onion, *Ann. Rep. Natl. Veg. Res. Stn. for 1986/87,* 86, 1987.
181. **Entwistle, A. R. and Munasinghe, H. L.,** Evaluation of prochloraz and nemacin for the control of white rot in spring-sown salad onions, *Tests of Agrochemicals and Cultivars,* 2, *(Ann. Appl Biol.* Suppl. 97), 28, 1981.
182. **Entwistle, A. R. and Marian, S. E.,** Evaluation of thiabendazole for the control of white rot in spring-sown salad onions, *Tests of Agrochemicals and Cultivars,* 4, *(Ann. Appl. Biol.* Suppl. 102), 66, 1983.
183. **Entwistle, A. R.,** Evaluation of the effect of propiconazole and penconazole on onion white rot, *Tests of Agrochemicals and Cultivars,* 8, *(Ann. Appl. Biol.* Suppl. 110), 60, 1987.
184. **Entwistle, A. R. and Marian, S. E.,** Evaluation of carboxin, oxycarboxin, triadimefon and tolclofos-methyl for the control of white rot in spring-sown salad onions, *Tests of Agrochemicals and Cultivars,* 4, *(Ann. Appl. Biol.* Suppl. 102), 64, 1983.

185. **Utkhede, R. S. and Rahe, J. E.,** Biological control of onion white rot, *Soil Biol. Biochem.,* 12, 101, 1980.
186. **Abd El Moity, T. H. and Shatla, M. N.,** Biological control of white rot disease of onion (*Sclerotium cepivorum*) by *Trichoderma harzianum, Phytopathol. Z.,* 100, 29, 1981.
187. **Rahe, J. E.,** A critical perspective on biological control of *Allium* white rot, in *Proc. 3rd. Int. Workshop on Allium White Rot,* Entwistle, A. R., Ed., T. W. Printing Associates Ltd, Leamington Spa, U.K., 1987, Section 4, 1.
188. **Ahmed, A. H. M. and Tribe, H. T.,** Biological control of white rot of onion (*Sclerotium cepivorum*) by *Coniothyrium minitans, Plant Pathol.,* 26, 75, 1977.
189. **Adams, P. B,** Comparison of isolates of *Sporidesmium sclerotivorum in vitro* and in soil for potential as active agents in microbial pesticides, *Phytopathology,* 77, 575, 1987.
190. **Adams, P. B. and Ayers, W. A.,** *Sporidesmium sclerotivorum:* distribution and function in natural biological control of sclerotial fungi, *Phytopathology,* 71, 90, 1981.
191. **Trutmann, P., Keane, P. J., and Merriman, P. R.,** Reduction of sclerotial inoculum of *Sclerotinia sclerotiorum* with *Coniothyrium minitans, Soil Biol. Biochem.,* 12, 461, 1980.
192. **Ghaffar, A.,** Biological control of white rot of onion I. Interactions of soil micro-organisms with *Sclerotium cepivorum Berk., Mycopathol. Mycol. Appl.,* 38, 101, 1969.
193. **Fisher, C. G., Conway, K. E., and Motes, J. E.,** Fluid drilling: A potential delivery system for fungal biological control agents with small-seeded vegetables, *Proc. Oklahoma Acad. Sci.,* 63, 100, 1983.
194. **Ghaffer, A.,** Biological control of white rot of onion II. Effectiveness of *Penicillium nigricans* (Bain.) Thom., *Mycopathol. Mycol. Appl.,* 38, 113, 1969.
195. **Abd El Moity, T. H.,** Biological control of white rot of onion in the field, in *Proc. 2nd Int. Workshop on Allium White Rot,* Adams, P. B. and Entwistle A. R., Eds., 1983, 81.
196. **Utkhede, R. S. and Rahe, J. E.,** Interactions of antagonist and pathogen in biological control of onion white rot, *Phytopathology,* 73, 890, 1983.
197. **Utkhede, R. S. and Rahe, J. E.,** Screening commercial onion cultivars for resistance to white rot, *Phytopathology,* 68, 1080, 1978.
198. **Abd El Moity, T. H.,** A new laboratory method to evaluate and screening *Allium* spp. and onion varieties against *Sclerotium cepivorum* Berk., the causal organism of white rot disease, *Agric. Res. Rev.,* 62, 131, 1984.
199. **Gabelman, W. H.,** White rot resistance from *Allium cepa cv.* Zittauer Gelb, in *Proc. 3rd. Int. Workshop on Allium White Rot,* Entwistle, A. R., Ed., T. W. Printing Associates Ltd, Leamington Spa, U.K., 1987, Section 5, 9.
200. **Utkhede, R. S., Rahe, J. E., Coley-Smith, J. R., van der Meer, Q. P., Brewer, J. G., and Criscola, V.,** Genotype-environment interactions for resistance to onion white rot, *Can. J. Plant Pathol.,* 4, 269, 1982.
201. **Esler, G. and Coley-Smith, J. R.,** Flavour and odour characteristics of species of *Allium* in relation to their capacity to stimulate germination of sclerotia of *Sclerotium cepivorum, Plant Pathol.,* 32, 13, 1983.
202. **Coley-Smith, J. R.,** Interactions between *Sclerotium cepivorum* and cultivars of onion, leek, garlic and *Allium fistulosum, Plant Pathol.,* 35, 362, 1986.
203. **Esler, G. and Coley-Smith, J. R.,** Resistance to *Sclerotium cepivorum* in *Allium* and other genera, *Plant Pathol.,* 33, 199, 1984.
204. **Coley-Smith, J. R., Reese, R. A., and Georgy, N. I.,** Differential stimulation of germination of sclerotia of *Sclerotium cepivorum* by cultivars of onion and its effect on white rot disease, *Plant Pathol.,* 36, 246, 1987.
205. **Entwistle, A. R., Riddle, S. L., Horobin, J. F., and Dowker, B. D.,** Genetic resistance in *A. cepa. Ann. Rep. Natl. Veg. Res. Stn. 1984,* 69, 1985.
206. **Rahe, J. E.,** Detection and selection for field resistance to onion white rot, in *Proc. 3rd. Int. Workshop on Allium White Rot,* Entwistle, A. R., Ed., T. W. Printing Associates Ltd, Leamington Spa, U.K., 1987, Section 5, 11.
207. **van der Meer, Q. P.,** Review on the availability of resistance to *Allium* white rot (*Sclerotium cepivorum* Berk.), in *Proc. 3rd. Int. Workshop on Allium White Rot,* Entwistle, A. R., Ed., T. W. Printing Associates Ltd, Leamington Spa, U.K., 1987, Section 5, 1.
208. **Taubenhaus, J. J. and Mally, F. W.,** Pink root disease of onions and its control in Texas, *Texas Agric. Exp. Stn. Bull.,* 273, 1921.
209. **Rengwalska, M. M. and Simon, P. W.,** Laboratory evaluation of pink root and *Fusarium* basal rot resistance in garlic, *Plant Dis.,* 70, 670, 1986.
210. **Klingner, A. and Pontis-Videla, R. E.,** Pink root disease of onions in Mendoza, Argentina, *Plant Dis. Rep.,* 45, 235, 1961.
211. **Valdivia Minaya, G.,** Onion pink root (*Pyrenochaeta terrestris*) in Peru, *Fitopatologia,* 9, 78, 1974.
212. **Yassin, A. M., Babiker, A. G. T., and Ahmed, M. K.,** First report of pink root of onion in Sudan, *Plant Dis.,* 66, 741, 1982.

213. **Kreutzer, W. A.,** Host-parasite relationships in pink root of *Allium cepa* II. The action of *Phoma terrestris* on *Allium cepa* and other hosts. *Phytopathology,* 31, 907, 1941.
214. **Sprague, R.,** *Phoma terrestris* on Gramineae in the Northern Great Plains, *Phytopathology,* 34, 129, 1944.
215. **Pfleger, F. L. and Vaughan, E. K.,** *Pyrenochaeta terrestris* as a naturally occurring component of the soil in Oregon, *Plant Dis. Res.,* 56, 180, 1972.
216. **Hansen, H. N.,** Etiology of the pink-root disease of onions, *Phytopathology,* 19, 691, 1929.
217. **Vaughan, E. K., Siemer, S. R., and Lu, M.,** Soil fumigation for onion pink root control, *Down to Earth,* 26, 1, 1971.
218. **McCoy, P. R.,** Fighting pink root, *Onion World,* 3, (September), 22, 1987.
219. **Rabinowitch, H. D., Katan, J., and Rotem, I.,** The response of onions to solar heating, agricultural practices and pink-root disease, *Sci. Hortic.,* 15, 331, 1981.
220. **Lacy, M. L. and Roberts, D. L.,** Yields of onion cultivars in midwestern organic soil infested with *Fusarium oxysporum* f.sp. *cepae* and *Pyrenochaeta terrestris, Plant Dis.,* 66, 1003, 1982.
221. **Gorenz, A. M., Walker, J. C., and Larson, R. H.,** Morphology and taxonomy of the onion pink-root fungus, *Phytopathology,* 38, 831, 1948.
222. **Watson, R. D.,** Rapid identification of the onion pink root fungus, *Plant Dis. Rep.,* 45, 289, 1961.
223. **Hess, W. M., Vaughan, E. K., and Leach, C. M.,** Use of ultraviolet-induced sporulation for identification of *Pyrenochaeta terrestris, Phytopathology,* 54, 113, 1964.
224. **Kreutzer, W. A.,** Host-parasite relationships in pink root of *Allium cepa* L. I. The pigment of *Phoma terrestris, Phytopathology,* 29, 629, 1939.
225. **Johann, H.,** *Phoma terrestris* in the roots of mature maize plants, *Phytopathology,* 33, 526, 1943.
226. **Punithalingam, E. and Holliday, P.,** *Pyrenochaeta terrestris, CMI Descriptions of Pathogenic Fungi and Bacteria,* Commonwealth Mycological Institute Kew, 1973, 397.
227. **Mention, M. and Bouhot, D.,** Role of *Fusarium roseum* in the symptom of pink root of leek, *Rev. Hortic.,* 179, 37, 1977.
228. **Katan, J., Rotem, I., Finkel, Y., and Daniel, J.,** Solar heating of the soil for the control of pink root and other soilborne diseases in onions, *Phytoparasitica,* 8, 39, 1980.
229. **Jenkins, J. M.,** Shallots — North Carolina's newest crop, *Res. Farm.,* Winter, 8, 1951.
230. **Siemer, S. R. and Vaughan, E. K.,** Bioassy of *Pyrenochaeta terrestris* inoculum in soil, *Phytopathology,* 61, 146, 1971.
231. **Davis, G. N. and Henderson, W. J.,** The interrelation of the pathogenicity of a *Phoma* and a *Fusarium* on onions, *Phytopathology,* 27, 763, 1937.
232. **Levy, D. and Gornik, A.,** Tolerance of onions (*Allium cepa* L.) to the pink root disease caused by *Pyrenochaeta terrestris, Phytoparasitica,* 9, 51, 1981.
233. **Netzer, D., Rabinowitch, H. D., and Weintal, Ch.,** Greenhouse technique to evaluate onion resistance to pink root, *Euphytica,* 34, 385, 1985.
234. **Nichols, C. G., Gabelman, W. H., Larson, R. H., and Walker, J. C.,** The expression and inheritance of resistance to pink root in onion seedlings, *Phytopathology,* 55, 752, 1965.
235. **Porter, D. R. and Jones, H. A.,** Resistance of some of the cultivated species of *Allium* to pink root *(Phoma terrestris), Phytopathology,* 23, 290, 1933.
236. **Link, G. K. K. and Bailey, A. A.,** *Fusaria* causing bulb rot of onions, *J. Agric. Res.,* 33, 929, 1926.
237. **Jones, H. A. and Perry, B. A.,** Inheritance of resistance to pink root in the onion, *J. Hered.,* 47, 33, 1956.
238. **Peters, R. J., Netzer, D., and Rabinowitch, H. D.,** A progress report: Pink root resistance in *Allium cepa* L. X *Allium fistulosum* L. hybrids and progeny, *Synopses of the 3rd Eucarpia Allium Symposium,* Wageningen, Netherlands, 1984, 70.
239. **Sparace, S. A., Shahin, E. A., and Mudd, J. B.,** Accumulation of pyrenocines in cultures of the onion pink root pathogen *Pyrenochaeta terrestris* and some observations on their phytotoxicities, *Phys. Mol. Plant Pathol.,* 28, 381, 1986.
240. **Sneh, B., Netzer, D., and Krikun, J.,** Isolation and identification of *Pyrenochaeta terrestris* from soil on dilution plates, *Phytopathology,* 64, 275, 1974.
241. **Walker, J. C. and Tims, E. C.,** A *Fusarium* bulb rot of onion and the relation of environment to its development, *J. Agric. Res.,* 28, 683, 1924.
242. **Snyder, W. C. and Hansen, H. N.,** The species concept in *Fusarium, Am. J. Bot.,* 27, 64, 1940.
243. **Sherf, A. F. and MacNab, A. A.,** Onion, garlic, leeks, and shallots, in *Vegetable Diseases and their Control,* John Wiley & Sons, New York, 1986, chap. 13.
244. **Palo, M. A.,** A *Fusarium* causing bulb rot of onion in the Philippines, *Philippine Agric.,* 17, 301, 1928.
245. **Booth, C.,** The genus *Fusarium, Commonwealth Mycological Institute,* Kew, Surrey, UK, 1971.
246. **Everts, K. L., Schwartz, H. F., Epsky, N. D., and Capinera, J. L.,** Effects of maggots and wounding on occurrence of *Fusarium* basal rot of onions in Colorado, *Plant Dis.,* 69, 878, 1985.
247. **Marlatt, R. B.,** Onion *Fusarium* basal rot in Arizona, *Plant Dis. Rep.,* 42, 667, 1958.

248. **Armstrong, G. M. and Armstrong, J. K.,** Nonsusceptible hosts as carriers of wilt *Fusaria, Phytopathology,* 38, 808, 1948.
249. **Price, D.,** *Fusarium* and plant pathology: the resevoir of infection, in *The Applied Mycology of Fusarium,* Moss, M. O. and Smith, J. E., Ed., British Mycological Society, 1984.
250. **Matuo, T., Tooyama, A., and Isaka, M.,** *Fusarium* basal rot of *Allium bakeri* Regel and its causal fungus, *Fusarium oxysporum* Schl. f.sp. *allii* n.f., *Ann. Phytopathol. Soc. Jpn.,* 45, 305, 1979.
251. **Abawi, G. S. and Lorbeer, J. W.,** Several aspects of the ecology and pathology of *Fusarium oxysporum* f.sp. *cepae, Phytopathology,* 62, 870, 1972.
252. **Chattopadhyay, S. B. and Mustafee, T. P.,** Outbreaks and new records, India. Occurrence of two new diseases on vegetable, *Plant Protection Bulletin, Food and Agriculture Organization,* 15, 40, 1967.
253. **Takakuwa, M., Ishizaka, N., Kodama, F., and Saito, I.,** Host range of *Fusarium oxysporum* f.sp. *cepae,* causal fungus of *Fusarium* basal rot of onion, *Ann. Phytopathol. Soc. Jpn.,* 43, 479, 1977.
254. **Jaworski, C. A., McCarter, S. M., Johnson, A. W., and Williamson, R. E.,** Response of onions grown for transplants to soil fumigation, *J. Am. Soc. Hortic. Sci.,* 103, 385, 1978.
255. **Joffe, A. Z., Ausher, R., and Palti, J.,** Distribution and pathogenicity of *Fusarium* species associated with onion in Israel, *Phytopathol. Mediterr.,* 11, 159, 1972.
256. **Davis, G. N. and Reddy, C. S.,** A seedling-blight stage of onion bulb rot, *Phytopathology,* 22, 8, 1932.
257. **Barnoczkine Stoilova, E.,** Possibilities to control *Fusarium* on onions, *Zoldsegtermesztesi Kutato Intezet Bulletinje,* 19, 33, 1986.
258. **Kodama, F.,** Studies on basal rot of onion caused by *Fusarium oxysporum* f.sp. *cepae* and its control, *Rep. Hokkaido Pref. Agric. Exp. Stn.,* 39, 65, 1983.
259. **Abawi, G. S. and Lorbeer, J. W.,** Pathological histology of four onion cultivars infected by *Fusarium oxysporum* f.sp. *cepae, Phytopathology,* 61, 1164, 1971.
260. **Abawi, G. S. and Lorbeer, J. W.,** Cultural variability and pathogenicity of *Fusarium oxysporum* f.sp. *cepae, Phytopathology,* 55, 1051, 1965.
261. **Abawi, G. S. and Lorbeer, J. W.,** Populations of *Fusarium oxysporum* f.sp. *cepae* in organic soils in New York, *Phytopathology,* 61, 1042, 1971.
262. **Srivastava, K. J. and Qadri, S. M. H.,** Some studies on damping-off disease of onion (*Allium cepa* L.), *Indian Bot. Rep.,* 3, 147, 1984.
263. **Lorbeer, J. W. and Stone, K. W.,** Reaction of onion to *Fusarium* basal rot, *Plant Dis. Rep.,* 49, 522, 1965.
264. **Geard, J. D.,** Diseases of onions, *Tasmanian J. Agric.,* 30, 165, 1959.
265. **Sokhi, S. S., Sohi, H. S., Singh, D. P., and Joshi, M. C.,** Sources of resistance to basal rot of onion caused by *Fusarium oxysporum, Indian J. Mycol. Plant Pathol.,* 4, 214, 1974.
266. **Robbs, C. F. and Ribeiro, R. D. L. D.,** Studies on the aetiology of 'seven-curls disease' of onion in the San Francisco Valley, *Anais Reuniao Fitossanit. Brasil,* 10, 85, 1967.
267. **Hess, W. M. and Vaughan, E. K.,** Pathogenicity of *Pyrenochaeta terrestris* and *Fusarium* sp. from onion roots, *Phytopathology,* 52, 735, 1962.
268. **Elewa, I. S., Ali, A. A., and Dabash, T. S.,** Histopathological studies on onion cultivars susceptible and tolerant to *Fusarium* basal rot disease, *Agric. Res. Rev.,* 57, 127, 1980.
269. **Holz, G. and Knox-Davies, P. S.,** Production of pectic enzymes by *Fusarium oxysporum* f.sp. *cepae* and its involvement in onion rot, *Phytopathol. Z.,* 112, 69, 1985.
270. **Ashour, W. A., Ali, M. D. H., Morsy, A. A., and Diab, M. M. M.,** Effect of temperature, moisture and aeration on the development of basal bulb rot of onion during storage, *Agric. Res. Rev.,* 51, 163, 1973.
271. **Retig, N., Kust, A. F., and Gabelman, W. H.,** Greenhouse and field tests for determining the resistance of onion lines to *Fusarium* basal rot, *J. Am. Soc. Hortic. Sci.,* 95, 422, 1970.
272. **Latham, A. J. and Watson, R. D.,** Effect of specific crop residues on soil fungi, onion infection, and bulb rotting, *Plant Dis. Rep.,* 50, 469, 1966.
273. **Takakuwa, M., Kodama, F., Saito, I., and Inoue, H.,** A survey on onion basal rot in Hokkaido, *Ann. Rep. Soc. Plant Prot. North Jpn.,* 32, 136, 1981.
274. **Ramsey, G. R. and Lorbeer, J. W.,** Seed borne fungi of onion in New York, *Phytopathology,* 70, 468, 1980.
275. **Katan, J., Abramski, M., and Levi, D.,** Transport and transmission of *Fusarium* pathogens by tomato and onion seeds, *Phytoparasitica,* 3, 74, 1975.
276. **Holz, G. and Knox-Davies, P. S.,** *Fusarium oxysporum* f. *cepae* on *Oxalis* species in the Western Cape Province, *Phytophylactica,* 8, 89, 1976.
277. **Kawamoto, S. O. and Lorbeer, J. W.,** Protection of onion seedling from *Fusarium oxysporum* f.sp. *cepae* by seed and soil infestation with *Pseudomonas cepacia, Plant Dis. Rep.,* 60, 189, 1976.
278. **Fantino, M. G. and Bazzi, C.,** Antagonistic effect of *Pseudomonas cepacia* on *Fusarium oxysporum* f.sp. *cepae, Informatore Fitopatologico,* 32, 55, 1982.
279. **Lopez, C. R. and Gonzalez, U. L. C.,** The effects of fungicides and fumigants applied to soil of onion seed beds, *Proc. Trop. Reg. Am. Soc. Hortic. Sci.,* 17, 277, 1973.

280. **Abawi, G. S. and Lorbeer, J. W.,** Reaction of selected onion varieties to infection by *Fusarium oxysporum* f.sp. *cepae, Plant Dis. Rep.*, 55, 1000, 1971.
281. **Fantino, M. G. and Badino, M.,** Evaluation of the tolerance of *Fusarium oxysporum* f.sp. *cepae* Sn. Hans. in onion (*Allium cepa* L.) lines and cultivars, *Informatore Fitopatologico,* 31, 3, 1981.
282. **Holz, G. and Knox-Davies, P. S.,** Resistance of onion selections to *Fusarium oxysporum* f.sp. *cepae, Phytophylactica,* 6, 153, 1974.
283. **Kehr, A. E., O'Brien, M. J., and Davis, E. W.,** Pathogenicity of *Fusarium oxysporum* f.sp. *cepae* and its interaction with *Pyrenochaeta terrestris* on onion, *Euphytica,* 11, 197, 1962.
284. **Ashour, W. A., Aly, A. A., Elewa, I. S., and Dabash, T.,** Interaction of soil microorganism and rhizosphere of different onion cultivars with *Fusarium oxysporum* f.sp. *cepae*, the cause of basal rot disease of onions, *Agric. Res. Rev.*, 58, 129, 1980.
285. **Holz, G. and Knox-Davies, P. S.,** Natural sugars present in different parts of onion at different growth stages in relation to pectic enzyme production by *Fusarium oxysporum* f.sp. *cepae, Phytophylactica,* 17, 157, 1985.
286. **Holz, G. and Knox-Davies, P. S.,** Pectic enzyme production by *Fusarium oxysporum* f.sp. *cepae*: induction by cell walls from different parts of onion bulbs at different growth stages, *Phytopathol. Z.*, 12, 81, 1985.
287. **Holz, G. and Knox-Davies, P. S.,** Possible involvement of apoplast sugar in endopectin-*trans*-eliminase synthesis and onion bulb rot caused by *Fusarium oxysporum* f.sp. *cepae., Phys. Molecular Plant Pathol.*, 28, 403, 1986.
288. **Naik, D. M. and Burden, O. J.,** Chemical control of basal rot of onion in Zambia, *Trop. Pest Manage.*, 27, 455, 1981.
289. **Tahvonen, R.,** Storage fungi of onion and their control, *J. Sci. Agric. Soc. Finland,* 53, 27, 1981.
290. **Gaur, R. B. and Agnihotri, J. P.,** Root rot of onion caused by *Fusarium solani* (Mart.) Sacc., *Madras Agric. J.*, 66, 837, 1979.
291. **Du Plessis, S. J.,** The parasitism, morphology and physiology of *Fusarium solani* (Mart.) Sacc. on onions, *Annale van die Universiteit van Stellenbosch,* 10, 1, 1932.
292. **Singh, J. P. and Bhatnagar, D. K.,** Note on the irrigation of onion as a factor predisposing its bulbs to infection by *Fusarium solani* (Martius)Appel.-Wollenweber in storage, *Indian J. Agric. Sci.*, 51, 686, 1981.
293. **Anon**, *Annual Report for 1980-81 of the Associated Agricultural Development Foundation.* New Delhi, India, 27, 1982.
294. **Vannacci, G.,** Reperimento di *Fusarium moniliforme* var. *subglutinans* Wr. et Reink. su seme di cipolla e sua patogenicita, *Phytopathol. Mediterr.*, 20, 144, 1981.
295. **Michail, S. H., Abd El Rehim, M. A., Elarosi, H., and Khairy, E. A.,** Dry rot of garlic cloves in U.A.R. (Egypt), *Phytopathol. Mediterr.*, 10, 202, 1971.
296. **Roelands, C. and Alofs, W.,** Is it possible to control *Fusarium culmorum* in leeks?, *Groenten en Fruit,* 34, 58, 1979.
297. **Dingley, J. M.,** *Records of Plant Diseases in New Zealand, Bulletin 192,* New Zealand Department of Scientific and Industrial Research, Wellington, N.Z., 18, 1969.
298. **Ellis, M. B. and Ellis, J. P.,** *Microfungi on Food Plants,* Croom Helm, London, 1985, 303, 307.
299. **Westcott, C.,** *Plant Disease Handbook,* 3rd Ed., Van Nostrand, Reinhold, New York, 1969, chaps. 4, 5.
300. **Fischer, F. W.,** *Manual of the North American Smut Fungi,* The Ronald Press Company, New York, 1953, 10.
301. **Kochman, J. and Macias, W.,** Effect of soil type, pH, moisture and temperature on the phytotoxicity and effectiveness of fungicides of the PCNB and thiram groups against onion smut *(Urocystis magica Pass.), Acta Agrobot.*, 27, 85, 1974.
302. **Lacy, M. L.,** Optimum conditions for germination of *Urocystis colchici* teliospores, *Phytopathology,* 58, 1460, 1968.
303. **Dow, R. L. and Lacy, M. L.,** Factors affecting growth of *Urocystis colchici* in culture, *Phytopathology,* 59, 1219, 1969.
304. **Walker, J. C.,** White rot of shallot, onion and garlic, in *Plant Pathology,* 3rd Ed., New York, McGraw-Hill Book Company, 1957, 819.
305. **Tachibana, H. and Duran, R.,** Comparative survival of teliospores and mycelia of the onion smut fungus in soil, *Phytopathology,* 56, 136, 1966.
315. **Aycock, R.,** Stemrot and other diseases caused by *Sclerotium rolfsii, North Carolina Agric. Exp. Stn. Tech. Bull.*, 174, 201, 1961.
316. **Zeidan, O., Elad, Y., Hadar, Y., and Chet, I.,** Integrating onion in crop rotation to control *Sclerotium rolfsii, Plant Dis.*, 70, 426, 1986.
317. **Robertson, G. I.,** *Pythium* species in market gardens and their pathogenicity to fourteen vegetable crops, *N.Z.J. Agric. Res.*, 19, 97, 1976.
318. **Suhag, L. S. and Rana, R. S.,** Studies on the chemical control of seedling mortality of kharif onions in nursery, *Indian J. Plant Pathol.*, 2, 13, 1984.

319. **Bruckart, W. L. and Lorbeer, J. W.,** *Pythium* species pathogenic to onion seedlings grown on organic soils in New York, *Phytopathology,* 72, 469, 1982.
320. **van der Plaats-Niterink, A. J.,** Monograph of the genus *Pythium, Studies in Mycology No. 21,* Centraalbureau voor Schimmelcultures.Baarn, 242, 1981.
321. **McKeen, C. D.,** Preliminary studies on a *Pythium* rootrot of Spanish onion seedlings, *Sci. Agric.,* 30, 125, 1950.
322. **Semeniuk, G. and Wadley, B. N.,** *Pythium* seedling rot and root necrosis of *Allium cepa, Phytopathology,* 38, 29, 1948.
323. **Anon.,** Crop protection, *Agric. Supply Ind.,* 17, Nov 6, 3, 1987.
324. **Walker, J. C.,** Onion smudge, *J. Agric. Res.,* 20, 685, 1921.
325. **Docampo, D. M. and Nome, S. F.,** New records of fungus and bacterial diseases of plants in Cordoba, Argentina, *Plant Dis. Rep.,* 54, 939, 1970.
326. **Yamamoto, H. and Naito, N.,** The perfect stage of an anthracnose of onion for seed production, *Kagawa Daigaku Nogakubu Gakuzyutu Hokoku,* 23, 57, 1971.
327. **Altstatt, G. E. and Smith, H. P.,** Production, diseases, and insects of garlic in Texas, *Circ. Tex. Agric. Exp. Stn.,* 98, 13, 1942.
328. **Chandra, S. and Tandon, R. N.,** Occurrence of *Macrophomina phaseoli* (Maubl.) Ashby on garlic bulbs, *Curr. Sci.,* 34, 56, 1965.
329. **Garg, R. P. and Chauhan, R. K. S.,** Occurrence of *Macrophomina phaseolina* on onion bulbs, *Indian Phytopathol.,* 34, 501, 1981.

Chapter 7

NEMATODE PESTS OF *ALLIUM* SPECIES

Chris D. Green

TABLE OF CONTENTS

I. INTRODUCTION

Any review of nematode pests of *Allium* species is bound to concentrate on the stem and bulb nematode *(Ditylenchus dipsaci)* in onion and garlic. This was first noticed as a pest and probable cause of ''Onion Bloat'' in Germany by Kuhn and later found in Holland[1] and in France.[2] However, in this review we look at the pests of the roots first and then those of the shoot.

II. ROOT PARASITES

Several root parasitic nematodes have been shown to attack onions and more have been associated with onion roots, but there are no specific parasites of onion. Less is known about root parasites of garlic and almost nothing about those of other *Allium* species. The root parasitic nematodes are conveniently divided into endoparasites and ectoparasites. Each group includes, sedentary species, which modify cells and are thus able to feed at one site for a long time, and migratory species that destroy tissue and move to feed ahead of the damaged area.

A. SEDENTARY ENDOPARASITES (FIGURE 1A)

1. Root Knot Nematodes

Meloidogyne arenaria, M. hapla,[3] *M. exigua,*[4] *M. incognita, M. javanica,* and *M. thamesi*[5] are the main sedentary endoparasites of onions. The larvae hatch from eggs in the soil and are attracted to areas of high metabolic activity, so they generally invade near the root tips. Larvae can also invade other tissues, even stems, leaves and flowers if these are in contact with soil. They migrate to the pericycle where they settle. As they penetrate, the root swells and forms a gall by hypertrophy and hyperplasia of the cells around the nematode. The pericycle cells enlarge into syncytia to form giant transfer cells on which the nematode feeds. The nematodes feed for 2 to 3 weeks before becoming adult. Females continue to feed and grow rapidly. About 1 week after becoming adult they begin laying eggs in a gelatinous matrix which they secrete. The egg masses usually extrude from the gall, and the larvae may hatch and reinvade nearby roots, so that the root system is successively attacked as it grows. In dry conditions the matrix may shrink and harden so protecting the eggs from dessication and preventing immediate hatching. Many species are parthenogenetic and females develop eggs without insemination. However, the number of larvae that become males depends on the environment, unfavorable nutritional conditions increasing the number of males. It has not been determined if in such conditions, the migration of males and insemination of distant females increases the genetic diversity and the chance of reselection for individuals more fitted to the environment.

Most damage to plants occurs through root growth inhibition and shows as poor growth. Experiments with *M. hapla* in sterilized soil showed that approximately 2000 nematode larvae per killigram of soil significantly reduced onion yield.[6] The damage would have been greater in some natural soils in which mycorrhizal fungi enhance onion growth, because *M. hapla* inhibited this effect of the mycorrhiza as well as causing direct damage to the plants.

In other plants root knot nematode attack may predispose roots to invasion by fungi, notably *Fusarium* spp., and bacteria. No such interactions however, were found in *Allium* spp. The increased susceptibility may occur because the resistance of the host is suppressed and systemic invaders can spread more easily throughout the plant, or because gall tissue is juvenile in character and this enables invasion of pathogens usually restricted to seedlings.

The host ranges of root knot nematodes are too great for effective control by rotation. No *Allium* spp. are known to have resistance to root knot nematodes. Therefore, control, if necessary, must be by soil pesticides. Methyl bromide, chloropicrin, or 1,2-dichloropro-

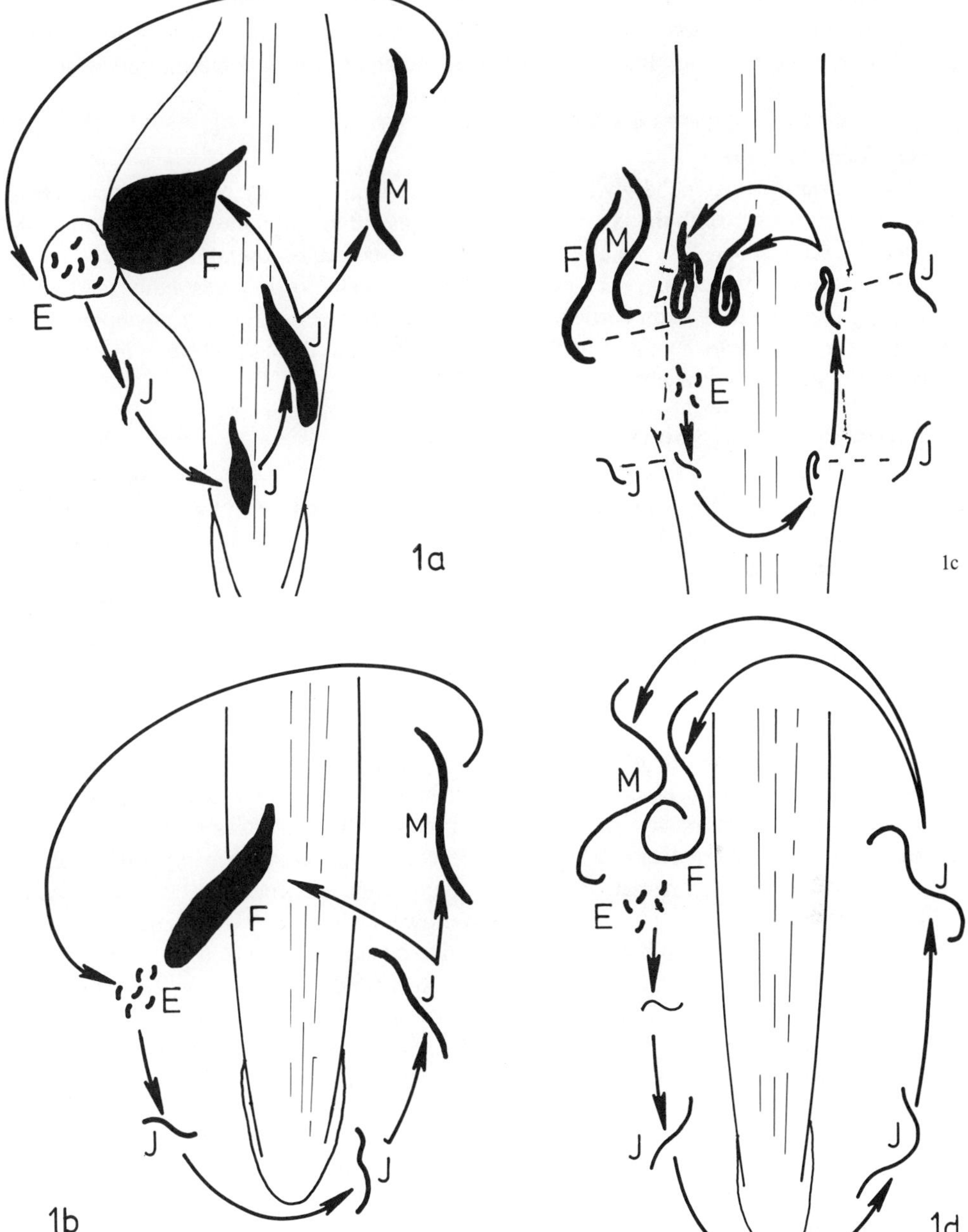

FIGURE 1. Diagramatic life cycles of ecological types of plant parasitic nematodes; (a) sedentary endoparasite, (b) sedentary ectoparasite, (c) migratory endosparasite, and (d) migratory ectoparasite. The nematodes, not to scale, are shown, in black within or outside the outline of a root. E — egg, J — juvenile, M — male, F — female.

pane with 1,3-dichloroporpene (D-D) are effective preplanting soil fumigants.[7,8] Some nematicides do not penetrate root tissue readily. Thus, infested roots of previous crops must be removed or rotted before treatments with dibromochloropropane or ethylene dibromide.[9,10] Fumigants differ in this respect and D-D and methyl bromide penetrate root galls better than chloropicrin.[11] A wide range of other compounds including N-methyl carbamates, alkane sulphorates, and organophosphates have been found to prevent root galling of various crops.[12] crops.[12]

Herbicide action on plants may affect root knot nematode invasion and damage. Fewer galls were formed in onion roots in soil treated with chlorthal-dimethyl than in untreated soil. This was attributed to changes in root structure and exudates inhibiting invasion.[13]

B. SEDENTARY ECTOPARASITES (FIGURE 1B)

1. Reniform Nematode

The reniform nematode, *Rotylenchus reniformis,* is a sedentary ectoparasite which inhibits onion growth.[14] Both the larval stages and males are found in the soil, but the females invaded the roots about 2 weeks after a hatching, just before becoming adult.

They penetrate the cortex, feeding and destroying cells to lie partly embedded in the root. They continue to feed on cortical cells which develop dense, granular cytoplasm and, after about a week, start laying 50 to 100 eggs per day into the soil. Developing roots are successively invaded by newly developing females.

C. MIGRATORY ENDOPARASITES (FIGURE 1C)

1. Lesion Nematodes

Lesion nematodes (*Pratylenchus* sp.) are the principle migratory endoparasites of onions, which can invade roots of all stages of development. They usually invade just behind the root tip. Eggs are laid in the cortex and colonies develop in the root as it matures. The feeding and movement of the nematodes cause necrosis and release of polyphenols from the plant cells. The lesions break the epidermis and may girdle the root. Feeding by the nematodes also causes granulation of the cytoplasm and changes in the walls of adjacent cells.

Most damage occurs when nematodes penetrate to the pericycle, a more frequent occurrence with some species, such as *P. penetrans* than others.

Differences in virulence between nematode genotypes are marked. One race of *P. penetrans* reduced growth of onions while another did not.[15] The effect of *P. penetrans* depends on the temperature so that 100/g of root reduced the growth of onions at 7 to 10°C as much as 40/g of root at 16 to 20°C.[16] This effect of temperature probably applies also to the other species of lesion nematodes because the damage was found to be variable and it is likely that most damage occurs when the roots grow slowly relative to the rate of nematode invasion. Winter death of onions in England could sometimes be related to the presence of lesion nematodes, and both *P. penetrans* and *P. neglectus* were associated with poor seedling establishment.[17] At three sites, more lesion nematodes (30 to 50/g of root) were found in roots of weakly growing onions than in roots, of nearby vigorous plants (5 to 10 nematodes/g of root). At two sites the plants were also attacked by migratory *Dorylaimid* nematodes, the dagger nematode *(Xiphinema diversicaudatum)* or the lance nematode *(Longidorus caespiticola).* At all sites the tops were stunted (Figure 2) and the invaded (Figure 3) roots were stained light brown. In Brazil, *P. brachyurus* appeared to cause the greatest damage to onions.[5] Such associations are difficult to establish unless root samples are examined because most of the active population resides within the root tissue.[18]

Economic control of lesion nematodes is not usually feasible as damage is unpredictable. Dazomet, D-D and ethoprophos were effective nematicides with no phytotoxicity when applied to onion fields.[19] D-D and 1,3-dichloropropene applied to the soil increased the yield of leeks in soil heavily infested with *P. penetrans* and a spiral nematode (*Rotylenchus robustus*).[20]

2. Burrowing Nematode

The burrowing nematode (*Radopholus similis*) behaves similarly to lesion nematodes but is restricted to tropical and subtropical countries. The banana race of this nematode invades onion roots destroying the cortex and leaving cavities,[21] but its effects on plant growth are not recorded. As with lesion nematodes the destroyed cortex may provide entry for secondary pathogens.

FIGURE 2. Onion plants attacked by *Pratylenchus neglectus* and *Longidorus caespiticola* are stunted and poor establishment and growth results.

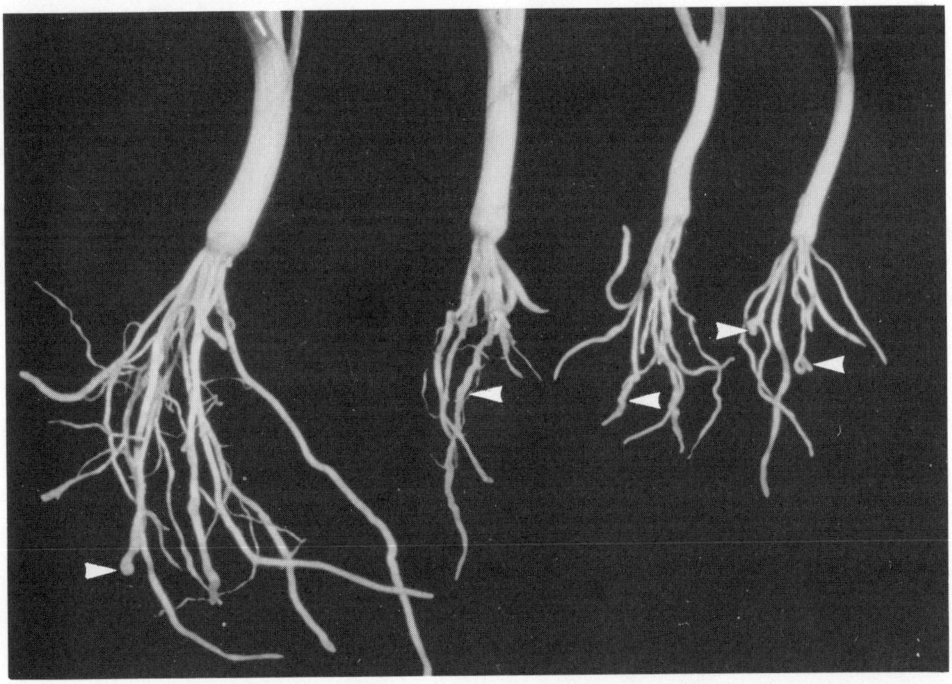

FIGURE 3. The effect of nematode density on root development, the plants on the right are more heavily attacked by *Longidorus caespiticola* and *Pratylenchus neglectus*. The roots are swollen, knobbed or curled where *L. caespiticola* has attacked. Root branching may also be a response to nematode attack.

D. MIGRATORY ECTOPARASITES (FIGURE 1D)

1. Stubby Root Nematodes

Stubby root nematodes (*Trichodorus* sp.) are migratory ectoparasites of roots which are consistently associated with stunted growth of onions and which make them subject to water stress. Stubby root nematodes feed on cell contents released after puncturing cells with a solid stylet. In contrast, the other nematode species suck out the contents through hollow stylets. The nematodes follow the extending root, congregating at the sites of lateral root emergence and the root tip. Their feeding weakens the young roots and stops their growth, so that the root systems become stubby. In onions they also cause a darkening of the attacked roots.[3] Eggs are laid in the soil around the roots so that as the eggs hatch the attack is continued and increased. Stubby root nematodes are restricted to light sandy loams.

In Oregon, *T. allius* causes important losses but can be controlled by crop rotation because it has a limited host range.[3] *T. christii,* which has a wide host range and cannot be so controlled, causes major damage in eastern U.S.[3] and Brazil.[22] In Sweden, various species of *Trichodorus* appeared to damage onions.[23]

Stubby root nematodes are vectors for tubular tobraviruses and often cause more damage by introducing viruses than by feeding. There are no reports of such problems in *Allium* sp. but the tobravirus, tobacco rattle virus, together with potyvirus, garlic mosaic virus, were isolated from poor growing plants.[26]

As with most ectoparasitic nematodes, *Trichodorus* spp. can be controlled by soil fumigation[24] but the treatments are often too expensive to be economic. The nematocide aldicarb applied to the rows at sowing, is more economic and enables seedlings to establish effective root systems.[25]

2. Lance and Dagger Nematodes

A lance nematode (*Longidorus vineacola*) was associated with damage to onions in Israel.[27] Other lance nematodes (*L. caespiticola* and *L. elongatus*), and a dagger nematode (*Xiphinema diversichaudatum*), were associated with stunted young plants when lesion nematodes were also present.[17] The characteristic damage is hyperplasia of the roots causing small galls, swellings, or curls (Figure 3). The nematodes select root tips for feeding, and the swollen tissue becomes a preferred feeding site for other individuals. The eggs are laid in the soil and the nematodes move around the root system.

Lance and dagger nematodes are vectors for polyhedral nepo viruses but no damage from these has been reported although leeks, chives, shallots, onion, and *Allium flavum* were found to be symptomless hosts of two such viruses Tomato Black Ring Virus and Arabis Mosaic Virus.[26]

III. CROWN PARASITES

In contrast to the root there is only one major parasite of the tops of *Allium,* the stem and bulb nematode (*Ditylenchus dipsaci*). Foliar nematodes, *Aphelenchoides* sp. may be found in the bulbs and leaves. *Ap. ritzemabosi* was recorded in *Allium sikkimense* and *Allium flavum.*[28] Hot water treatment of *Allium giganteum* bulbs was found to control *Ap. subtenuis.*[29] Another *Aphelenchoides,* probably *Ap. saprophilus,* was regularly associated with rotting bulbs in one property in central England. I found the nematodes in the scale leaves and the rot was associated with bacteria and not fungi. These parasites are all occasional and cause only local problems.

A. STEM AND BULB NEMATODE

Ditylenchus dipsaci, or a very similar nematode called *D. allii* in the USSR[30] is a potential major problem of onion, and garlic wherever they are grown. The nematode is readily distributed and introduced into cultivated land (see Section III.A.5). In certain soils it persists for long periods and spreads within the crop. It has a short life cycle, about 3 weeks, and a high multiplication rate.[30,31]

Chemical control is effective only for crops for immediate consumption and not for bulbs to be stored (see Section III.A.8). Some chemicals cannot be used at sowing of green salad onions because of their short growth period and the likelihood of residues in the crop.

1. Behavior of Stem and Bulb Nematode

The nematode is well adapted as a parasite of leaves and crowns but can feed and invade roots. This occurred readily in aseptic culture[32] but is rare in field soils. Onion roots are frequently colonized by mycorrhizal and other fungi and these may be expected to alter the root surface and inhibit nematode invasion.[33] Invasion of the bulb and leaves may occur at any stage in the nematode life cycle but invasion by the preadult 4th stage, is the most likely to occur in the soil.[34,35] The nematode invades garlic through cracks in the scale leaves[36] and damage to onion seedlings increases the likelihood of invasion.[37] The preadult stage is particularly able to withstand dessication and be distributed in dry seed, dry plant debris, and dry soil. This stage also stores food reserves as oil droplets and is able to survive long periods without feeding.[35] After invading plant tissue and feeding on the cells, the preadults molt to become adults which mate and lay eggs in the plant tissue.[38] The life cycle may be completed in the tissue or emigration may occur to spread the infestation.

2. Host Response

The nematodes are attracted to the cotyledon of a germinating seed and may invade it. The cotyledon swells around the nematodes and may collapse (Figure 4a). Seedling death of onions in eastern England was attributed to early invasion of stem nematodes from the soil. Later invasion caused deformed leaves and split stems 10 d after invasion.[39] Multiplication of the nematode in the leaves causes the classical bloat symptoms (Figure 4b) which is the major problem of green salad onions. The infected leaves become swollen and distorted as localized extension of the tissue is slowed. Axiliary buds may sprout creating multistemmed plants. The plants may be swollen and stubby (Figure 4c). The inner leaves may coil in the sheathing leaves and fail to emerge completely. In *Allium* spp. with laminate leaves the symptoms are similar; garlic plants are shortened and distorted.[40] Leek crops rarely show symptoms, possibly because distorted seedlings are removed or rejected during thinning and transplanting. However, when infected, leek shoots are stunted and distorted by curling of the leaves and also axiliary buds grow out (Figure 5). These symptoms on the leaves could be associated with localized release of plant hormones at the invasion sites. Plant growth regulators that would cause cell multiplication and hypertrophy were found in extracts of stem nematodes[41] and they were also found to secrete an auxin inactivator that stopped elongation of the stem apex.[42] Symptoms very similar to those in stem and bulb nematode invasion may be caused by pesticides or herbicides that affect leaf growth and false bloat, where no nematodes have invaded, has been seen in onion crops.[43]

Pectolytic enzymes can be secreted by the nematode[44,45] and these cause the cells to separate, facilitating nematode movement in the tissues. Infestation of bulb scales loosens the cells and softens the tissue. The infested areas become brown, probably by the release of polyphenols from the damaged cells (Figure 6). The infestations spread more readily within a scale, so that bands of damaged tissue may be seen in cut bulbs. Infested onion bulbs usually develop bacterial rot. In garlic the ''Cafe au lait'' damage was associated with stem and bulb nematode or the bacterium *Pseudomonas fluorescens*. Inoculation with both organisms did not affect nematode invasion but it doubled the proportion of bulbs invaded by bacteria.[46]

Growing onion plants may be killed during development when the leaf bases become

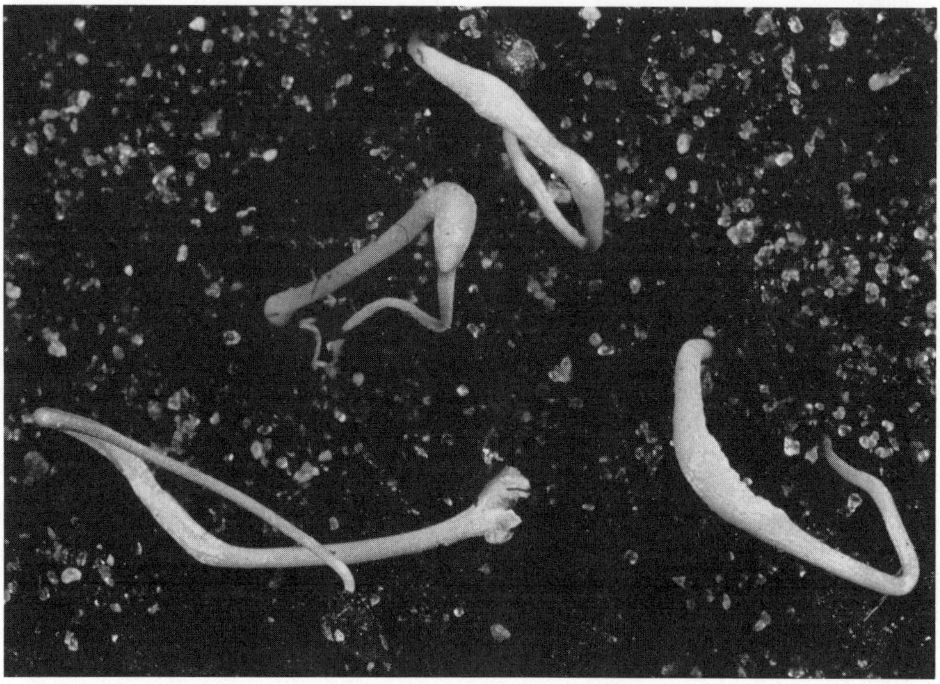

A

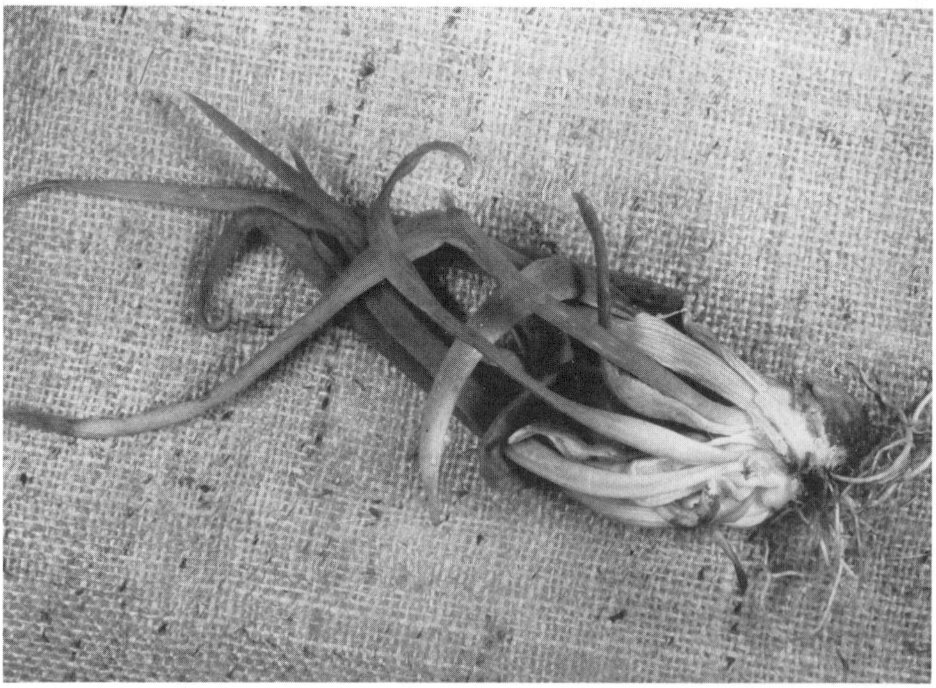

B

FIGURE 4. Onion damage by stem and bulb nematodes. (a) Cotyledon distortion and collapse after invasion. Field symptoms of infection of onion plants by *Ditylenchus dipsaci*. (b) A bloated plant showing multiheading and distortion. (c) Short, smaller and distorted leaves. (d) Death of plant due to breaking of the leaf bases.

Figure 4C.

Figure 4D.

FIGURE 5. Distorted shoot of a leek plant invaded by *Ditylenchus dipsaci*.

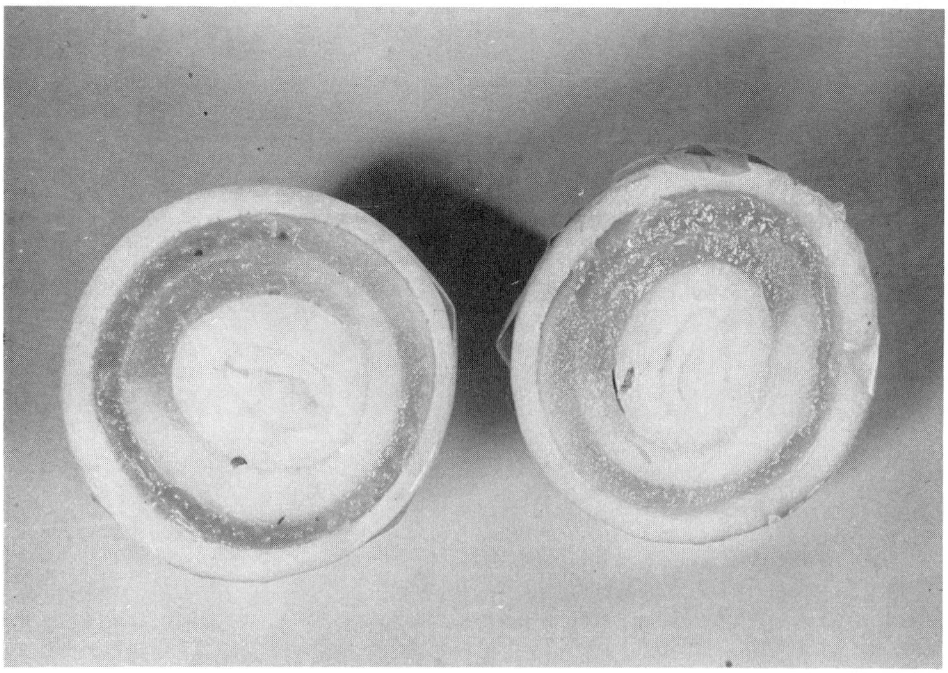

FIGURE 6. Pectolytic enzymes from *D. dispaci* infected scales, the tissue browns and is vulnerable to bacterial infection.

heavily infested and break away from the basal plate (Figure 4d). This is most frequent in late spring or early summer in the U.K. when nematode attack can devastate large areas of crop. The damage occurs at the same time in both spring- and autumn-sown crops, and may be associated with nematodes spread across the surface by water run-off. Massive invasion of stem and bulb nematode occurred in onions in Foggia (Italy) after the autumn rains occurred.[47]

3. Sources of Infestation

Stem and bulb nematode is particularly troublesome because infestations can be started in many ways. It is an internal parasite which invades all *Allium* spp. tissues and therefore can be present in any propagating or transplanting material. These distribute the nematode and introduce it to new sites. Furthermore, enough plants may be infested initially to cause economic damage without any further spread.

Although the nematode invades onion seed,[48-50] usually only a few seeds are infested and the numbers of nematodes are too few to cause economic damage.[51] Further infestation after field spread may cause more loss of plants at harvest and is likely to cause post harvest bulb damage. The introduction of the nematode to new sites is the most important effect of seed infestation. In this respect nematodes carried in seed of other hosts, e.g., carrot, beet, broad bean,[52] and field bean[53,54] are of as much concern as sources of infection as those in onion seed. These hosts are usually infested by the field race (see Section II.A.6) of stem and bulb nematode, but *Allium* spp. are susceptible to invasion by most races of the nematode. When nematodes are introduced in crops preceeding onion or garlic, they can multiply in those crops and are likely to be present in large numbers when onions or garlic are planted. The host crops, unlike onion, are likely to show few or no symptoms of attack.

Nematodes emigrate from infected plants into the soil, and debris from an infested crop that remains in the field is a major source of nematodes. Such debris, either wind blown or dropped from carted or stored material, distributes the nematode. Sufficient nematodes may be so introduced to cause economic damage, for example a severely damaged onion crop in eastern England was most probably infested from bulb skins blowing from a nearby pile of refuse outside a vegetable pack house.

A less obvious introduction occurred from a native plant community in grazed pasture in southwest England. There was a small, stable population of stem nematodes in this pasture which was on hills above strawberry fields.[55] Flood water from the hills ran across the cultivated land and would be a continual source of nematodes in those fields. Similar stable populations occurred in hedgerow vegetation in central England but the importance of such sources of nematodes is unknown.

4. Survival

Stem and bulb nematodes survive in seed and debris because they can withstand dessication. Their survival in dry soil depends on the same ability and may be important in some circumstances, particularly in semidesert conditions. However, the most important need for survival occurs in moist soil when host plants are not present. In temperate regions such survival is necessary to overwinter and, in all regions, to persist when non- or poor hosts are grown. When stem and bulb nematodes are buried deeply by cultivation, they migrate upwards to growing host plants.[34] Adult and young larval stages of the nematode may be found in soil, around host crops or soon after harvest, but the preadult stage is the one most commonly found.[34] The incidence of stem and bulb nematode may often be correlated with soil type as the nematode persists longer in clay soils,[56,57] and is frequently found in organic soils used for onion growing in U.S.[58] In a sandy loam, preadult nematodes survived better at 5°C than at 15 or 25°C. In a clay loam, they survived as well at 15 or 25°C as at 5°C in sandy or clay loam. Soil moisture content had no effect on survival in either soil.[58] Depletion of food reserves was not the cause of death of most preadult nematodes, but many were parasitized by *Arthrobotrys* fungi.[35] However the frequency of such parasitism has not been related to soil type.

Rotations may be an effective control of the nematode in soils in which they do not survive. Then it is particularly important that reintroductions do not occur.

5. Spread Within a Crop

Often too few nematodes are present, or their distribution is too aggregated, for more than a few plants to be invaded so that there is no major crop damage. However, the risk of crop damage is greatly increased by conditions which enable the nematode to spread rapidly from infested plants. Migration of the nematode in soil is most rapid when the soil is moist but not waterlogged.[60] However, even under optimal conditions spread is not more than, a meter or so in a year.[61] Such a rate of movement would not give rise to epidemic situations. Emigration from infested plants will be greatly enhanced if the foliage is wet for several hours. Movement across a field is extended by surface water movement. This enables areas of plants to be infested from a single source, either crop plant or weed, many of which are good hosts.[62] Capped and heavy soils into which surface water does not easily drain are most likely to provide conditions for such crop damage. Early spread can cause the death of the plants, later spread will infest bulbs and cause post harvest damage in store or infestation of propagating material.

6. Races of Stem and Bulb Nematodes

The stem and bulb nematode occurs as diploid, and tetraploid (giant) forms as well as host selected races.[63-65] The giant forms have been recorded on specific hosts but their host ranges have not been fully investigated. The usual form is the diploid and this is the one always found in *Alliums*. One race or another of this form can invade plants from 45 angiosperm plant families.[55] Host selected races seem most likely when the crop is grown and propagated so that the nematode is cycled continuously in the one host species. This occurs with red clover and lucerne in which seed infestation is likely to be the major source of reinfection of the new crop. Races with limited host ranges occur in bulbous plants such as tulip and narcissus in which, unlike onion, the nematodes are transmitted in the bulbs. Garlic and shallot nematode races have not been described, and races may only develop when selection pressure for particular genes is engendered by the presence of particular crop species. *Allium* spp. appear to be unselective and they host all diploid races.

A nematode field race with a wide host range is found in many crops which are frequently invaded by populations of stem and bulb nematode in soil. Multiplication within the crop may select certain host specific genes but, unless transmission occurs directly to new plants of the same species, frequent mixing with the residual population restricts the development of specific races and maintains genetic diversity.

Most often endemic populations in natural plant communities are likely to be a field race. The populations seem to be stable as the nematodes are common but few occur in any plant. It was postulated that the stability is caused by frequent migration between hosts.[55] Constant migration slows reproduction directly and also indirectly because of (1) invasion of unsuitable hosts in which no reproduction occurs, and (2) very susceptible hosts which may die as seedlings, causing the nematodes to migrate before completing their life cycle.

Weed free monoculture probably provides greater opportunity for plant damage than do plant communities and it is notable that plants, even *Allium* spp. or *Narcissus* spp. in natural plant communities have rarely or never been recorded to have nematode damage. The lack of major damage to wild hosts and the lack of impairment to seed production, makes natural selection of tolerance and resistance unimportant, and infrequent except in plants which naturally form monocultures or are subject to other restraints which make nematode invasion critical. There is a need for more information on the behavior of the nematode populations in plant communities to substantiate the above suggestions and to establish an alternative basis for finding resistance to the nematode in sensitive crops.

7. Resistance to Stem and Bulb Nematode

Resistance to stem and bulb nematodes has effectively controlled their damage in several host plants of fodder and seed crops. Host resistance has been reported in garlic[66,67] and onion[67] but has not yet been shown to be effective for nematode control. Reduced nematode multiplication and apparent resistance in some varieties of onion early in the season did not affect the final numbers of nematodes.[68]

Host tolerance in garlic,[69,70] onion and *A. fistulosum*[71] is not likely to be useful except in green salad onions, as invasion of bulbs is likely to increase bacterial rot after harvest.

8. Control of Nematode Invasion from Soil

Stem and bulb nematodes may be killed in soil by soil fumigants thereby increasing the yields of onions. Dazomet was effective[72] but could be phytotoxic[73] and sometimes only partly controlled the nematodes.[74] Ethylene dibromide,[75] 1,2-dichloropropane with 1,3-dichloropropene and methyl isothiocyanate, 1,3-dichoropropene alone, Di-Trapex,[75,76] D-D,[76,77] and Telone[77] gave inconsistent or partial control. The lack of complete control was because the few nematodes that escaped the treatment usually multiplied rapidly and rotted bulbs after harvest or, if conditions were suitable, spread through the crop. D-D followed by 4-5 foliar treatments of oxamyl were necessary to keep plants free of nematodes.[78] This would not be considered practical except perhaps for seed crops. Methyl bromide or solar heating gave excellent control[75] however the cost of such treatments may limit them to valuable crops.

Granular nematocides, thionazin,[79] oxamyl,[73,74,80] aldicarb,[72,73] and fenamiphos,[81] applied at sowing prevented early invasion. Increases in yields were measured in most cases but it was noted that treatments only partly controlled the nematodes[74] and did not decrease the spread of nematodes in stored bulbs.[80,82] The failure to prevent invasion later in the season makes any soil treatment ineffective in controlling damage to autumn sown onions in the U.K. since these have a long growing period. No field treatments were effective in eradicating nematodes that had invaded plants.[80]

9. Eradicating Nematodes from Seeds, Sets and Cloves

Methyl bromide fumigation killed nematodes in onion[83] and shallot[84] seed. Germination was not greatly reduced if the seed was sufficiently dry.[85] Onion seed is now usually treated, or else is produced from nematode free crops and seed rarely causes nematode infestations in the U.K.[52] Distribution of the nematode with other vegetable seeds is still a likely source of infestation and methyl bromide fumigation was too lethal to *faba* bean seed to be acceptable.[86]

Soaking shallot sets in hot water at 43.5°C for 2 h,[87] onion sets at 45°C for 3 h,[88] and garlic cloves at 49°C for 20 min[89,90] eradicated stem and bulb nematodes in them. The garlic cloves were presoaked at 10°C for 30 min and treated in formalin solution. Other work did not confirm that the nematodes were eradicated from garlic cloves by hot water treatment alone but showed that dipping in calcium polysulfide was effective[91] Dipping in thionazin,[79,91,92] ethoprophos, and oxamyl[75] have also been shown to be effective and metham-sodium is effective in fumigation treatments.[93]

There is little doubt that the most efficient means of distributing nematode free seeds and sets is to produce them from nematode free crops. Alliums are sensitive to stem and bulb nematodes and infestations in the growing crop can be seen. Crop inspections and the elimination of infested crops from seed and set production should be feasible. Seed, set, and clove examination for nematodes is feasible and infested material should be discarded.

10. Hygiene, Soil Conditions, and Rotation

The control of stem and bulb nematode requires an overall approach, and an understanding where introduction of, and invasion by, stem nematode occurs. In some soils for example clay loams and fine silts there is little that can be done once they are infested except to protect the crop from early invasion and market the crop for immediate use after harvest. On other soils, namely sandy loams and organic soils, the problem can best be alleviated by taking care in cultivations, preventing any contamination with debris and by using nematode free propagating material for the *Allium* crop and preceeding crops.

One of the main limitations to the effectiveness of chemical treatment of the soil is the later spread from a few surviving foci of infestation. This is particularly true for fumigants. Precautions to prevent this later spread occurring would probably be effective in improving the control. Free draining conditions and the prevention of soil capping would restrict field spread, and this may be one of the reasons for the disappearance of bloat as a major crop problem of salad onions in central England. The cultivation of the crop was transferred from heavy soil in the valleys to lighter more open soil on the slopes and bloat no longer occurred. Similarly, large scale losses are rarely seen on deep organic soils or well cultivated beds on small holdings, even when these are on a heavy soil normally conducive to damge.

The nematode survives for too long and *Allium* spp. are host to too many races for rotation to be an effective control on soils in which the nematode persists for long periods. There is also a considerable risk of introducing nematodes with the seed of some crops that may be grown in rotation with alliums and this may be why onion beds were never used for other crops in traditional vegetable garden planning. However, on soils in which the nematodes do not persist barley, wheat, brassicas and lettuce may lessen the risk of damage to following *Allium* crops.

IV. CONCLUSION

Bulb and foliage distortion or destruction by stem and bulb nematode is the most important problem caused by nematodes in cultivated *Allium* spp. In most circumstances stem and bulb nematode can be controlled only by a comprehensive program preventing introduction, survival and spread of the nematode with or without soil pesticide application. In contrast, soil nematocide treatments are effective protection against nematode root parasites. Root parasites slow the growth of the plants due to root damage and early protection is sufficient to enable plants to become established. The major problem is predicting if and where sufficient damage will occur to make soil treatment economic. Damage from root knot and stubby root nematodes is usually sufficiently predictable to justify treatments.

REFERENCES

1. **Beijerinck, M. W.,** De oorzak derkroefziekte van de jonge ajuinplanten, *Maanblad uitgegeven van wege de Hollandsche maatschappij van Landlouw, V,* 1883.
2. **Chatin, J.,** Reeherches sur l'anguillule de l'oignon, Paris, 1884.
3. **Jensen, H. J.,** Nematode pests of vegetable and related crops, in *Economic Nematology,* Webster, J. M., Ed. Academic Press, London, 1972, chapt. 16.
4. **Moraes, M. V., de, Lordello, L. G. E., Piccinin, O. A., and Lordello, R. R. A.,** Pesquisas sobre plantas hospedeiras do nematoide do cafeeiro, *Meloidogyne exigua* Goeldi, 1887, *Cien. Cult.,* 24, 658, 1972.
5. **Lordello, L. G. E. and Mello Filho, A. de T.,** Ocorrencia de nematoides em culturas de cebola, *Pesqui. Agropecu. Bras.,* 6, 127, 1971.
6. **Olthof, T. H. A. and Potter, J. W.,** Relating nematode populations to crop losses, *Can. Agric.,* 17, 18, 1972.

7. **MacGuidwin, A. E., Bird, G. W., and Safir, G. R.,** Influence of *Glomas fasciculatum on Meloidogyne hapla* infecting *Allium caps, J. Nematol.,* 17, 389, 1985.
8. **Jaworski, C. A., McCarter, S. M., Johnson, A. W., and Williamson, R. E.,** Response of onions grown for transplants to soil fumigation, *J. Am. Soc. Hortic. Sci.,* 103, 385, 1978.
9. **Lear, B. and Raski, D. J.,** Survival of root-knot nematodes in grape and tomato roots recovered from soils fumigated with Nemagon, *Phytopathology,* 46, 18, 1956.
10. **Lear, B. and Raski, D. J.,** Survival of root-knot nematodes in excised grape roots in moist soil fumigated with ethylene dibromide, *Phytopathology,* 52, 1309, 1962.
11. **Stark, F. L., Jr. and Lear, B.,** Miscellaneous greenhouse tests with various soil fumigants for the control of fungi and nematodes, *Phytopathology,* 37, 698, 1947.
12. **Lamberti, F.,** Chemical and cultural methods of control in *Root Knot Nematodes (Meloidogyne species) Systematics, Biology and Control,* Lamberti, F. and Taylor, C. E., Eds., Academic Press, London, 1979, Chapter, 26.
13. **Romney, R. K., Anderson, J. L., and Griffin, G. D.,** Effects of DCPA on seedling infection by root-knot nematode, *Weed Sci.,* 22, 51, 1974.
14. **El-Hameed, S. H. A., Kheir, A. M., and Shafiee, M. F.,** Host-parasite relations of *Rotylenchulus reinformis* on onion *(Allium cepa* L.), *Phytopathol. Mediterr.,* 16, 11, 1977.
15. **Townshend, J. L., Tarte, R., and Mai, W. F.,** Growth response of three vegetables on smooth- and crenate-tailed females of three species of *Pratylenchus, J. Nematol.,* 10, 259, 1978.
16. **Ferris, J. M.,** Soil temperature effects on onion seedling injury by *Pratylenchus penetrans, J. Nematol.,* 2, 248, 1970.
17. **Green, C. D.,** Root-feeding nematodes in onions, *24th Annual Report 1973, National Vegetable Research Station,* 93, 1974.
18. **Sidam, A. K. Bin.,** *Pratylenchus* spp. attack on onions and lettuce, *29th Annual Report 1978, National Vegetable Research Station,* 52, 1979.
19. **Mancini, G., Rameri, G., and Villani, R.,** Prove di lotta contro Nematodi parassiti della cipolla, *Ann. Faco. Sci. Agrar. Univ. Studi. Napoli, Portici,* 6, 3, 1972.
20. **Wuyts, F., Vanachter, A., and Assche, C.,** van., Grondontsmetting in openluchtteelten, *Meded. Fac. Landbouwwet. Rijksuniv. Gent,* 36, 1027, 1971.
21. **Koshy, P. K. and Sosamma, V. K.,** Host range of the burrowing nematode *Radopholus similis* (Cobb, 1893), Thorne, 1949, *Indian J. Nematol.,* 5, 255, 1977.
22. **Zem, A. C.,** Nematoides associados a algumas plantas nativas e cultivadas do cerrado de Itirapina, Sao Paulo, *Rev. Agric., (Piracicaba, Bras.),* 52, 112, 1977.
23. **Banck, A. and Andersson, S.,** Frittlevande nematodes orsaker svara skadori matloken, *Vaxtskyddsnotiser,* 40, 93, 1976, *Trichodorus similis chapters.*
24. **Banck, A.,** Inventering av rotnematoder i matlok, *Vaxtskyddsnotiser,* 41, 111, 1977.
25. **Graichen, K.,** Allium-Arten als naturliche Wirte nematodenubertragbarer Viren, *Archiv. Phytopathol. Pflanzenschutz,* 11, 399, 1975.
26. **Cooper, J. I. and Thomas, P. R.,** Chemical treatment of soil to prevent transmission of tobacco rattle virus to potatoes by *Trichodorus* sp., *Ann. Appl. Biol.,* 69, 23, 1971.
27. **Cooke, D. A., Dunning, R. A., and Winder, G. H.,** The effect of nematicides applied to the seed rows in spring, on growth and yield of sugar beet in Docking-disorder-affected fields, *Ann. Appl. Biol.,* 76, 289, 1974.
28. **Cohn, E., Yisraeli, U., and Krikun, J.,** A new disease of onion caused by an ectoparasitic nematode, and its control, *Ktavim (Engl. Ed.),* 17, 147, 1967.
29. **Southey, J. F.,** New or unusual host-plant records for plant-parasitic nematodes, 1971-1973, *Plant Pathol.,* 23, 45, 1974.
30. **Muller, P. J., Hollinger, T. C., and Coijn, C. G. M.,** Diagnostisch onderzoek bij bol -en knolgewassen, *Jaarverslag 1978 Laboratorium voor Bloembollenonderzoek, Lisse,* 34, 100, 1979.
31. **Petrova, Z. I.,** Observations on the biology of the onion stem nematode *Ditylenchus allii* Beijerinck, 1883, *Tr. Vses. Inst. Zashch. Rast.,* No. 16, 174, 1961.
32. **Yuksel, H. S.,** Observations on the life cycle of *Ditylenchus dipsaci* on onion seedling, *Nematologica,* 5, 289, 1960.
33. **Green, C. D.,** Interactions between nematodes and *Rhizobium* in relation to root nodulation of pea plants, in *The Pea Crop A Basis for Improvement,* Hebblethwaite, P. D., Heath, M. C., and Dawkins, T. C. K., Eds., Butterworths, London, 1985, 37.
34. **Green, C. D., Ghumra, M. F., and Salt, G. A.,** Interaction of root- lesion nematodes, *Pratylenchus thornei* and *P. crenatus,* with the fungus *Thielaviopsis basicola* and a grey sterile fungus on roots of pea *(Pisum sativum), Plant Pathol.,* 32, 281, 1983.
35. **Caubel, G.,** Etude des populations de *Ditylenchus dipsaci* (Kuhn) Filiipjev distribution et fluctuations dans les sols de l'ouest de la France en cultures legumieres et fourrageres, *Ann. Zool. Ecol. Animale,* 5, 309, 1973.

36. **Clayden, I. J.,** Factors Influencing the Survival of *Ditylenchus dipsaci* in Soil, *Ph.D. thesis, University of Leeds,* 1985.
37. **Alfaro, A. and Martinez Beringola, M. L.,** *Ditylenchus dipsaci* on garlic, *Nematologia Mediterr.,* 1, 69, 1973.
38. **Norwood, J. M. and Green, C. D.,** Stem eelworm invasion of onion, *26th Annual Report 1975,* National Vegetable Research Station, 1976, 94.
39. **Windrich, W. A.,** Attraction of immature female nematodes by males of the tulip race of *Ditylenchus dipsaci* in leaf tissue on tulip, *Nematologica,* 19, 570, 1970.
40. **Artukhova, G. A.,** The reaction of young *Allium cepa* plants to the penetration of the stem nematode, *Byul. Vsesoyuznogo Inst. Gel'mintologii K.I. Skryabina,* 5, 1978.
41. **Salvarredi, A. U.,** Sintomatologia del ataque del nematodo *Ditylenchus dipsaci* (Kuhn) Goodey y su distribucion en la planta de ajo, *Rev. Invest. Agropecu. INTA Buenos Aires,* 9, 1, 1972.
42. **Viglierchio, D. R.,** Nematodes and other kpasthogens in auxin-related plant growth disorders, *Bot. Rev.,* 37, 1, 1971.
43. **Viglierchio, D. R. and Pyung, K. Yu.,** Plant parasitic nematodes: a new mechanism for injury of hosts, *Science,* 147, 1301, 1965.
44. **Chitwood, D. J. and Krusberg, L. R.,** Pectolytic enzymes in three populations of *Ditylenchus dipsaci, J. Nematol.,* 9, 187, 1977.
45. **Riedel, R. M. and Mai, W. F.,** A comparison of pectinases from *Ditylenchus dipsaci* and *Allium cepa* callus tissue, *J. Nematol.,* 3, 174, 1971.
46. **Caubel, G. and Samson, R.,** Influence du nematode des tiges, *Ditylenchus dipsaci* (Kuhn) Fil. dans le developpement de la bacteriose "cafe au lait" de l'ail *(Allium sativum* L.) occasionnee par un biovar de *Pseudomonas fluorescens* Migula, *Agronomie,* 4, 311, 1984.
47. **Greco, N., Lamberti, F., and Brandonisio, A.,** Indagini su biologia ed epidemiologia de *Ditylenchus dipsaci* (Kuhn) Filipjev su cipolla in Puglia, *Laboratorio di Nematologia Agraria del C.N.R.,* Bari, 149, 1974.
48. **Goodey, T.,** *Anguillulina dipsaci* in the inflorescence of onions and in samples of onion seed, *J. Helminthol.,* 3, 22, 1943.
49. **Southey, J. F.,** The incidence and location of stem eelworms on onion seed, *Plant Pathol.,* 14, 55, 1965.
50. **Brzeski, M. W. and Rajewski, M.,** Badania nad wystepowaniem I rozmieszczeniem wegorka niszczyka *(Ditylenchus dipsaci)* w pedach nasiennych i w nasionach cebuli, *Biul. Warzywniczy,* 7, 196, 1966.
51. **Sienhorst, J. W. and Koert, J. L.,** Stengelaaltjes op uienzaad. *Overdruk uit Gewasbescherming,* 2, 1979.
52. **Green, C. D. and Sime, S.,** The dispersal of *Ditylenchus* dipsaci with vegetable seeds, *Ann. Appl. Biol.,* 92, 263, 1979.
53. **Hooper, D. J.,** Stem eelworm *(Ditylenchus dipsaci),* a seed and soil-borne pathogen of field beans *(Vicia faba), Plant Pathol.,* 20, 25, 1971.
54. **Pereira, M. C. L. S. and Santos, M. S. N. De A.,** Seasonal changes in the numbers and stages of *Ditylenchus dipsaci* (Kuhn) Filipjev, in soil and host plants in Portugal, *Nematologia Mediterr.,* 2, 91, 1974.
55. **Green, C. D.,** The effects of weeds and wild plants on the reinfestation of land by *Ditylenchus dipsaci* (stem and bulb nematode) and on the stability of its populations, in *Pests Pathogens and Vegetation,* Thresh, J. M., Ed., 1981, 217.
56. **Green, C. D. and McNee, H.,** Stem-eelworm persistence, *28th Annual Report 1977, National Vegetable Research Station,* 1978.
57. **Seinhorst, J. W.,** De betekenis van de toestand van de grond voor het optreden van aantasting door het stengelaaltji *(Dithylenchus dipsaci)* (Kuhn) Filipjev), *Tijdschri. Plantenziekten,* 56, 289, 1950.
58. **Lewis, G. D. and Mai, W. F.,** Overwintering and migration of *Ditylenchus dipsaci* in organic soils of southern New York, *Phytopathology,* 50, 341, 1960.
59. **Green, C. D. and Wilson, A.,** Stem nematode survival, *30th Annual Report 1979, National Vegetable Research Station,* 46, 1980.
60. **Wallace, H. R.,** Movement of eelworms, II, *Ann. Appl. Biol.,* 46, 89, 1958b.
61. **Webster, J. M. and Greet, D. N.,** The effect of host crop and cultivation on the rate that *Ditylenchus dipsaci* reinfested a partially sterilized area of land, *Nematologica,* 13, 295, 1967.
62. **Gentzsch, D.,** Die Vermehrung de Zwiebeln schadigenden Stengelnematoden *(Ditylenchus dipsaci)* in Unkrautern und Schlussfolgerungen fur den Herbizideinsatz, *Vortrage der 12 Tagung ube Probleme der Phytonematologie, Gross Lusewitz, I Juni 1973.... Gros Lusewitz Akademie der Landwirtschaften der Deutsch Demokratischa Republik,* 109, 1973.
63. **Sturhan, D.,** Das Rassen problem bei *Ditylenchus dipsaci, Mitt. Biol. Bundesanst. Land Forstwirsch., Berlin-Dahlem,* 136, 87, 1969.
64. **Ladygina, N. M. and Barbaschova, V. N.,** Genetic and physiological compatibility and karyotypes of stem eelworms, *Parazitologiya,* (Leningrad) 10, 449, 1976.

65. **Hesling, J. J.,** Biological races of stem eelworm., *Annual Report, Glasshouse Crops Res. Inst. Ann. Rep. for 1965,* 132, 1966.
66. **Shubina, L. V.,** Steblevye nematody sel'skokhozyaistvennykh kul'tur i mery bor'by s nimi. (Materialy simpoziuma, Voronezh, 27-29 Sentyabrya, 1983 g). *Voronezh USSR, Vserossiiskii NII Zashchity Rastenii,* 147, 1983.
67. **Shavrov, G. N.,** Steblevye nematody sel'skokhozyaistvennykh kul'tur i mery bor'by s nimi, (Materialy simpoziuma, Voronezh, 27-29 Sentyabrya 1983 g), *Vormezh USSR Vserossiiskii NII Zashchity Rastenii* 137, 1983.
68. **Greco, N. and Lamberti, F.,** Riproduzione di *Ditylenchus* dipsaci in varieta di cipolla, *Nematologia Mediterr.,* 4, 95, 1976.
69. **Romascu, E. and Lemeni, V.,** Nematodul *Ditylenchus dipsaci* Kuhn — un daunator periculos al culturilor de usturoi si ceapa, *Rev. Hortic. Vitic.,* 21, 73, 1972.
70. **Curi, S. M., Silveira, S. G. P. da, Bona, A. de, Siqueira, W. J., and Prates, H. S.,** Ocorrencia e sintomatologia do nematoidee do alho, *Ditylenchus dipsaci,* no estado de Sao Paulo, *Biologico,* 50, 187, 1984.
71. **Bergquist, R. R. and Riedel, R. M.,** Screening onion *(Allium* cepa) in a controlled environment for resistance to *Ditylenchus dipsaci, Plant Dis. Rep.,* 56, 329, 1972.
72. **Whitehead, A. G. and Tite, D. J.,** Control of stem nematode *(Ditylenchus dipsaci)* attacking onions by Aldicarb and Dazomet in sandy loam, *Plant Pathol.,* 21, 89, 1972.
73. **Winfield, A. L., Murdoch, G., and John, M. E.,** Chemical control of stem nematode *(Ditylenchus dipsaci)* of bulb onions, *Proc. 6th British Insecticide Fungicide Conference,* 1971, 141.
74. **Siti, E., Cohn, E., Katan, J., and Mordechia, M.,** Control of *Ditylenchus dipsaci* in garlic by bulb and soil treatments, *Phytoparasitica,* 10, 93, 1982.
75. **Inserra, R. and Lamberti, F.,** Preliminary trials of the chemical control of the stem nematode on onion in Apulia, *International Symposium of Nematology (11th),* 31, 1972.
76. **Inserra, R., Lamberti, F., and Greco, N.,** Prove di lotta chimica contro *Ditylenchus dipsaci* (Kuhn) Filipjev su cipolla in Puglia. Primi risultati, *Nematologia Mediterr.,* 2, 29, 1974.
77. **Bird, G. W. and Smith, H. A.,** Factors affecting control of onion bloat by fumigants containing 1,3-dichloropropene in organic soils in southern New York, *Plant Dis. Rep.,* 49, 33, 1965.
78. **Greco, N., Lamberti, F., and Brandonisio, A.,** Produzione di semenzali di cipolla esenti da *Ditylenchus dipsaci, Nematologia Mediterr.,* 4, 71, 1976.
79. **Brzeski, M. W.,** Wegorek niszczyk na cebuliiczosnku, *Skierniewice, Instytut Warzynictwa,* 16, 1974.
80. **Allison, D. A., Sinclair, C., and Smith, M. R.,** Oxamyl — a nematicide for potatoes, onions and sugar beet, *Proceedings 7th British Insecticide and Fungicide Conference,* 1973, 393.
81. **Homeyer, B.,** Results with fenamiphos in potatoes and vegetable crops, *Proceedings 7th British Insecticide and Fungicide Conference,* 1973, 415.
82. **Williamson, K. and Davies, P. C.,** Stem-nematode, occurrence in bulb stores, *30th Annual Report 1979 National Vegetable Research Station,* 47, 1980.
83. **Goodey, T.,** *Anguillulina dipsaci* on onion seed and its control by fumigation with methyl bromide, *J. Helminthol.,* 21, 45, 1945.
84. **Hague, N. G. M.,** Fumigation of shallot seed with methyl bromide to control stem eelworm, *Ditylenchus dipsaci, Plant Pathol.,* 17, 127, 1968.
85. **Powell, D. F.,** The effect of methyl bromide fumigation on the germination of onion seed and vigor of the seedlings *Ditylenchus dipsaci, Plant Pathol.,* 24, 237, 1975.
86. **Powell, D. F.,** Fumigation of field beans against *Ditylenchus dipsaci, Plant Pathol.,* 23, 110, 1974.
87. **Bruinsma, F. and Seinhorst, J. W.,** Warmwaterbehandeling van sjalotten tegen aantasing door stengelaaltjes *(Ditylenchus dipsaci* (Kuhn) Filipjev), *Meded. Dir. Tuinbuw. (Neth.),* 17, 437, 1954.
88. **Meijneke, C. A. R. and Scheele, J.,** Plantuitjes stengelaaltjes en warmwaterbehandelingen, *Zaadbelangen,* 24, 248, 1970.
89. **Lear, B. and Johnson, D. E.,** Treatments for eradication of *Ditylenchus dipsaci* in cloves of garlic, *Plant Dis. Rep.,* 46, 635, 1962.
90. **Johnson, D. E. and Lear, B.,** Additional information regarding the hot water treatment of seed garlic cloves for the control of the stem and bulb nematode, *(Ditylenchus dipsaci), Plant Dis. Rep.,* 49, 898, 1965.
91. **Romascu, E., Lemeni, V., Popescu, I., Vasile, I., and Jobanu, E.,** Rezultate experimentale privind combaterea nematodului *Ditylenchus dipsaci* Kuhn din culturile de usturoi, I. Tratamente aplicate la materialul vegetal pentru saminta, *An. Inst. Cercet. Prot. Plant, Acad. Stiinte Agric. Silvice,* 12, 399, 1974.
92. **Brzeski, M. W.,** Zwalczanie wegorka niszczyka *Ditylenchus* dipsaci (Kuhn) na cebuli, *Materialy Ogolnoposkiego Zjazdu Warzywniczego, Skierniewice,* 78, 1973.
93. **Saly, A.,** Ochrana cesnaku pred hadatkem cibulovym *Ditylenchus allii* (Biejerinck) fumigovanim sadby nematocidmi, *Pol'nohospodarstvo,* 10, 125, 1964.

Chapter 8

LEAF DISEASES OF ONIONS

R. B. Maude

TABLE OF CONTENTS

I. INTRODUCTION

There are some 12 leaf pathogens, mainly fungal, of onion crops throughout the world. About 6 of these cause major disease problems. Excluded from this account are the fungal and bacterial storage pathogens (see chapter on Storage Diseases of Onions) which also have a leaf-borne phase.

The pathogens considered in order of importance, include *Peronospora destructor* (Berk.) Casp. (downy mildew), *Alternaria porri* (Ellis) Cif. (purple blotch), *Sclerotinia squamosa* (Vienn. Bourg.) Dennis (leaf rot or blast), *Botrytis cinerea* Rud. (leaf spot, collar rot), *Cladosporium allii-cepae* (Ranojevic) M. B. Ellis (leaf blotch), *Puccinia allii* (DC) Rud. (rust), *Pleospora herbarum* (Pers. ex Fr.) Rabert. (black stalk rot), *Stemphylium vesicarium* (Wallr.) Simmons (leaf blight), *Glomerella cingulata* (Stonem.) Spauld. and Schrenk (seven curls disease), and *Cercospora duddiae* Welles (leaf spot). Other problems include onion yellow dwarf, the main virus disease and occasional physical and pollutant disorders which produce effects reminiscent of disease.

The effects of pathogens listed range from slight to severe leaf blemishes, which may necessitate leaf trimming in preparation for market thereby increasing costs to the grower,[1] to leaf kill resulting in yield depression[2,3] or even total crop loss. Foliar fungi cause greatest damage in high density crops such as green (salad) or pickling onions particularly where these are produced as a series of successional and often adjoining crops. Onions grown for seed production are also susceptible. Bulb onions grown at lower densities on wide rotations may escape severe infection unless they are in close proximity to potent sources of inoculum such as affected onion residues or diseased crops. There are recent accounts of the disease cycles of many of these pathogens[4-7] and of their distribution throughout the world.[8]

II. ONION LEAF PATHOGENS

A. *PERONOSPORA DESTRUCTOR* (DOWNY MILDEW)

1. Disease Distribution and Host Range

The disease is widely distributed on all continents (Commonwealth Mycological Institute (CMI) map no. 76)[8] particularly temperate growing areas of the world. Considerable losses have occurred in bulb and seed production onion crops[3,9,10] despite the use of prophylactic fungicides.

The pathogen was first named and the disease described by Berkeley in the U.K. in the latter half of the 19th century.[11,12] A comprehensive review of research progress up to 1981 has been published.[3]

P. destructor infects the common onion *(Allium cepa),* tree onion (*A. cepa* var. *viviparum*), potato multiplier onion (*A. cepa* var. *solanina*), shallots (*A. cepa* var. *ascalonicum*), and Welsh onion (*A. fistulosum*).[13]

2. Disease Symptoms and Infection Sources

Infections which occurred in the autumn in overwintering onion crops develop into sporulating lesions during mild humid weather in the following spring. Leaves of affected plants curve outward. Leaf lesions are long and have a pale yellow, glazed appearance; sporulation on these areas produces a thick grey to purple felt of conidiophores.[13] Under less favorable climatic conditions the lesions are restricted to nonsporulating white spots on the foliage.[4]

The main infection source for seed production crops is systemically affected onion bulbs.[4,14,15] Leaves of overwintering onions also carry the disease, on which sporangia are produced and these disseminate to infect spring-sown onions. Decaying onion tissues may release oospores into the soil thereby increasing the inoculum potential.[4] It is reported that mycelium may invade onion seeds, creating a potent source of the disease.[16,17] Others have considered seed transmission improbable.[3]

3. The Pathogen and Disease Biology

The fungus is a Phycomycete with aseptate mycelium, and asexual monopodially branched sporangiophores which bear fusiform to pyriform sporangia. It has sexual oogonia and antheridia which fuse to produce numerous thick-walled oospores.[4,13] Sporangia are produced over a temperature range 3 to 25°C[3] at saturation relative humidity (RH).[18] The mature sporangia are released in daylight (between 08.00 to 09.00 h in Canadian onion fields[10]) as the leaves dry and RH is reduced.[19] Reductions in humidity to below 59% were most effective in stimulating spore release.[19] In these conditions the vibration of leaves triggered the forcible discharge of the sporangia, probably under the control of an electrostatic mechanism.[20] The sporangia subsequently infected plants only on days when there was cloud cover and persistent moisture on the leaves and not on dry and sunny days.[10] Sporangia germinate only in free water in the temperature range 6 to 27°C (optimum 10 to 12°C) producing germ tubes in 2 to 4 h.[21] Appressoria formed over the stomata penetrate the host plant and produce substomatal vesicles from which mycelium penetrates cells followed by the development of haustoria. Invasion is progressive with the infection cycle in leaf tissues taking 11 to 14 d,[13,22] but longer when temperatures are higher.[3] The sporangia so produced are wind disseminated.[22] The pathogen may also develop systemically in onion tissues with invasion of the necks of plants leading to penetration of the bulb storage tissues, where it may survive for many months.[18,23] In inoculated sets *P. destructor* remained viable for five months (the maximum duration of the test) at 1 to 3°C and for almost 2 months in sets stored at 20 to 22°C.[24]

Under certain conditions,[3] planting of systemically affected onion bulbs may result in infection, sporulation of the pathogen on emergent leaves and the development of the disease in seed production crops resulting in the destruction of seed stems and inflorescences.[13,14]

Oospores which may be produced in abundance in diseased onion tissues are able to survive for at least 26 years[13] but their role in perpetuating and re-initiating the disease is not well proven.[3]

4. Disease Control

There is little data on resistance in onion to the downy mildew pathogen. In the U.S., physiological resistance under certain inoculum and environmental conditions was claimed for a breeding line of *A. cepa.*[25] Cuticular wax[26] and lignified leaf cells[27] correlated with resistance to the pathogen in *Allium* spp. Elsewhere, some resistance was detected in cultivars of onion under strong disease pressure.[28] Resistance and immunity of the flower bearing stem is reported.[29,30] Cultural control methods have included such measures as the burning of onion debris, elmination of volunteer plants, long rotations, and spatial separation of overwintered and spring planted crops.[13]

Eradication of mycelium in bulbs destined for seed production, by direct heat treatment for 4 h at 41°C was proposed by Yarwood.[18] A similar effect may be achieved naturally, in areas with hot climates, by field drying onions for 12 d at temperatures of 40°C for 4 h and longer.[31] Preplanting heating of onion mother bulbs (43 to 45°C for 8 h) followed by fungicide treatments are reported to be effective.[32] Frequent foliar applications of the dithiocarbamate fungicides, zineb,[33,34] mancozeb,[35] and maneb/zineb mixtures[36] have controlled onion downy mildew over the last 30 years. Further improvements in disease control were obtained with the introduction of the acylalanine fungicides in the late 1970s; these chemicals were systemic in action, selective for Phycomycete fungi and required only 1 to 3 applications.[37,38] Unfortunately, insensitivity to metalaxyl has developed in many *Phytophthora, Pythium* and *Peronospora* fungi.[39] Hence, the fungicide is now formulated with mancozeb to preserve its selective action; this product is approved for the control of onion downy mildew in the U.K.

An integrated control strategy, based on a bioclimatological model developed to forecast the onset of downy mildew outbreaks linked with threshold values for levels of inoculum, has been proposed.[40]

B. *ALTERNARIA PORRI* (PURPLE BLOTCH, SCALD)

1. Disease Distribution and Host Range

The pathogen is widespread (CMI map no. 350)[8] but because it requires temperatures of 21 to 30°C for development it is most serious in hot humid climates where severe crop losses, including reductions of 50% in seed crop yield, have been reported.[7] The greatest losses often occur in store where symptoms may not be expressed for up to 5 months.[41]

The pathogen attacks members of the *Allium* family including onion *(A. cepa)*, Egyptian onion (*A. proliferum*), Welsh onion (*A. fistulosum*), leek (*A. ampeloprasum* var. *porrum*), and possibly others.[5] The name, "purple blotch" was proposed by Nolla[42] who named the fungus *Alternaria allii* sp. nov.; this was later amended to *A. porri.*

2. Disease Symptoms and Infection Sources

Initially, small white sunken spots develop on the leaves. These enlarge, become zonate and under moist conditions, turn purple, and are surrounded by a broad chlorotic margin. These are very prominent on the inflorescence stalks.[43] Lesions may girdle leaves and stalks causing them to drop after 2 to 3 weeks. Infection can cause a semiwatery rot of necks of bulbs which turn yellow-red in color; bulb tissues eventually become papery.[7,41] The fungus is seed-borne[44,45] but the relevance of this phase in initiating disease outbreaks in hot climates is not well documented. Infected onion debris was implicated as an infection source in the U.S.[42] and India.[44] In the latter country conidia present on debris on the soil surface survived for 8 months, but this was reduced to less than 2 months if the debris was buried.[44]

3. The Pathogen and Disease Biology

Alternaria porri is a dematiaceous Hyphomycete. Conidia are brown, septate 100 to 300 μm long × 15 to 20 μm thick with a long flexuous beak. They occur singly on short conidiophores borne on a septate mycelium.[46] There is no known sexual state. Conidia are produced at night on lesions and are released after 07.00 h[47,48] as vapor pressure decreases. Maximum sporulation on affected leaves in Indian crops occurred at 08.00 h after rain. In the U.S., peak concentrations of spores occurred above onion crops between 08.00 and 14.00.[47] Crop disturbance by rain, sprinkler irrigation, spraying, or wind, increased release of conidia[47] which were subsequently wind dispersed.[41]

The fungus is primarily a wound pathogen[41] although inoculation studies have indicated that it can invade intact onion leaves.[49] Maximum germination of conidia occurred in free water after 6 h at 25°C.[50] Germ tubes penetrated stomata or wounds made mechanically by wind blown sand, hail, or by insects, particularly thrips (*Thrips tabaci*).[41,51,52] Old leaves are more susceptible to infection than young leaves, and emerging leaves are more susceptible the closer the bulbs are to maturity.[53] Infection of the leaves leads to invasion of the developing bulbs.[42] Infected bulbs transmit the disease to the flower stalks on the debris of which conidia survive to produce an infection source for other crops.[44]

4. Disease Control

Sweet Spanish onion types appear to be more susceptible to infection than Yellow Globe types.[5,7] Reduced infection has been reported in hybrid breeding material in Brazil[54] and in certain lines of the cv. 'Red Creole' in Cuba.[55] Resistance is reported for the cv. 'Taliana Red' in Hungary.[56] In the latter cultivar, the high content of phenolic compounds and oxidase activity is reported to inhibit the macerating enzymes released by virulent isolates of the fungus.[56] In experimental systems, development of mycelium and consequent infection of onion leaves by *A. porri* was reduced by the addition of cells of the leaf saprophyte *Aureobasidium pullulans* (De Bary) Arnand.[57] The role of this naturally occurring component of the leaf mycoflora, as a potential biological control organism for onion leaf pathogens, is not known.

Cultural control methods include long rotations with nonrelated crops and good drainage.[5,7] Lowering the density of transplanted crops reduced infection, as did the application of high rates of calcium superphosphate and potassium fertilizers.[58,59] Nitrogen fertilizers at low and high rates increased disease.[58,59] Routine (7-d interval) field sprays with dithiocarbamate fungicides, particularly mancozeb, and also with iprodione and chlorothalonil are reported to be effective.[7,60] Surprisingly, the benzimidazole-based fungicides which are not highly toxic to *Alternaria* spp. are reported to reduce the disease.[58,61]

C. *SCLEROTINIA SQUAMOSA* (LEAF ROT, LEAF BLIGHT, BLAST)

1. Disease Distribution and Host Range

Botrytis squamosa, the imperfect form, causes disease in temperate growing areas of the world (CMI map no. 164)[8] but is less widely distributed than neck rot *B. allii* (CMI map no. 169).[8]

The conidial form was identified as *B. squamosa* by Walker,[62] later the perfect form, a Discomycete, was found and placed in the family Sclerotinacea.

The fungus, *B. squamosa,* is pathogenic to *A. cepa* cultivars; there is resistance in certain cultivars of *A. fistulosum* (Welsh onion) and immunity in *A. schoenoprasum* (chives) and *A. bouddhae.*[63,64]

2. Disease Symptoms and Infection Sources

The disease is most prevalent in densely sown crops such as green (salad) and pickling onions, particularly where crops have been grown in succession and close to the preceding onions. Severe symptoms include the die-back of onion leaf tips and leaves on whose necrotic tissues conidiophores and conidia are produced. More usually leaf lesions develop in the form of grayish-white desiccated spots, circular to elliptical in shape (1 to 10 mm long × 1 to 4 mm wide).[65,66]

B. squamosa was shown to be seed-borne[67] but disease transmission was not demonstrated.[68] Sclerotia bearing conidia of the fungus occurred on onion bulbs and leaf debris in cull piles and seed production fields in New York,[68] and these represent a primary source of inoculum for the infection of commercial onion fields.[68] Additionally, apothecia, produced in the spring on sclerotia on the surface of soil which had previously borne onions, released ascospores which were capable of infecting onions.[68] Inoculum bridging occurs, through conidia or ascospores moving from crop to crop, in successionally grown onions in a single season, or over a span of at least a year, where overwintered crops (green, ware or seed production) effectively link the seasons into a continuous cycle of onions.

3. The Pathogen and Disease Biology

Taxonomically the perfect state of the fungus was placed in the genus *Sclerotinia* by Cronshey in 1946.[69] Later (1952) and independently, the sexual form was identified as a *Botryotinia* by Viennot-Bourgin.[70] The question of the generic status of the perfect form recently has been reviewed.[71]

Apothecia are produced from sclerotia and are stalked (3 to 4 mm tall) and cupulate producing asci (162.5 to 200.0 μm × 13.8 to 16.5 μm) containing 8 bi-guttulate single-celled, ovoid, hyaline ascospores (15.0 to 17.5 μm × 10.0 to 12.5 μm).[72]

The common asexual *B. squamosa* form produces monopodial, short-stemmed conidiophores which, in contrast to other onion *Botrytis* spp., bear heads of very large conidia. Also, and unlike the other spp. when the conidia are released, the spore bearing branches (ampullae) of the conidiophore head collapse in a characteristic concertina-like fashion.[73] The comparative characteristics, including measurements for conidia, of onion *Botrytis* spp. have been given in Table 1 of the chapter on Storage Diseases of Onions.

Sporulation is promoted by persistent leaf surface wetness (≥13 h) and temperatures of 14 to 20°C; it is restricted by shorter wet periods (≤12 h) and temperatures of ≤12°C.[74] Spore release is triggered by increasing or decreasing relative humidity and light rain.[74] Conidia germinate between 6 and 33°C but more rapidly at 20 to 28°C.[75] Infection occurs over a temperature range of 6 to 28°C and requires 9 h of leaf wetness at 15 to 18°C.[75] Leaf penetration is direct and localized leaf spot lesions develop.[76] Temperatures of approximately 27°C and persistent wetness favor leaf dieback when extensive tissue invasion may occur.[65,77] Growth of invasive hyphae within senescent tissues is much faster than in nonsenescent leaf tissues.[27] The greatest production of spores (conidia) occurs on necrotic tissues of leaves and leaf tips (i.e., leaf rot), but not on the bleached spots on the green parts of leaves; these lesions represent a host hypersensitivity reaction to a weak pathogen.[78] Leaf spotting also may be incited nonbiologically by ozone pollution of the atmosphere in onion crops.[79,80] Increased leaf spotting occurred where *Botrytis squamosa* inoculum was added to the ozone contaminating onions.[80] Where ozone was removed experimentally by filtration, onion yields were increased and *Botrytis* was reduced.[80] However, it is also reported that ozone, applied only at acute doses (0.25 ppm for 4 h), induced increases in *B. squamosa* lesions on nonsenescing onion leaves, whereas chronic doses (0.12 ppm for 5 h/d for 5 d) predisposed onion leaves to enhanced infection by *B. cinerea* (leaf fleck).[81,82]

White spots may also be produced by hail. These occur on the exposed side of leaves and are variable in shape and alignment relating to the force of impact of the hail and direction of the wind.

4. Disease Control

There is no consistent source of resistance within *A. cepa* to *B. squamosa*.[63,64] However, it may be possible to breed for resistance using interspecific crosses, with possibly *A. fistulosum* as the donor.[63,64] Although the system has been shown to be effective experimentally[63,64] there is no breeding programme, at present, for incorporating this source of resistance into *A. cepa*. Fungicides, particularly broad-based protectant dithiocarbamates, e.g., mancozeb, have been used routinely.[83,84] and as applications timed to coincide with increases in inoculum levels, to control the disease in summer crops of onions in the U.S.[85] Following the introduction of selective systemic fungicides, based on benzimidazole, *Botrytis* leaf pathogens were reduced to extremely low inocula potentials.[83] However, field insensitivity to the benzimidazole-based fungicides soon developed with subsequent loss of disease control.[86]

In the U.K., a program of four sprays, applied monthly between October and January, has been proposed for control of *B. squamosa* in overwintered green (salad) onions.[87] This involves two applications of thiram and two of the dicarboximide fungicide iprodione[86] or four applications of iprodione.[87] More frequent applications of iprodione were necessary to control the disease in summer-grown salad onions. Low volume applications (280 l/ha) and the use of an oil-based formulation increased leaf retention of the fungicide and gave an overall improved performance.[87] Although insensitivity in laboratory isolates of *B. squamosa* to iprodione has been reported,[88] it has not been observed in field isolates of the fungus and hence the disease continues to be controlled by foliar applications of iprodione.

An integrated approach to disease control, based on disease forecasting, has been developed in Canada utilizing quantitative relationships between weather variables and the pathogen infection cycle.[89] Using this system it is possible to predict the onset and potential severity of disease. Such information can be used to reduce the number of fungicide applications needed for control.[89] The system known as Botcast can be operated by farm staff with reasonable management skills.[89]

D. *BOTRYTIS CINEREA* (LEAF SPOT, LEAF FLECK, COLLAR ROT, BROWN STAIN)

1. Disease Distribution and Host Range

Botrytis cinerea is the common gray mold fungus which is ubiquitous in temperate areas of the world.[90] The types grouped under *B. cinerea* can infect a wide range of plants including field and glasshouse vegetables, small berry fruits including grapes, ornamentals, bulb and corm producing monocotyledons and forest tree seedlings.[90] The host range of the types of *B. cinerea* which can infect onions under certain conditions is not known.

2. Disease Symptoms and Infection Sources

B. cinerea has been reported to have caused flecking of leaves,[66] collar rotting of false stems,[91] and brown stain of bulbs of onions.[92]

Leaf spotting is expressed as small superficial lesions (0.5 × 1 to 1.5 mm) as distinct from the larger deeper lesions caused by *B. squamosa* (1 × 1 to 3 mm). *B. cinerea* was generally less frequent than *B. squamosa* in onion fields in the U.S. where it occupied a subordinate and mainly saprophytic role.[66] To distinguish between the field symptoms of the two fungi, Hancock and Lorbeer[66] proposed the name, "leaf fleck" for *B. cinerea* lesions and "leaf blight" for those produced by *B. squamosa*. The imposition of stress, for example exposure to ozone (see Section II.C.3) on onion crops, may enhance their vulnerability to fungi such as *B. cinerea*.[81,82]

Collar rotting caused by the fungus is a further example of infection occurring under conditions of stress during the winter.[91] Symptoms are not obvious, the leaf bases of plants which form the false stem or neck of onions are attacked at the soil surface, and when penetrated the plants collapse and die. Sometimes a typical gray mold develops around the false stem at the soil surface. More usually in high density crops a proportion of plants may collape, die and disappear unnoticed.

B. cinerea causes a brown discoloration of the outer dry scales of onion bulbs.[92] Staining occurs on bulbs in the field prior to harvest and is located on the neck to shoulder or on the middle regions of bulbs.[92]

The fungus overwinters as sclerotia in the soil or on onion debris and as mycelium in debris from diseased onions.[5] Other sources are similar to those for *B. squamosa* (see Section II.C.2).

3. The Pathogen and Disease Biology

Some isolates of *Botrytis* of the *cinerea* type have *Botryotinia fuckeliana* as the perfect form.[71] It is not known if this is true for the *B. cinerea* type or types affecting onions and only the asexual form will be treated in this account. Information on the generic sexual state is presented in reviews of the genus *Botrytis*.[71,90]

The fungus frequently produces a conspicuous grey mycelium on which many branched conidiophores develop.[46] Heads of conidia are produced on the ends of colorless swollen branches. The colourless to pale brown conidia are more ellipsoidal than those of *B. allii* (see chapter on Storage Diseases of Onions) and usually are 8 to 14 μm × 6 to 9 μm in dimension.[46] Sclerotia are larger than those of *B. allii* but smaller than those of *S. sclerotiorum*.

B. cinerea is better adapted as a saprophyte of onion foliage than *B. squamosa*.[66] It releases pectolytic enzymes which produce a slight hypersensitivity reaction, e.g., flecking but without fungal penetration.[76]

Penetration of leaf tissues can occur, with the greatest incidence of infection occurring when winter temperatures in the U.K. are low (about 6°C) and leaf growth had ceased. *B. cinerea* grew and sporulated at 5°C and almost 30% of leaves were invaded within 48 h of inoculation at that temperature.[91] Collar rotting ceased and some infected plants recovered when temperatures rose in the spring and existing leaves resumed growth and new leaves were produced.[91,93]

B. cinerea causing brown stain also infects the necks of maturing bulbs and affects yellow onions (e.g., especially *cv.* 'Elba Globe') more seriously than red and white bulb types.[94]

4. Disease Control

The procedures of rotation, removal of onion debris, avoidance of infection bridging by overlapping onion crops, effective in the control of *B. squamosa,* will also be effective against *B. cinerea*. Fungicide programs devised to control *B. squamosa* in overwintered and summer onions will also control *B. cinerea*.

E. *CLADOSPORIUM ALLII-CEPAE* N.SP. (ONION LEAF BLOTCH)

1. Distribution and Host Range

The pathogen occurs as the anamorphic form *Cladosporium allii-cepae* (syn. *Heterosporium allii cepae* Ran.). Recently, the teleomorph has been described and named *Mycosphaerella allii-cepae* n.sp.[95] The disease has occurred infrequently in the past, mainly in temperate growing areas.[96-101] In recent years, however, there have been more frequent and severe outbreaks of leaf blotch in Ireland[102,103] and, to a lesser extent, in the U.K.[104]

The name "leaf blotch" was given by Moore in 1946 who described the disease on salad onions in the U.K.[100] It has also been called "leaf blight"[101] and "leaf spot".[105]

There are recent accounts of the taxonomy, pathology and host ranges of *C. allii-cepae* (on onion) and *C. allii* (leaf blotch of leek).[106,107] The species are distinct.[106,107] *C. allii-cepae* affects a range of horticultural *Allium* spp. including *A. cepa* (onion), *A. cepa* var. *ascalonicum* (shallot), and *A. fistulosum* (Welsh onion) and the wild spp., *A. altaicum, A. galanthum, A. pskemense,* and *A. vavilovii*. In contrast, *C. allii* infects only leek (*A. ampeloprasum* var. *porrum*) and great-headed garlic (*A. ampeloprasum* var. *holmense*).[107]

2. Disease Symptoms and Infection Sources

Lesions of *C. allii-cepae* develop from small bleached depressed blotches into distinct bleached elliptical eye-spots (up to 1.5 cm × 0.5 cm[106] or 4 cm × 1 cm[108]) running parallel to the leaf veins. Powdery olivaceous conidiophores bearing conidia develop on the centers of the lesions which darken with age as pseudothecia form and become more diffused and blotchy in appearance.[108] Lesions coalesce causing withering and death of leaves. The disease becomes aggressive as the foliage senesces[105] and severe lesioning may result in widespread bleaching of the foliage making crops appear as if they had been sprayed with a contact herbicide, e.g., paraquat. Reduced yields may result.

The random distribution of disease outbreaks is reminiscent of infection initiated by a seed-borne source.[109] However, the causal organism has not been found on commercially produced onion seeds.[108,109] Inoculation of onion inflorescences produced a low incidence of superficially infected seeds but transmission of disease from this source was not proved.[108] It is more likely that the disease progresses from crop to crop where these overlap in time. In Ireland, leaf blotch was most severe in spring sown bulb onion crops planted in areas where there were infected overwintered bulb crops.[109] *C. allii-cepae* persisted for 3 months on surface onion debris; conidia of the fungus in unsterile soil lost viability in 10 d.[108] Similarly pseudothecia of the fungus could not be recovered from unsterile soil after 28 d. The asexual and sexual states of the pathogen present on infected debris on the soil surface may act as infection sources in crop overlap situations, but they appear unlikely to function as direct sources of crop infection unless one crop follows another in very close succession in the same soil.

3. The Pathogen and Disease Biology

Conidiophores of *C. allii-cepae* on leaf lesions are solitary or in groups. They are brown, generally 80 to 160 μm long × 8 to 11 μm wide; conidia are usually borne singly and are cylindrical with rounded ends, 1-septate, pale brown echinulate 65 to 95 μm long × 13 to 17 μm wide.[106] Conidial dimensions may vary on different *Allium* hosts.[107]

Pseudothecia of *M. allii-cepae* n. sp. measure 140 to 310 μm × 130 to 300 μm and contain hyaline 8-spored asci (62.5 to 80 μm × 20 to 30 μm). Ascospores are two celled, slightly curved, with one larger cell and measure 25 to 40 μm × 10 to 20 μm.

In nature pseudothecia are produced readily on mature lesions. In tests under controlled conditions they tend to be sterile, or after a period of development, produce conidiophores which protrude from their ostioles. Onion material inoculated in November and overwintered out-of-doors in a protected situation produced pseudothecial initials in January and mature pseudothecia in March and April.[110] Thus pseudothecia (producing conidia and ascospores) are potential survival structures possibly in situations where debris is sheltered, e.g., as in dumps not subjected to rapid decay.

Conidiophores of the fungus develop in light and dark conditions in relative humidities (RH) of $<95\%$; at RHs $>95\%$, mature conidia develop only after 8 h of darkness. Production of conidia is inhibited by light intensities of 0.3 mw/cm^2. Although in high latitudes, 8 h darkness rarely occurs in summer months (June—August), the light intensities (0.1 mw/cm^2) which occur at night may only prevent optimal spore production.[108] Release of conidia occurs under conditions of declining RH.[111] Profuse sporulation occurred in March and April with plant infection developing in May and June in the U.K. Although sporulation occurred over a wide temperature range (2 to 20°C), spore germination was slow on onion leaves (36% in 24 h at 100% RH) and bait plants became infected only after 17 h exposure to aerial inoculum at 100% RH. Thus continuous periods of rainfall to maintain prolonged periods of high RHs in onion crops may be necessary to induce high levels of onion leaf blotch.[108,112] This may account for the spasmodic nature of the disease and also for its prevalence in Ireland where rainfall is usually frequent in the spring and summer.

4. Disease Control

C. allii-cepae is able to infect a wide range of *Allium* spp. in the *Cepa* and *Phyllodolon* groups of the genus and it seems unlikely therefore that resistance will provide a potential disease control strategy. As infected debris is the most likely method of carry-over of the pathogen, crop rotations of two to three years should be practiced and overlapping of onion crops in space and time should be avoided. Spore trapping tests indicated that *C. allii-cepae* conidia traveled relatively short distances downwind, i.e., 15 m from a diseased crop[107] therefore it is only necessary to locate overwintered crops at a reasonable distance from those planted in the spring. The greatest release of conidia occurs when a crop is harvested and crop-to-crop transference of the pathogen is most likely at that time. Prophylactic fungicides such as chlorothalonil are approved for use against leaf blotch of onion in the U.K. This fungicide and iprodione were most effective in disease control trials in the U.K. and Ireland.[107,108]

F. *PUCCINIA ALLII* (RUST)

1. Disease Distribution and Host Range

Puccinia allii is a widely distributed but sporadic pathogen of temperate areas (CMI map no. 400).[8] *A. cepa, A. fistulosum, A. sativum, A. cepa* var. *ascalonicum, A. schoenoprasum,* and other *Allium* species are listed as host of the fungus.[8] In the U.K., however, (*P. allii*) is rarely seen on onions (*A. cepa*) but occurs frequently on leek.[113] Evidence indicates that U.K. onion cultivars are resistant to *P. allii* from leek.[113,114] *P. allii* on leek in the U.K. is the same sp. as occurs on European leeks, garlic and certain wild *Allium* spp. from the Mediterranean.[113] Rust on chives (*A. shoenoprasum*) in the U.K. is caused by *P. mixta* Fuckc; rust on Babington's leek (*A. babingtonii*) is caused by *Uromyces ambiguus* (DC) Lév.;[113] *A. fistulosum* which is probably an intermediate species can be infected by *P. allii* and *P. mixta.*[113]

2. Disease Symptoms and Infection Sources

Typical orange uredial pustules develop from small white flecks which are indicative of an initial invasion of the leaves. Fewer pustules, which also take longer to develop, occur on onion by comparison with those in leek.[113,114] Where field development of the disease was compared on 38 cultivars of leek and 4 cultivars of onion, all onion cultivars had few rust pustules in contrast to severe infection on most leek cultivars.[114]

It appears that there is no known onion pathotype of *P. allii* and the disease is rare on onions in the U.K. It is presumed (author) that when rust occasionally appears in onion crops, it has arisen as a result of contamination from nearby infected leek crops or debris.

3. The Pathogen and Disease Biology

Puccinia allii is a Basidiomycete of the order Uredinales (Rusts). It is probably an autoecious rust. The uredial and the telial stage have been reported on onion and leek.[4,5,114] Uredospores are globoid to ellipsoid, yellowish 23 to 29 μm × 20 to 24 μm, with a thick echinulate (1 to 2 μm) wall with 5 to 10 pores. Uredospores are contained in scattered uredia which are initially covered by the swollen epidermis which soon ruptures revealing the yellowish to reddish yellow spore mass.[115] Telia are scattered, lead colored with few or many paraphyses. Teleutospores are ellipsoid to obovoid, slightly constricted 28 to 45 μm × 20 to 26 μm with a smooth 1 to 2 μm thick wall and colorless short pedicels.[115]

On leeks, most severe infections occur in crops following prolonged periods of leaf wetness coinciding with daily temperatures of 15 to 20°C.[116] Temperatures above 24°C and below 10°C were inhibitory to infection. Immersion of uredospores in water, as might occur in heavy rainfall, reduced their viability but spores which floated remained viable.[117]

In controlled environments, uredospores needed unbroken RH of 97% for at least 4 h to germinate and infect.[118] At 100% RH, most infections occurred between 10 to 15°C.[119]

4. Disease Control

Rotation of crops, and the separation of onion and leek crops should ensure freedom from disease. There is no spray fungicide available for the control of rust of onions but the disease is eradicated from leek crops by application of fenpropimorph.

G. *PLEOSPORA HERBARUM* (BLACK STALK MOLD, BLACK MOLD, LEAF SPOT)

1. Disease Distribution and Host Range

The anamorphic form of *P. herbarum* (Pers. ex Fr.) Rabenh., *Stemphylium botryosum* causes the disease variously called black stalk mold,[5] black mold,[13] and leaf spot[7] of onion leaves. It occurs on onion crops grown in temperate and subtropical parts of the world. The pathogen is normally a secondary invader of damaged tissues and has a wide plant host range;[120] it is known to attack onion (*A. cepa*)[13] and leek (*A. ampeloprasum* var. *porrum*)[121] and presumably, given the correct environmental conditions would infect other members of the genus *Allium*.

2. Disease Symptoms, Biology and Control

Infection starts mainly at the apex of leaves which turn brown and collapse. As the infection progresses down leaves, they become flattened and covered with a black mold. There is a demarcation, almost a constriction, between healthy and diseased parts of leaves.[13] When seed stalks are attacked, they collapse and the seed umbels are lost. Although infection sometimes follows downy mildew attack, *S. botryosum* can be strongly parasitic,[13] sometimes killing onions in the absence of any other pathogen.[122] The fungus is seed-borne[45] but transmission from this source is not proven. It is more likely that onion debris acts as a source of inoculum.

The fungus is an Ascomycete of the order Pseudosphaeriales. The teleomorph is readily produced in culture, the pseudothecia containing an indefinite number of asci, with each ascus containing eight brown muriform ascospores (31 to 39 μm × 13.5 to 18.0 μm).[123] The conidiophores arise from a submerged stroma and are about 80 μm long × 4 to 7 μm wide with successive swellings where conidia have been borne.[46] Each bears a single olive brown echinulate conidium which is multicellular with 3 to 10 transverse septa and 1 to 10 or more vertical septa.[13] Conidia which measure 27 to 42 μm × 24 to 30 μm are variable in shape being roundish-square to rectangular with larger spores having a central transverse constricting septum.[13] Differential susceptibility to the disease in certain onion cultivars has been reported.[13] The fungus becomes aggressive under cool moist conditions, onion leaves are infected and abundant conidial inoculum is produced. Cultural hygiene is important, thus damage to the field crop should be avoided and prophylactic fungicide programs used to prevent infection by downy mildew and other foliar pathogens. Such fungicides, in controlling *P. destructor* and *B. squamosa,* prevent the development of suitable infection points for *S. botryosum,* they also may be toxic to conidia of that fungus.

H. *STEMPHYLIUM VESICARIUM (STEMPHYLIUM* LEAF BLIGHT)

1. Disease, Distribution, and Host Range

The fungus has been recorded from dead herbaceous stems in Europe, Africa, North, and South America.[46] *S. vesicarium* however caused a disease of onions in Texas in 1976.[124] The fungus produced significant damage alone and in a complex with *Alternaria porri.* In some fields, foliage losses of 80 to 90% were recorded.

2. Disease Symptoms, Biology, and Control

Lesions are nondelineated, light yellow to brown, water-soaked, and progress from the tip to the base of leaves as symptoms develop. Lesions turn light brown to tan and blacken when the conidia develop.[124]

Conidia of *S. vesicarium* have up to six transverse septa and are constricted at the three major transverse septa. There are several vertical septa. Conidia measure 20 to 50 μm × 15 to 26 μm, measurements similar to those given for this species from other sources.[46] The perfect state may be *Pleospora allii* (Rab.) Ces and de Not.[125]

The fungus normally invades dead and dying onion tissue for example, leaf tips, purple blotch lesions, etc., causing minimal foliage damage. When wet conditions persist for more than 24 h, conidial inoculum may reach high enough concentrations to cause minute lesions which later coalesce to form large lesions which may cover the entire leaf.

I. *GLOMERELLA CINGULATA* (TWISTER DISEASE, ANTHRACNOSE, SEVEN CURLS DISEASE)

1. Distribution and Host Range

Onion twister disease, anthracnose, or seven curls disease is reported to be widespread throughout the world but more usual in the tropics and subtropics.[7]

G. cingulata is a facultative parasite with a wide host range.[7] It causes a serious disease of onions in Brazil[126] and in Nigeria[127] where there have been 50 to 100% losses in yield.

2. Disease Symptoms, Biology, and Control

The most characteristic symptom of the disease is a curling, twisting, and chlorosis of the leaves and abnormal elongation of the neck (false stem). White oval sunken lesions also occur on the leaf blades and, in advanced states of infection, roots become sparse, and the plant may die. Bulbs are smaller than usual; some may rot before harvest while others rot in store.[127] *G. cingulata* is the perfect state of the fungus. The symptoms described bear the asexual *Colletotrichum* form which taxonomically is very similar to *C. gloeosporoides* var. *minor* var. nov.[127]

Acervuli do not always develop, but when they do, they appear on lesions on the leaf blades and leaf sheaths.[127] Pinkish masses of conidia are produced which are dispersed by rain and infect under conditions of high humidity and temperatures of 23 to 30°C. The application of fungicides, particularly zineb, will give some control. The pathogen survives on onion debris and the practice of crop hygiene, including the destruction of crop residues, will reduce this source of infection.[7]

J. *CERCOSPORA DUDDIAE* (LEAF SPOT)

1. Disease, Distribution, and Host Range

This *Cercospora* leaf spot is a hot climate disease which has been reported in Brazil, Central America,[5] Brunei, Cuba, Europe, Ghana, Kenya, Malawi, Zimbabwe, Zambia, Hong Kong,[128] and India.[129] The pathogen was first recorded in the Philippines in 1923 and described by Welles.[130] It attacks shallot (*A. cepa* var. *ascalonicum*),[131] onion (*A. cepa*), garlic (*A. sativum*), and other species of *Allium*.[128]

2. Disease Symptoms, Biology, and Control

Leaf spots are circular to elongate, at first chlorotic with brown or red borders turning brown with sporulation.[5,128] The centers of lesions may fall out leaving a hole.[7] Extensive lesioning results in coalescence causing leaf death and defoliation.

C. duddiae conidiophores are produced in bunches on the lesions. Each is brown single stemmed (50 to 170 μm × 5 to 8 μm) and bears a single hyaline 3 to 19 septate whip-like conidium (50 to 190 μm × 4 to 7 μm).[128] Conidia can be either wind blown, water-borne, or transported with onion parts.[5] Conidia survive in crop debris.[7] *C. duddiae* is probably seed-borne.[5]

High temperatures and wet conditions favor infection.[5,7] The disease is of minor importance, but crop hygiene, rotation, and routine sprays with protectant fungicides, such as Bordeaux mixture, maneb or with the systemic fungicide thiabendazole, assist disease reduction.

III. CONCLUSIONS

Foliar pathogens infect and sporulate on the leaves of onions and are therefore ideally located for subsequent dispersal by rain and wind. This capacity for random movement of the infection propagules (spores, conidia, sporangia, uredospores) over long distances confers considerable advantages over the relatively immobile soil-borne pathogens and therefore presents a different problem in respect of disease control strategies.

In the majority of cases a rotation of 1 to 4 years, would, with certain exceptions, for example *Peronospora destructor,* eliminate pathogens from soil. However, the influx of inoculum from other sources, such as neighboring farms, wild spp., and alternative hosts, can present problems. Additionally, successional sowings of crops such as salad (green) onions provide a continuing substrate for the development of the epidemic and attendant increase of inoculum. Similarly, the inclusion of an extra crop (bulb or seed production) to complete an all-year-round cycle of onions ensures the carry over and increase of existing foliar diseases from year to year. Such growing systems probably provide conditions suitable for the expression of rarer diseases, for example, leaf blotch (*Cladosporium allii-cepae*) which otherwise might have remained at a subthreshold incidence in the U.K. Other crop systems, such as onion sets and modular onion transplants, which are planted early in the growing season may provide substrates for the production of inoculum which may infect the main bulb onion crop which is grown from seeds sown at a later date.

In addition to crop rotational practices, other approaches are required including that of disposal of plant debris, where possible, by deep ploughing to increase the rate of degradation of the refuse and the use of artificial or natural heat (the sun), particularly in hot climates, to eliminate bulb-borne fungi. In some cases, plant resistance provides an effective control measure. With most foliar pathogens, however, the use of fungicides remains the dominant strategy for disease control. Recent advanced research, linking disease forecasting systems (identified from studies of the epidemiology of a pathogen), to timed application of fungicides, offers the prospect for a more effective use, and possible reduction in the amounts of chemicals applied. Nevertheless, the genetic plasticity of foliar fungi facilitates an increasing insensitivity to selective fungicides and requires that these chemicals should be used carefully, in conjunction with broad-spectrum compounds, in an effort to preserve and prolong their useful life.

REFERENCES

1. **Presly, A. H. and Maude, R. B.,** Spray programme for *Botrytis, The Grower,* 93(3), 31, 1980.
2. **Ryan, E. W.,** Leaf spot of onions caused by *Cladosporium allii-cepae, Plant Pathol.,* 27, 200, 1978.
3. **Viranyi, F.,** Downy mildew of onions, in *The Downy Mildews,* Spencer, D. M., Ed., Academic Press, London, 1981.
4. **Dixon, G. R.,** *Vegetable Crops Diseases,* Macmillan Publishers, London, 1981.
5. **Sherf, A. F. and Macnab, A. A.,** *Vegetable Diseases and their Control,* John Wiley & Sons, New York, 1986.
6. **Smith, I. M., Archer, S. A., Jimez, J., Lelliot, R., and Phillips, D. H.,** *European Handbook of Plant Diseases,* Blackwell Scientific Publications Ltd., Oxford, 1987.
7. **Anon.,** Pest control in tropical onions, *Trop. Develop. Res. Inst., London,* p. 109, 1986.
8. **Anon.,** Distribution Maps of Plant Diseases, C.M.I., Kew, Surrey, 1969.
9. **Van Doorn, A. M.,** Onderzoekingen over het opreden en de bestrijding van valse meeldauw *(Peronospora destructor* (Berk.) Fries) bij uien, *Tijdschr. Plantenziekten,* 65, 193, 1959.
10. **Hildebrand, P. O. and Sutton, J. C.,** Weather variables in relation to an epidemic of onion downy mildew, *Phytopathology,* 72, 219, 1982.
11. **Berkley, M. J.,** *Botrytis destructor* n.sp. in *Notices of British Fungi, Ann. Mag. Nat. Hist.,* (Ser. 1), 6, 431, 1841.

12. **Berkley, M. J.,** *Gard. Chron.,* 22, 689, 1884.
13. **McKay, R.,** *Onion Diseases,* The Agricultural Institute, Dublin, 1959.
14. **Viranyi, F.,** Studies on the biology of onion downy mildew *(Peronospora destructor* (Berk). Fries) in Hungary, I. Overwintering of the pathogen in onion bulbs, *Acta. Phytopathol. Acad. Sci. Hung.,* 9, 311, 1974.
15. **Rondomanski, W.,** The sources of primary infection of onion downy mildew *(Peronospora destructor* (Berk.) Fries), *Inst. Hortic. Sci., Poland,* 157, 1973.
16. **Glushchenko, V. I.,** The seed transmission of downy mildew of onion, *Zashch. Rast.,* 8, 31, 1980.
17. **Popkova, K. V., Palilov, N. A., and Kir'yanova, E. V.,** The sources of infection of onion by downy mildew, *Miklol. Fitopatol.,* 14, 435, 1980.
18. **Yarwood, C. E.,** Onion downy mildew, *Hilgardia,* 14, 595, 1943.
19. **Leach, C. M., Hildebrand, P. D., and Sutton, J. C.,** Sporangium discharge by *Peronospora destructor:* influence of humidity, red-infrared radiation, and vibration, *Phytopathology,* 72, 1052, 1982.
20. **Leach, C. M.,** Active sporangium discharge by *Peronospora destructor, Phytopathology,* 72, 881, 1982.
21. **Viranyi, F.,** Studies on the biology and ecology of onion downy mildew *(Peronospora destructor* (Berk.) Fries) in Hungary II. Factors influencing sporulation and conidium germination, *Acta Phytopathol. Acad. Sci. Hung.,* 10, 321, 1975.
22. **Viranyi, F.,** Studies on the biology and ecology of onion downy mildew *(Peronospora destructor* (Berk.) Fries) in Hungary, III. Epidemiology of the disease, *Acta Phytopathol. Acad. Sci. Hung.,* 10, 321, 1975.
23. **Walker, J. C.,** *Plant Pathology,* 3 ed., McGraw-Hill Book Company, New York, 1969, 819.
24. **Hildebrand, P. D. and Sutton, J. C.,** Maintenance of *Peronospora destructor* in onion sets, *Can. J. Plant Pathol.,* 2, 239, 1980.
25. **Berry, S. Z.,** Resistance in onion to downy mildew, *Phytopathology,* 49, 486, 1959.
26. **Talieva, M.,** Role of cuticular wax in the resistance of onion to peronosporosis, in *Physiology and Immunity of Cultivated Plants,* Tsitsin, N. V., Ed., Nauka, Moscow, 1976, 36.
27. **Furst, G.,** Anatomical and histochemical characters of resistance in onion to downy mildew, in *Physiology and Immunity of Cultivated Plants,* Tsitsin, N. V., Ed., Nauka, Moscow, 1976, 36.
28. **Khaisin, M. F.,** The resistance of varieties of onion *(Allium cepa* L.) to downy mildew — *Peronospora destructor* (Berk.), *Zushchita ovoshch. kul'tur ot boleznei i sornyakov.,* Kishinev, USSR, 21, 1978.
29. **Vitanov, M.,** Reaction of some onion varieties to *P. destructor,* cause of onion downy mildew, *Gradinar. Lozar. Nauka,* 7, 123, 1970.
30. **Angelov, D., Buchvarov, S., and Vitanov, M.,** Possibilities to breed onion *(Allium cepa)* lines immune to downy mildew, *Hortic. Vitic. Sci.,* 14, 91, 1977.
31. **Vitanov, M. and Angelov, D.,** Possible control of *Peronospora destructor* (Berk.) Casp. by insolating onion bulbs for planting, *Hortic. Vitic. Sci.,* 11, 94, 1977.
32. **Ledyear, A. A., Grishin, V. A., and Machneva, N. P.,** Pre-planting heating of onion mother bulbs, *Zashch. Rast.,* No. 3, 31, 1980.
33. **Newhall, A. G. and Rawlins, W. A.,** Control of onion blast and mildew with carbamates, *Phytopathology,* 42, 212, 1952.
34. **Woolliams, G. E.,** Downy mildew of onions and its control in the British Columbia interior, *Can. J. Plant. Sci.,* 37, 237, 1957.
35. **Smith, R. W., Lorbeer, J. W., and Abd-Elrazik, A. A.,** Reappearance and control of onion downy mildew epidemics in New York, *Plant Dis.,* 69, 703, 1985.
36. **Issa, E., Ramos, R. S., and Maia, J. B. G.,** Control of mildew *(Peronospora destructor* (Berk.) Casp. of onion *(Allium cepa* L.)), *Biologica,* 45, 273, 1979.
37. **Wilson, G. H.,** Downy mildew control in onions, *N.Z. Commer. Grower,* 35, 27, 1980.
38. **Teviotdale, B. L., May, D. M., Harper, D., and Jorde, D.,** New fungicide apparently controls onion downy mildew, *Calif. Agric.,* 34, 22, 1980.
39. **Davidse, L. C.,** Acylalines: resistance in downy mildew, *Phythium and Phytophthora* spp., in *Fungicide Resistance in Crop Protection,* Dekker, J. and Georgopoulos, S. G., Eds., Pudoc, Wageningen, 1982, 118.
40. **Weille, G. A., de,** Approach to the possibilities of forecasting downy mildew infection in onion crops, *Mededelingen en Verhandelingen, Koninklijk Nederlands Meteorologisch Instituut,* 97, 83, 1975.
41. **Skiles, R. L.,** Purple and brown blotch of onions, *Phytopathology,* 43, 409, 1953.
42. **Nolla, J. A. B.,** A new *Alternaria* disease of onions *(Allium cepa* L.), *Phytopathology,* 17, 115, 1927.
43. **Pandotra, V. R.,** Purple blotch disease of onion in Punjab. 1. Its occurrence, pathogenicity and host range, *Indian Acad. Sci. Sect. B., Proc.,* 60, 336, 1964.
44. **Pandotra, V. R.,** Purple blotch disease of onion in Punjab. II. Studies on the life-history, viability and infectivity of the causal organism *Alternaria porri, Indian Acad. Sci. Sect. B, Proc.,* 61, 326, 1965.
45. **Richardson, M. J.,** An annotated list of seed-borne diseases, C.M.I., Kew, Surrey, 1979, 220.
46. **Ellis, M. B.,** Dematiaceous Hyphomycetes, C.M.I., Kew, Surrey, 1971, 608.

47. **Meredith, D. S.,** Spore dispersal in *Alternaria porri* (Ellis) Neerg. on onions in Nebraska, *Ann. Appl. Biol.*, 57, 67, 1966.
48. **Khare, U. K. and Nema, K. G.,** Studies on purple blotch of onion-sporulation on host and dispersal of conidia, *Indian Phytopathol.*, 34, 214, 1981.
49. **Fahim, M. M. and El-Shehedi, A.,** The mode of penetration of *Alternaria porri* into onion leaves, *Trans. Br. Mycol. Soc.*, 49, 79, 1966.
50. **Khare, U. K. and Nema, K. G.,** Factors affecting germination of spores of *Alternaria porri in vitro* and *in vivo, Indian Phytopathol.*, 35, 100, 1982.
51. **Bhangale, G. T. and Joi, M. B.,** Role of thrips in development of purple blotch of onion, *J. Maharashtra Agric. Univ.*, 8, 299, 1983.
52. **Thind, T. S. and Jhooty, J. S.,** Association of thrips with purple blotch infection on onion plants caused by *Alternaria porri, Indian Phytopathol.*, 35, 696, 1982.
53. **Miller, M. E.,** Relationships between onion leaf age and susceptibility to *Alternaria porri, Plant Dis.*, 67, 284, 1983.
54. **Alves, M. L. B., Paiva, W. O., and Assis, L. A. G.,** Incidence of purple spot *(Alternaria porri* Ell. Cif.) on onion *(Allium cepa* L.) cultivars and hybrids in Manaus, Amazonia, *Acta Amazonica*, 12, 673, 1982.
55. **Castellanos Linares, J. J. and Gonzales, A. M.,** Behaviour of different onion lines towards the pathogen *Alternaria porri* (Ell.) Cif., *Informe Cientifico-Tecnico No. 78*, 5 pp. 1978.
56. **Michail, S. H. and Salem, M. A.,** Reaction of onion cultivars to scald disease incited by *Alternaria porri, Acta Phytopathol. Acad. Sci. Hung.*, 16, 299, 1981.
57. **Fokkema, N. J. and Lorbeer, J. W.,** Interactions between *Alternaria porri* and the saprophytic mycoflora on onion leaves, *Phytopathology*, 64, 1128, 1974.
58. **Awad, M. A., El-Shenawy, Z., Omran, A. F., and Shatla, M. N.,** Cultural practices in relation to purple blotch disease of onion, *Sci. Hortic.*, 9, 237, 1978.
59. **Sandu, K. S., Gill, S. S., and Singh, H.,** Effect of cultural practices in purple blotch disease in onion seed crop, *J. Res. Punjab Agric. Univ.*, 19, 118, 1982.
60. **Padule, D. N. and Utikar, P. G.,** Evaluation of fungicides for the control of *Alternaria* blight on onions, *Madras Agric. J.*, 64, 693, 1977.
61. **Ashrafuzzaman, M. H. and Ahmad, M. W.,** Control of foliage diseases of onions by spray fungicides, *Bangladesh Hortic.*, 4, 25, 1976.
62. **Walker, J. C.,** Two undescribed species of *Botrytis* associated with the neck rot diseases of onion bulbs, *Phytopathology*, 15, 708, 1925.
63. **Bergquist, R. R. and Lorbeer, J. W.,** Reaction of *Allium* spp. and *Allium cepa* to *Botryotinia (Botrytis squamosa), Plant Dis. Rep.*, 55, 394, 1971.
64. **Currah, L. and Maude, R. B.,** Laboratory tests for leaf resistance to *Botrytis squamosa* in onions, *Ann. Appl. Biol.*, 105, 277, 1984.
65. **Page, O. T.,** *Botrytis* leaf spot on onions and its control, *Can. J. Agric. Sci.*, 35, 358, 1955.
66. **Hancock, J. G. and Lorbeer, J. W.,** Pathogenesis of *Botrytis cinerea, B. squamosa* and *B. allii* on onion leaves, *Phytopathology*, 53, 669, 1963.
67. **Ellerbrock, L. A. and Lorbeer, J. W.,** Survival of sclerotia and conidia of *Botrytis squamosa, Phytopathology*, 67, 219, 1977.
68. **Ellerbrock, L. A. and Lorbeer, J. W.,** Sources of primary inoculum of *Botrytis squamosa, Phytopathology*, 67, 363, 1977.
69. **Cronshey, J. F. H.,** The perfect stage of *Botrytis squamosa* Walker, *Nature, (London)*, 158, 379, 1946.
70. **Viennot-Bourgin, G.,** Un parasite nouveau de l'oignon en France: *Botrytis squamosa* Walker et sa forme parfaite, *Botryotinia squamosa* sp. nov., *Anales des Epiphytics Phylogenelique*, 4, 23, 1953.
71. **Jarvis, W. R.,** Taxonomy, in *The Biology of Botrytis*, Coley-Smith, J. R., Verhoeff, K., and Jarvis, W. R., Eds., Academic Press, London, 1980, 318.
72. **McLean, D. M.,** The apothecial stage of *Botrytis squamosa* cause of tip and leaf blight of onions, *Plant Dis. Rep.*, 44, 585, 1960.
73. **Hickman, C. J. and Ashworth, D.,** The occurrence of *Botrytis* spp. on onion leaves with special reference to *B. squamosa, Trans. Br. Mycol. Soc.*, 26, 153, 1943.
74. **Sutton, J. C., Swanton, C. J., and Gillespie, T. J.,** Relation of weather variables and host factors to incidence of airborne spores of *Botrytis squamosa, Can. J. Bot.*, 56, 2460, 1978.
75. **Sutton, J. C., James, T. D. W., and Rowell, P. M.,** Relation of weather and host factors to an epidemic of *Botrytis* leaf blight in onions, *Can. J. Plant Pathol.*, 5, 256, 1983.
76. **Clark, C. A. and Lorbeer, J. W.,** Comparative histopathology of *Botrytis squamosa* and *B. cinerea* on onion leaves, *Phytopathology*, 66, 1279, 1976.
77. **Shoemaker, P. B. and Lorbeer, J. W.,** The role of dew and temperature in the epidemiology of *Botrytis* leaf blight of onion, *Phytopathology*, 67, 1267, 1977.

78. **Alderman, S. C. and Lacy, M. L.,** Influence of leaf position and maturity on development of *Botrytis squamosa* on onion leaves, *Phytopathology,* 74, 1461, 1984.
79. **Wukasch, R. T. and Hofstra, G.,** Ozone and *Botrytis* spp. interactions in onion leaf dieback: Field Studies. *J. Am. Soc. Hortic. Sci.,* 102, 543, 1977.
80. **Wukasch, R. T. and Hofstra, G.,** Ozone and *Botrytis* interactions in onion-leaf dieback: open-top chamber studies, *Phytopathology,* 67, 1080, 1977.
81. **Rist, D. L. and Lorbeer, J. W.,** Moderate dosages of ozone enhance infection of onion leaves by *Botrytis cinerea* but not by *B. squamosa, Phytopathology,* 74, 761, 1984.
82. **Rist, D. L. and Lorbeer, J. W.,** Ozone-enhanced leaching of onion leaves in relation to lesion production by *Botrytis cinerea, Phytopathology,* 74, 1217, 1984.
83. **Maude, R. B.,** Disease control, in: *The Biology of Botrytis'* Coley-Smith, J. R., Verhoeff, K. and Jarvis, W. R., Eds., Academic Press, London, UK, 1980, 318.
84. **Shoemaker, P. B. and Lorbeer, J. W.,** Spray volume, interval and fungicide rate for control of *Botrytis* leaf blight of onion, *Plant Dis. Rep.,* 55, 565, 1971.
85. **Shoemaker, P. B. and Lorbeer, J. W.,** Timing initial fungicide application to control *Botrytis* leaf epidemics on onions, *Phytopathology,* 67, 409, 1977.
86. **Presly, A. H. and Maude, R. B.,** Control of *Botrytis cinerea* and *Botrytis squamosa* in overwintered salad onions by fungicide sprays, *Ann. Appl. Biol.,* 94, 197, 1980.
87. **Presly, A. H.,** Retention of iprodione on salad onion leaves in relation to the control of *Botrytis* spp., *Plant Pathol.,* 33, 571, 1984.
88. **Presly, A. H. and Maude, R. B.,** Tolerance in *Botrytis squamosa* to iprodione, *Ann. Appl. Biol.,* 100, 117, 1982.
89. **Sutton, J. C., James, T. D. W., and Rowell, P. M.,** Botcast: A forecasting system to time the initial fungicide spray for managing *Botrytis* leaf blight of onions, *Agriculture, Ecosystems and Environment,* 18, 123, 1986.
90. **Coley-Smith, J. R.,** Introduction, in *The Biology of Botrytis,* Coley-Smith, J. R., Verhoeff, K., and Jarvis, W. R., Eds., Academic Press, London, 1980, 318.
91. **Maude, R. B. and Presly, A. H.,** Studies on the biology of *Botrytis* species on overwintered salad onions, *Ann. Appl. Biol.,* 94, 175, 1980.
92. **Clark, C. A.,** The *Botrytis* Brown Stain Disease of Onion *(Allium cepa* L.) Caused by *Botrytis cinerea* Pers., M.Sc. Thesis, Cornell University, Ithaca, New York, 1973, 63.
93. **Brewster, J. L., Salter, P. J., and Darby, R. J.,** Analysis of the growth and yield of overwintered onions, *J. Hortic. Sci.,* 52, 335, 1977.
94. **Clark, C. A. and Lorbeer, J. W.,** Reaction of onion cultivars of *Botrytis* brown stain, *Plant Dis. Rep.,* 57, 210, 1973.
95. **Jordan, M. M., Maude, R. B., and Burchill, R. T.,** Development of the teleomorph *(Mycosphaerella allii-cepae* sp. nov.) of *Cladosporium allii-cepae* (leaf blotch of onion), *Trans. Br. Mycol. Soc.,* 86, 387, 1986.
96. **Ranojevic, N.,** Zweiter Beitrag zur Pilzflora Serbiens, *Ann. Mycologici.,* 8, 347, 1910.
97. **Micholas, G. and Aggery, B.,** Sur un *Heterosporium* parasite de l'oignon, *Revue de Pathologie Vegetale et d'Entomologie Agricole,* 14, 195, 1927.
98. **Salmon, E. S. and Ware, W. M.,** Department of Mycology, *J. Southeast. Agric. Coll., Wye, Engl.,* 29, 17, 1932.
99. **Moore, W. C.,** Report on fungal, bacterial and other diseases of crops in England and Wales for years 1933-42, *Bulletin of the Ministry of Agriculture, Fisheries and Food,* 126, 50, 1943.
100. **Moore, W. C.,** New and interesting plant diseases, *Trans. Br. Mycol. Soc.,* 29, 90, 1946.
101. **Kaul, T. N.,** Occurrence of *Heterosporium* blight of onions in India, *Curr. Sci.,* 29, 31, 1960.
102. **Anon.,** Disease threat to onion crops in Ireland, *The Grower,* 90, 621, 1978.
103. **Ryan, E. W.,** New disease threatens onion production, *Farm Food Res.,* 9, 139, 1978.
104. **Maude, R. B. and Presly, A. H.,** Onion leaf blotch, *Rep. Natl. Veg. Res. Stn. 1981,* 64, 1982.
105. **Ryan, E. W., Doyle, J. A., and O'Donnell, T. B.,** Control of *Cladosporium* leaf spot in spring onions and overwintered bulb onions, *Proceedings 1981 British Crop Protection Conference: Pests and Diseases,* 349, 1981.
106. **Kirk, P. M. and Crompton, J. G.,** Pathology and taxonomy of *Cladosporium* leaf blotch of onion *(Allium cepa)* and leek *(A. porrum), Plant Pathol.,* 33, 317, 1984.
107. **Jordan, M. M., Maude, R. B., and Burchill, R. T.,** Investigation of the host ranges of the leaf blotch pathogens of onion *(Cladosporium allii cepae* and leek *(C. allii), Plant Pathol.,* 36, 394, 1987.
108. **Jordan, M. M.,** Investigation of the Biology, Epidemiology and Control of Leaf Blotch Diseases of *Allium* Species, Ph.D. thesis, University of Birmingham, UK, 1985, 233.
109. **Ryan, E. W.,** Leaf spot of onions caused by *Cladosporium allii-cepae, Plant Pathol.,* 27, 200, 1978.
110. **Maude, R. B. and Peach L.,** Leaf blotch of *Allium* spp., *Rep. Natl. Veg. Res. Stn. 1986/87,* 62, 1987.

111. **Burchill, R. T. and Jennens, A. P.,** Onion leaf blotch epidemiology, *Rep. Natl. Veg. Res. Stn. 1981,* 64, 1982.
112. **Jordan, M. M., Maude, R. B., and Burchill, R. T.,** Leaf blotch of *Allium* spp., *Rep. Natl. Veg. Res. Stn. 1985,* 66, 1986.
113. **Jennings, D. M.,** Taxonomy and Pathogenicity of Rusts from *Allium* species, Ph.D. thesis, Birmingham University, UK, 1987.
114. **Norwood, J. M.,** Leek rust — identification of race specific resistance, *Rep. Natl. Veg. Res. Stn. 1984,* 72, 1985.
115. **Wilson, M. and Henderson, D. M.,** *British Rust Fungi,* Cambridge University Press, 1966.
116. **Burchill, R. T. and Everett, B. A.,** Leek rust — forecasting disease outbreaks, *Rep. Natl. Veg. Res. Stn. 1983,* 69, 1984.
117. **Burchill, R. T. and Everett, B. A.,** Leek rust — epidemiology, *Rep. Natl. Veg. Res. Stn. 1984,* 71, 1985.
118. **Burchill, R. T., Norwood, J. M., and Everett, B. A.,** Epidemiology of leek rust, *Rep. Natl. Veg. Res. Stn. 1985,* 66, 1986.
119. **Burchill, R. T. and Everett, B. A.,** Epidemiology of leek rust, *Rep. Natl. Veg. Res. Stn. 1986/87,* 62, 1987.
120. **Moore, W. C.,** British Parasitic Fungi, Cambridge University Press, UK, 1959, 430.
121. **Bakkar, M.,** A leaf spot disease of leek caused by *Pleospora herbarum* (Pers.) Rabenh., *Tijdschr. Plantenziekten,* 59, 25, 1953.
122. **McKay, R. and Loughnane, J. B.,** Plant diseases in Ireland in 1956 and 1957, *J. Dep. Agric. Eire,* 54, 181, 1959.
123. **Brooks, F. T.,** Plant Diseases, Oxford University Press, 1953, 457.
124. **Miller, M. E., Taber, R. A., and Amador, J. M.,** *Stemphylium* blight of onion in South Texas, *Plant Dis. Resp.,* 62, 851, 1978.
125. **Simmons, E. G.,** Perfect states of *Stemphylium, Mycologia,* 61, 1, 1969.
126. **Aquino, M. L. N. and Da Wanderley, L. J.,** O ''mal das sete voltas'' nos cebolais de Sao Francisco, *Inst. Agron. Pernamb. Bol. Tech.,* 16, 42, 1966.
127. **Ebenebe, A. C.,** Onion twister disease caused by *Glomerella cingulata* in North Nigeria, *Plant Dis.,* 64, 1030, 1980.
128. **Ellis, M. B.,** More Dematiaceous Hyphomycetes, C.M.I., Kew, Surrey, 1976, 507.
129. **Raju, K. S. and Raj, R. B.,** A note on some foliar diseases of onion from Andhra Pradesh, Haryana, *J. Hortic. Sci.,* 8, 95, 1979.
130. **Welles, C. B.,** A new leaf spot of onion and garlic, *Phytopathology,* 13, 362, 1923.
131. **Deighton, F. C.,** Mycological work, *Ann. Rep. Agric. Dep. Sierra Leone* for 1931, 20, 1932.

Chapter 9

VIRUS DISEASES

David G. A. Walkey

TABLE OF CONTENTS

I. INTRODUCTION

Of the viruses that infect *Allium* crops, those of onion and garlic are economically the most important and widespread on a world scale. Infection in other *Allium* species, such as leeks and shallots while internationally unimportant, may be of economic significance locally. In general, it is the viruses of the vegetatively propagated *Allium* species, such as garlic and shallot, that are the most difficult to control and worldwide reports indicate, that all garlic clones tested were originally infected with one or more viruses.[1-4] The various viruses affecting *Allium* spp. have been reviewed by Bos.[2] Those so far reported as infecting economically important and other *Allium* species are listed in Tables 1 and 2.

The most important *Allium* viruses, onion yellow dwarf (OYDV), leek yellow stripe (LYSV), and shallot latent (SLV), are not seed transmitted. Therefore, seed sown crops of onion and leek are initially free of virus infection when established in the field and subsequent infection must result from transmission of the virus from outside sources. Unfortunately, garlic cannot be grown from seed, so infected clones remain infected unless meristem-tip culture and thermotherapy are used to eradicate the virus and produce healthy lines.[3-6] The vegetatively propagated shallot may also be freed of virus by meristem-tip culture[4,7] and this method could probably be used to eradicate viruses from other similarly propagated *Allium* species. The Welsh onion (*A. fistulosum*) may be regenerated from meristem-tips by tissue culture[8] but although it is reported to be a host of OYDV,[9] there are no reports of tip-culture being used to produce virus-free clones.

Virus diseases of *Allium* have been known for a long time and as early as 1916 a mosaic disease of onion was reported from West Virginia in the U.S.[10] and Brierley and Smith[11] also working in the U.S., obtained virus isolates from onion, garlic, and shallot that had similar properties. Progress on the diagnosis of viruses isolated from *Allium* species was slow however, due to difficulties in handling the viruses and their host plants.

The most important *Allium* viruses have a narrow host range and are mainly confined to *Allium* species. This makes their isolation, propagation and subculture difficult, furthermore the mucilaginous sap present in *Allium* species is a further hinderance in virus purification. Diagnosis of the main *Allium* viruses, OYDV, LYSV, and SLV was also complicated by their having somewhat similar basic properties. In addition to their restricted host range, the three viruses have flexuous rod-shaped particles which look similar in the electron microscope, and are transmitted by aphids in a nonpersistent manner, consequently, they were difficult to separate and identify. It was not until the last decade that advances in virological techniques were sufficient to permit separation and critical identification of these viruses. Major advances were then made by Bos and his colleagues in the Netherlands when they recognized that LYSV[12] and SLV[13] were distinct from OYDV. The three viruses were purified[12,14] and antisera prepared. Subsequently, the introduction of the advanced serological techniques of enzyme-linked immunosorbent assay (ELISA)[15] and immunosorbent electron microscopy[16] (ISEM) has enabled such antisera to be extensively used for identification purposes.

TABLE 1
Viruses Isolated from Onion, Leek, Shallot and Garlic

Allium species	Virus	Group type	Ref.
A. cepa (onion)	Shallot latent	Carlavirus	13
	Onion yellow dwarf	Potyvirus	2
	Onion mosaic	Unknown	2, 20
	Tobacco mosaic	Tobamovirus	96
	Tobacco rattle	Tobravirus	111, 112
	Tomato black ring	Nepovirus	17
	Tomato spotted wilt	Tomato spotted wilt	113, 114
	Aster yellows[a]	Mycoplasma	104
A. ampeloprasum var. *porrum* (leek)	Leek yellow stripe	Potyvirus	12
	Shallot latent	Carlavirus	13, 47
	Tomato black ring	Nepovirus	17
	Turnip mosaic	Potyvirus	63
A. cepa var. *ascalonicum* (shallot)	Shallot latent	Carlavirus	13, 65, 66
	Onion yellow dwarf	Potyvirus	4, 66
A. sativum (garlic)	Aster yellows[a]	Mycoplasma	115, 116
	Onion yellow dwarf	Potyvirus	1, 4
	Leek yellow stripe	Potyvirus	1, 4
	Shallot latent	Carlavirus	1, 4
	Cucumber mosaic	Cucumovirus	86
	Tobacco mosaic	Tobamovirus	96
	Tobacco rattle	Tobravirus	95

[a] A mycoplasma disease that causes virus-like symptoms

TABLE 2
Viruses Isolated from Other Minor *Allium* Types

Allium species	Virus	Ref.
A. cepa var. *solanina* (multiplier or potato onion)	Onion yellow dwarf	11
A. cepa var. *viviparum*	Arabis mosaic (nepovirus)	95
(syn. *Allium* x *proliferum*)	Onion yellow dwarf	11
(tree onion)	Tomato black ring	95
A. flavum	Tomato black ring	95
A. fistulosum (Welsh onion)	Onion yellow dwarf	9
A. moly	Tobacco rattle	117
A. roseum	Turnip mosaic	63
A. schoenoprasum (chive)	Tomato black ring	95
A. scorodoprasum (serpent garlic)	Shallot latent	66
A. ursinum	Tobacco rattle	2

Unfortunately, the failure to diagnose correctly *Allium* viruses prior to these developments, has left a legacy of considerable confusion in the earlier literature as to the causal agents of the various *Allium* diseases. This is particularly true of garlic which may be infected with various combinations of different rod-shaped poty- and carlaviruses. Some of the individual viruses in these garlic complexes have been identified as OYDV, LYSV, or SLV,[4] but others remain to be identified. Diseases of leek and shallot may also result from infection by more than one virus. In some complexes, an individual virus may be entirely responsible for the disease symptoms and the additional virus or viruses present may be latent in respect of disease expression (see Section II.D.2). In other complexes, however, the combined, synergistic effect of two or more viruses may increase the severity of the disease symptoms compared with infection by one virus alone (see Section II.B.2.a).

Besides the mainly *Allium* specific viruses so far identified (e.g., OYDV, LYSV, and SLV), other viruses which cause diseases in other vegetable crops may also infect *Allium* species (Tables 1 and 2). These viruses are generally of little or no economic importance in the *Allium* crop concerned, but occasional outbreaks may be of local importance in a limited area. The nematode transmitted tomato blackring virus for instance, caused problems in onions and leeks in Northern Ireland in the early 1960s[17] but has not been reported subsequently.

In addition to virus diseases, mycoplasma may cause virus-like symptoms in some *Allium* species. The various reports of mycoplasma diseases are discussed in Section IV.

II. VIRUSES OF ECONOMICALLY IMPORTANT ALLIUM CROPS

A. ONION *(ALLIUM CEPA L.)*

1. Introduction

Of the various viruses that have been isolated from onion, OYDV is the most important and widespread. The name OYDV has led to some confusion in the past, for although stunting and yellow streaking of the leaves are essential features of the disease symptoms, some workers have referred to the yellow streaking as a mosaic and called the disease onion mosaic.[18,19] Other workers in Russia have, however, described a mosaic disease of onions which now appears to be caused by a separate, mite-transmitted virus.[20,22] The relationships of these two viruses are discussed later in Section II.A.3.

2. Onion Yellow Dwarf Virus (OYDV)

a. Disease Symptoms

Leaves of infected onion plants lose their cylindrical shape and become flattened and crinkled when infected with OYDV. The leaves then bend over near the base causing the foliage to lie on the ground, so that the plant looks stunted (Figure 1). The leaves also develop yellowish striping giving the appearance of a parallel mosaic pattern and some leaves may turn almost entirely yellow.[23-25] The bulbs of infected plants are usually reduced in size and crop yield reductions of up to 63% have been reported.[26] The storage life of infected bulbs is also reduced resulting in premature sprouting.[27,28] Infection can also affect plants grown for seed production, causing smaller flowers and reduced seed yield and quality.[29] The germination of seed collected from infected plants has been reported to be reduced from 91 to 52%.[26]

b. Occurrence and Economic Importance of OYDV

Since the disease was first recorded in the U.S. in 1916,[10] it has been reported from most countries where onions and shallots are grown (all References 30, unless otherwise stated) including Argentina, Australia, Austria, Brazil,[9] Canada, Czechoslovakia, Denmark, England, Egypt,[26] Finland, France, Germany, Hungary, New Zealand, Poland, Portugal, Morocco,[29] Scotland, Soviet Union, Switzerland, and Yugoslavia.[31] Disease incidence in individual outbreaks can be very high, with up to 95% of the plants infected in some crops in the U.S.[23] Similar high levels have been reported from Germany[27] and in the 1938—1939 season in New Zealand up to 50% of commercial bulb crops and 45% seed crops were affected.[32]

Following the introduction of successful control measures (see Section II.A.2.d) and the progressive replacement of sets with directly drilled seed, the incidence of the virus quickly fell until today it is no longer of economic importance in the major onion growing areas of North America and Western Europe. In England, despite the introduction of autumn sown onions whose maturation period overlaps with the spring sown crop, the disease is

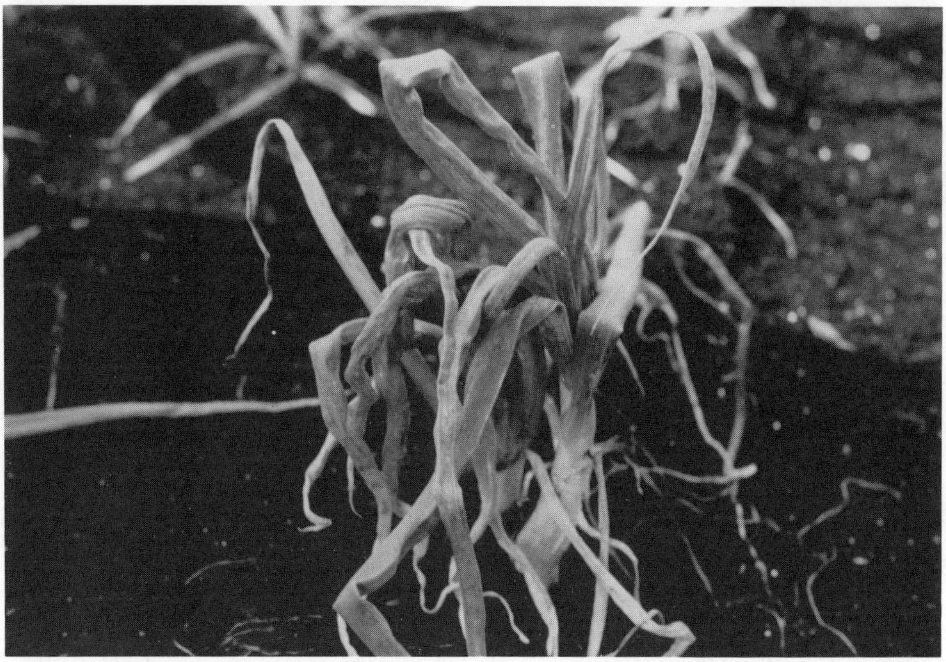

FIGURE 1. Onion infected with onion yellow dwarf virus showing chlorotic striping, flattening and crinkling of leaves (× 1/4) (Photo courtesy T.A. Zitter).

rarely seen. The isolated severe occurrence of the disease in countries such as Morocco,[29] however, emphasizes the importance of maintaining rigid control measures to minimize the risk of further outbreaks.

c. *Properties of OYDV*

OYDV is a typical member of subgroup 1 of the potyvirus group. It is sap-transmitted by mechanical inoculation and transmitted by aphids in a nonpersistent manner. Aphids do not normally colonize onions, but during migration, they may briefly land on a plant for feeding probes.[33,34] Although there are reports of OYDV being seed-transmitted,[35,36] these observations were based on field observations and were not substantiated in critical experiments carried out by other workers.[23,24,28] The virus may also be mechanically transmitted on contaminated knives when the leaf bases of the bulbs are cut, a process that is used in breeding programs to select single-centered bulbs before replanting.[37]

OYDV has a flexuous, rod-shaped particle reported to be between 720 nm[38] and 833 nm[12] in length (Figure 2A). It has been purified from infected onion leaf material using the method developed for leek yellow stripe virus.[14] Like other members of this potyvirus subgroup, it produces pinwheel and scroll-shaped inclusion bodies that may be seen in ultrathin sections in the electron microscope.[39]

This virus is very restricted in its host range. It can infect onion and shallot and has also been isolated from tree onion (*A. cepa* var. *viviparum* syn. *Allium* × *proliferum*) and multiplier onion (*A. cepa* var. *solanina*).[11] It may also infect garlic (*A. sativum*)[1,4,11] and the Welsh onion (*A. fistulosum*).[9] Although OYDV has been reported in *Narcissus* species[11,24] there is doubt as to the validity of the evidence and to whether these reports relate to typical OYDV strains.[2,40] Some isolates of OYDV are reported to produce local lesion symptoms both on *Chenopodium amaranticolor* and *C. quinoa,* but others do not.[41] Similarly, typical OYDV isolates, although they infect, do not cause symptoms in leek (*A. ampeloprasum*

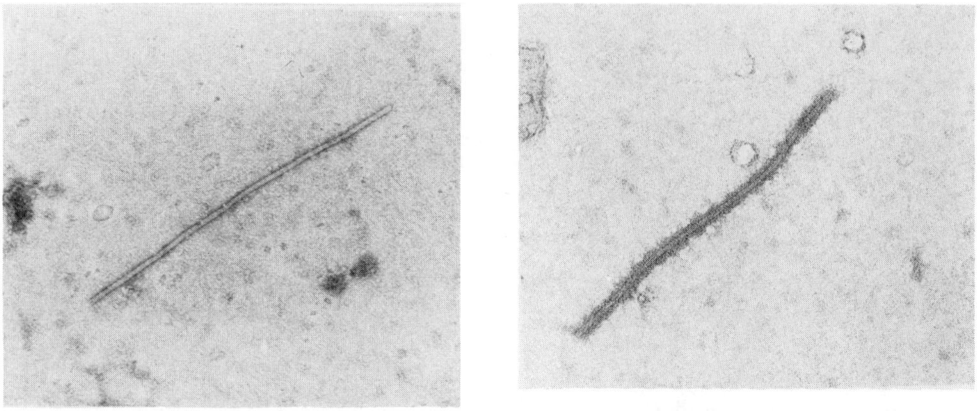

A B

FIGURE 2. (A) Particle of onion yellow dwarf virus approximately 760 nm in length (×96,400) (Electron micrograph courtesy M.J.W. Webb). (B) Particle of onion yellow dwarf virus decorated by antibodies to OYDV (×112,000) (Electron micrograph courtesy M.J.W. Webb).

var. *porrum*).[12] Early reports of OYDV infection of leek[27,42] probably resulted from a failure to identify correctly leek yellow stripe virus, which induces symptoms in leek similar to those produced by OYDV in onion (see Section II.B.2.a).

Strains of OYDV are known to exist, but there are no reports of critical studies on strain identification and properties. Brierley and Smith[11] working with OYDV isolates from onion, shallot, and garlic, reported that the shallot isolate was more virulent than the onion isolate when inoculated to onion, but that the garlic isolate infected onion only with difficulty and induced mild symptoms. Other properties of OYDV have been described by Bos.[41]

d. Control Measures

As OYDV is not seed transmitted and can only infect onions and closely related *Allium* species, preventative measures can be relatively easily taken to prevent outbreaks of the disease in commercial crops. These preventative measures are listed below.

1. A period free of onion and shallot crop culture may be used to prevent overlapping of diseased crops with new plantings.
2. Strict virus indexing and roguing of onion mother bulbs used for seed production, and their cultivation in areas isolated from commercial plantings of onion and shallot crops.
3. The removal of all volunteer onion plants remaining from previous onion crops.
4. Indexing of onion sets for virus-freedom and their production in areas isolated from infected commercial bulb crops.

The adoption of these measures quickly reduced the incidence of OYDV in Iowa from 40% in 1928 to a negligible level by 1933[24] and similar measures taken in New Zealand were also successful.[32] In recent years, the widespread use of direct seed drilling in the U.S. and onion growing areas of Western Europe in preference to onion sets, has resulted in a very low incidence of OYDV in these areas. In northern and northeastern Europe, however, significant amounts of onions are still grown from sets, and growers there must be particularly reliant on the provision and maintenance of healthy planting material.

The importance of excluding OYDV from planting sets is also relevant in respect of current onion production in countries such as the U.K., where in recent years, there has been a trend back to growing the commercial crop from sets (up to 15% of the crop in

1987). The use of sets combined with year-round cropping of winter and spring-sown onions, could be a potential disease threat should OYDV establish itself in the sets.

Insecticidal sprays to control the aphid vectors of OYDV are of little use in preventing virus spread for as with most viruses that are transmitted in a nonpersistent manner, the spray cannot kill the aphid before it probe-feeds and transmits the virus. Recently, certain pyrethroid chemicals have been effective against non-persistently transmitted viruses such as potato virus Y,[43] but it is not yet known if they are effective against OYDV. Early studies indicated that some onion cultivars were resistant to infection by some strains of OYDV,[11,24] but so far resistance has not been widely adopted as a control measure.

3. Other Viruses Infecting Onion

An onion disease called onion mosaic has been widely reported from Russia[2] and the relationship between this disease and OYDV is confused. Probably some of the reports of onion mosaic in Russia are referring to a disease caused by OYDV,[18,19] for the particle length of the virus observed 722 nm, was similar to that for OYDV. Other Russian workers have, however, described the causal virus of onion mosaic as being transmitted by an eriophid mite, *Aceria tulipae,* rather than by an aphid, and the virus particle being shorter, around 675 nm in length.[20-22,44] This suggests that the mosaic diseases observed in onions in Russia may be caused by two different viruses with the latter virus possibly related to the mite-transmitted subgroup of potyviruses, which infect monocotyledonous hosts of the family Gramineae. Members of the subgroup have particles which are generally shorter than aphid transmitted potyviruses (about 700 to 720 nm compared with about 750 to 850 nm). *Aceria tulipae* has also been reported to be the vector of a mosaic disease of garlic in India.[45] This disease is associated with tangle-top malady in garlic, which is caused by mite feeding damage.

There are other vegetable viruses that may occasionally infect onion. None of these viruses (see Table 1) are widespread in their occurrence in onion or of economic importance. Tomato black ring virus (TBRV) did, however, cause an isolated, but serious local disease in spring onions in Northern Ireland in 1961,[17] resulting in yellow blotching, striping, distortion, and stunting of the leaves. TBRV is transmitted by the eelworm *Longidorus elongatus* and can be avoided by cropping land not infested by viruliferous eelworms.

B. LEEK (*ALLIUM AMPELOPRASUM* VAR. *PORRUM*)

1. Introduction

Leek yellow stripe virus (LYSV) is the most important virus affecting leeks. For many years, the chlorotic streak disease observed in leeks was thought to be caused by OYDV, but critical diagnostic studies by Bos and his colleagues in the Netherlands in the late 1970s showed that the virus was distinct and hence they named it LYSV.[12] They found that although LYSV could infect onion, it was not pathogenic in onion (*A. cepa*) and that typical OYDV strains do not produce symptoms in leek and only infect leek with difficulty. Recently LYSV has been found to infect some garlic cultivars.[4]

2. Leek Yellow Stripe Virus (LYSV)

a. Disease Symptoms

LYSV causes irregular chlorotic stripe symptoms (Figure 3A) on the leaf blade from its base upward. In some plants the entire leaf may become chlorotic and flaccid[12,46,47] (Figure 3B); disease symptoms may be more severe if the plants are jointly infected with LYSV and shallot latent virus.[47] LYSV may cause yield reductions of up to 54%[28,48] and diseased plants may fail to withstand winter frost and have a reduced shelf-life.[28,46,47]

b. Occurrence and Economic Importance of the Disease

Symptoms of yellow stripe in leeks were first reported from Germany in 1937,[27] and

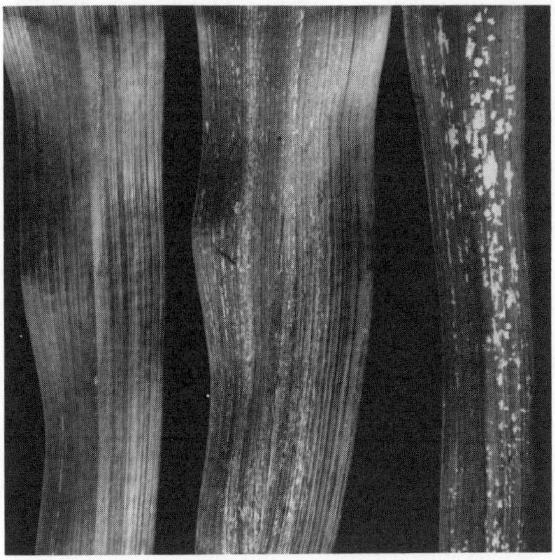

A

B

FIGURE 3. Leaves of leek *cv.* ''Odin'' showing different types of irregular chlorotic stripe symptoms caused by leek yellow stripe virus (×1/4) (Photo courtesy N. Paludan). (B) A leek plant *cv.* ''Sisu'' infected with leek yellow stripe virus showing flaccid and chlorotic leaves (×1/3) (Photo courtesy N. Paludan).

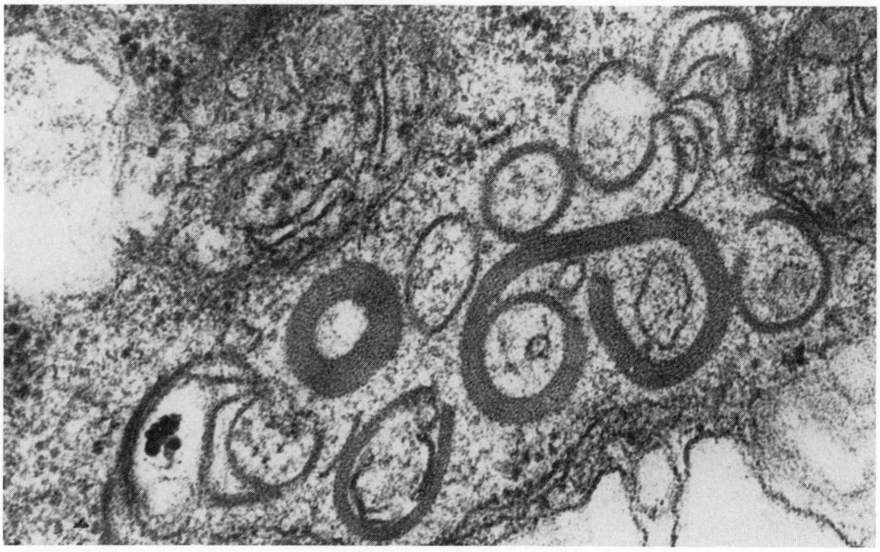

FIGURE 4. Ultrathin section of leek leaf infected with leek yellow stripe virus, showing pinwheel inclusion body structures typical of viruses belonging to the potyvirus group (ca. ×40,000) (Photo courtesy N. Paludan).

later from Belgium,[49] Czechoslovakia,[50] Denmark,[51] East Germany,[52] France,[53] Germany,[42,54,55] and Italy.[56] In all reports the disease was thought to be caused by OYDV. In the 1970s severe outbreaks of the disease occurred in the Netherlands, Belgium, Denmark, and neighboring parts of Germany which resulted in more intensive research in the Netherlands and the discovery that LYSV was the causal virus.[12]

In outbreaks in the southeast of the Netherlands many autumn and winter crops were totally infected. The incidence of the virus was also high in Belgium, with a rise in incidence from 40% in 1971 to 64% in 1972.[57] In Denmark the virus was present in 20% of the crops in 1976, 12% in 1977, and 43% in 1978,[47] and in New Zealand 100% of the crop was infected on one holding.[46] LYSV has also been reported to be widespread in vegetatively propagated "pot-leeks" grown mainly for exhibition purposes in the U.K.[58]

c. Properties of LYSV

LYSV belongs to subgroup 1 of the potyvirus group. It is transmitted by aphids in a nonpersistent manner and has been experimentally transmitted by *Myzus persicae*[12] and *Aphis fabae*.[57] The virus is also sap transmitted by mechanical inoculation but is not seed transmitted. The particle is a flexuous rod about 820 nm in length,[12] although an isolate from New Zealand is reported to have a particle length of about 700 nm.[46] Electron microscopic examination of ultrathin sections of diseased leek leaves showed tubular, lamella, and pinwheel structures characteristic of those formed by viruses of the potyvirus group[47,59] (Figure 4).

LYSV has a very restricted host range among *Allium* species. It infects leek, and although it will infect onion and shallot the latter two species have a high degree of resistance.[12] Recently, LYSV has been shown to occur naturally in numerous garlic cultivars,[4] but the infectivity of garlic isolates of LYSV on leek is not known. *Allium fistulosum* is reported to be immune to LYSV infection, whereas it is infected by OYDV.[12,60] The virus causes only local lesions on the virus assay plants *Chenopodium amaranticolor, C. quinoa, C. album,*[12,60] and *C. murale*.[46] It should be noted that unlike LYSV, OYDV does not usually infect *C. amaranticolor* and that the local lesions induced by shallot latent virus on *C.*

amaranticolor are smaller, more necrotic and develop earlier than those caused by LYSV. The virus may be maintained and propagated for purification in leek plants.

In ISEM serology decoration tests, antibodies to LYSV did not decorate OYDV particles and vice versa, but both viruses were decorated by their homologous antibodies.[4]

d. Control Measures

Studies on the epidemiology of LYSV have shown infected, overwintering leeks to be the main, if not the only source of infection for the following season's crops.[12] The recent discovery that LYSV occurs naturally in some garlic cultivars[4] however, suggests that garlic crops growing near leeks may also be a source of LYSV infection.

The major outbreaks of LYSV disease in the low-countries of Western Europe in the early to mid 1970s, coincided with an increase in the all-year-round cultivation of commercial leek crops in some areas. A break in these overlapping cycles of culture to provide a leek-free period, would almost certainly prevent further serious outbreaks of the disease. In addition, leek seed beds should not be located near overwintering crops, or close to private gardens that contain leeks or garlic. Finally, leek crops should be "rogued" in spring and summer to remove any diseased plants that might provide foci of infection for subsequent spread by aphid transmission.[12] As LYSV is transmitted by aphids in a nonpersistent manner, the virus in common with OYDV is unlikely to be controlled by insecticidal sprays.

All leek cultivars tested are reported to be susceptible to LYSV, although they vary in sensitivity.[12,47,61] Of 20 cultivars tested, Paludan[47] found the *cvs.* "Autumn Giant Triumphatur", "Ligina", "Siegfriend Frost", and "Winta" to be the most resistant. In addition, immunity to LYSV has been found in Kurrat (*A. ampeloprasum*) which is cross-fertile with leek.[62] This resistance could be used in a breeding progam to produce leek cultivars with increased resistance to LYSV.

In the U.K., "pot-leeks" which are vegetatively propagated and grown mainly for exhibition purposes, are known to be widely infected with LYSV.[58] Viruses may be eradicated from "pot-leeks" by meristem-tip tissue culture and healthy plants regenerated.

3. Other Leek Viruses

Shallot latent virus (SLV) (see Section II.C) has been isolated from leek crops in Denmark.[47] Leeks infected with only SLV developed mild chlorotic streak symptoms on the leaves, but when jointly infected with SLV and LYSV the plants developed severe chlorosis and white streaking, followed by death in some cultivars.

The nematode transmitted tomato black ring virus, has been reported to cause mild chlorotic leaf symptoms and stunting in an isolated outbreak in leeks in Northern Ireland,[17] but since this is the only known occurrence of the virus in leeks, TBRV is not considered to be of economic importance. In addition, turnip mosaic virus, an aphid-transmitted potyvirus, has been isolated from leeks in Yugoslavia[63] but again is economically unimportant.

C. SHALLOTS (*ALLIUM CEPA* VAR. *ASCALONICUM*)

1. Introduction

Although some cultivars of shallot flower and produce seed, most growers wish to maintain vegetatively propagated selected clones. Consequently, as with vegetatively propagated garlic, most shallots are universally infected with virus[13] unless virus-free clones have been produced by meristem-tip culture.[4,64] Shallots are frequently infected with a complex of two or more viruses. Shallot latent virus (SLV), belonging to the carlavirus group, was isolated and described by Bos et al.,[13] and appears to be universally present in most commercial cultivars[4,13,65-67] but causes no, or only very mild symptoms.[65] Onion yellow dwarf virus, a potyvirus, also infects shallots and is believed to be responsible for the main disease symptoms[65] (see Section II.C.3). Recently, in addition to SLV and OYDV, a second and so far unidentified potyvirus has been observed in the important Dutch commercial cultivar "Noordhollandse Strogele".[4]

2. Shallot Latent Virus (SLV)

a. Disease Symptoms

SLV generally causes no symptoms in shallots[13,66-67] but there is a report from France of a carlavirus, which may have been SLV, causing very mild symptoms in shallot.[65] If, however, OYDV is present as well as SLV, then severe symptoms similar to those caused by OYDV in onions are induced.[65]

b. Occurrence and Economic Importance

The virus is reported to be widespread if not universal in Dutch, Belgian and French cultivars of shallot[13,65-66] and has also been reported in the U.K.[4,68] The economic importance of SLV in shallot is not known, as trials to compare the bulb yield of infected and healthy crops have not been made. It is possible that the virus may increase disease symptoms when shallot bulbs are jointly infected with SLV and OYDV, as is the case with SLV and LYSV infected leeks.[47]

c. Properties of the Virus

Shallot latent virus is a typical member of the carlavirus group. It is sap transmitted by mechanical inoculation and transmitted by the aphid *Myzus ascalonicus* in a nonpersistent manner. It is not seed transmitted.[13] The virus particle is not as flexuous as that of a potyvirus, and is approximately 640[67] to 650 nm[13] in length. There are no inclusion bodies in the host associated with SLV infection.[13]

In common with OYDV and LYSV, SLV has a limited host range, although not as narrow as that of LYSV. It infects shallot, onion, and leek naturally[13,47] and has been isolated from field-infected garlic[4] and *Allium scorodoprasum* (the serpent garlic).[66] SLV has been experimentally transmitted to *A. fistulosum* and *A. jailae* in which it causes systemic infection, and to *Chenopodium amaranticolor, C. album,* and *C. quinoa* in which it only induces local lesions.[13]

The virus may be maintained and multiplied for purification in leek *cvs.* "Goliath" and "Winterreuzen". SLV may be separated from joint infection with LYSV by passage through *A. fistulosum* which is not a host of LYSV[13] and from OYDV by passage through leek.[66]

d. Control Measures

Shallot viruses may be controlled by the use of virus-free planting material. SLV, OYDV, and the other unidentified potyvirus may be eradicated from shallot by meristem-tip culture.[4] The details of the procedure are given in Section II.D.3. In these studies with the Dutch *cv.* "Noordhollandse Strogele", 25% of the plantlets regenerated from meristem-tips were virus-free. Heat treatment of the infected parent shallot bulbs at 33°C did not increase the percentage of virus-free plants obtained, and the survival of meristem-tips taken from parent bulbs heat-treated at 36°C was very low.

A virus-free certification scheme is required for the multiplication and distribution of healthy shallot bulbs produced by tip culture. The subsequent cultivation of commercial crops planted with healthy bulbs should be at sites isolated from other *Allium* crops that might be potential sources of SLV or OYDV.

3. Onion Yellow Dwarf Virus

As mentioned in the introduction to this section, OYDV may cause severe symptoms in shallot, including chlorotic stripes on the leaves, leaf curling, and stunting of the plant[11,65,69] and bulb yield may be reduced by 20 to 30%.[64] The first report of OYDV causing a disease in shallots was from Louisiana;[11] later the virus was found in shallots in France,[65,70] Belgium,[66] and the U.K.[4] A virus was also isolated by Hollings in the U.K.[68] from shallots showing yellowing symptoms; electron microscope examination showed virus particles about 620 to 640 nm in length. It seems likely in retrospect that these plants were infected with a mixture of OYDV and SLV. The particles observed were the correct length for SLV and the symptoms would have been caused by the OYDV.

D. GARLIC (*ALLIUM SATIVUM*)

1. Introduction

Garlic is grown throughout the world and is vegetatively propagated, although there is a recent report of fertile clones collected in Soviet Central Asia.[71] The widespread vegetative propagation of garlic over many centuries has resulted in it being infected with various virus mixtures. The identification of these viruses is complex and confusing with many reports of incompletely diagnosed viruses relating to the virus content of different local garlic cultivars from many countries.[1-4,65]

2. Virus Diseases of Garlic

a. Occurrence and Disease Symptoms

A mosaic disease of garlic was first reported from Oregon and California in 1946[11] and subsequently from Belgium,[49] Canada,[72] Czechoslovakia,[6] China,[73] France,[74,75] Japan,[76,77] Korea,[78,79] Morocco,[80] New Zealand,[93] North Yemen,[81] Phillipines,[45] Spain,[3] Soviet Union,[82] Thailand,[83] Taiwan,[84] U.K.,[4] Venezuela,[85] and Yugoslavia.[86]

The most frequently described virus symptom in garlic is a mosaic consisting of light and dark green longitudinal stripes along the leaves (Figure 5). Symptoms are often difficult to see early in the season on the young leaves, but intensify as the season progresses until near maturity when the mosaic develops into vivid chlorotic stripes. Infection may also cause leaf curling, yellowing and stunting.[77] Some viruses are reported to be latent and to cause no visual symptoms in garlic.[1] Little data are available on the effect of virus infection on bulb yield, but a field trial in France indicated a reduction of 25 to 50% depending on the cultivar grown.[87] The number of cloves and bulbils produced by infected plants may also be reduced by up to 45% compared with healthy plants.[88]

b. Identification of Garlic Viruses

Initially, the mosaic disease of garlic was thought to be caused by a single virus which various workers referred to as garlic mosaic virus.[75,85] The restricted host range of *Allium* viruses resulted in difficulties in handling garlic viruses and contributed to the slow progress in separating and identifying individual component viruses. Electron microscope observations by Pena-Iglesias and Ayuso[89] and Cadilhac et al.,[65] using different garlic cultivars, revealed the presence of two different length rod-shaped particles. One of these was a potyvirus about 750 nm and the other a carlavirus about 625 to 650 nm. Further progress on the identity of the viruses in French garlic was made by Delecolle and Lot[1] using the ISEM "decoration" procedure[16] (Figure 2B). They showed that the *cv.* "Fructidor" contained two potyviruses and one carlavirus, and that one of the potyviruses was decorated with OYDV antibodies. The second potyvirus was not identified. It was not decorated by OYDV antibodies or by antibodies to LYSV (a potyvirus) and in the same tests, the carlavirus was not decorated by antibodies to shallot latent virus (SLV). Poty- and carlavirus particles were also observed in another French *cv.* "Blanc de la Drome", but these were not decorated by OYDV, LYSV, or SLV antibodies. They also found that severe symptoms were only present in plants that contained OYDV, and that plants containing only the second potyvirus and the carlaviruses were symptomless. It is not known, however, if OYDV alone induces the symptoms in this complex, or whether OYDV interacts with one or other of the other viruses to cause them.

FIGURE 5. Mosaic symptoms in garlic, infected leaf left and healthy leaf right (× 1).

Pena-Iglesias and Ayuso[3] also believed that the potyvirus observed in Spanish garlic was responsible for disease symptoms, and that the carlavirus they observed of about 625 nm in length, was latent. Earlier the Spanish potyvirus had been called garlic degeneration virus[89,90] and the carlavirus, garlic latent virus. Complexes consisting of a potyvirus and a carlavirus had also been reported to infect garlic in Japan,[76] U.S.,[91] and Canada.[92] In Japan mosaic symptoms were induced in virus-free garlic by inoculation with a 750 nm virus (presumably a potyvirus) isolated from diseased garlic, but this virus did not react with OYDV antiserum.[77] A potyvirus was also responsible for mosaic symptoms in garlic in Venezuela.[85] In New Zealand a potyvirus, referred to as garlic yellow streak virus, was responsible for field symptoms in garlic. This virus also occurred in a complex with a carlavirus which was thought to be latent.[93] The New Zealand potyvirus was reported to be related, but not identical to OYDV and LYSV as a result of microprecipitin tube serology tests, but as the reciprocal control tests aginst OYDV and LYSV were not carried out, the true relationship of the potyvirus to OYDV and LYSV is doubtful.

Recent studies in the U.K. using ISEM electron microscopy on a wider range of garlic cultivars[94] emphasizes the present state of confusion in respect of garlic virus diagnosis. These results (Table 3) confirmed the presence of a complex of one or more potyviruses and carlaviruses in each cultivar examined, but individual cultivars and even clones of the

TABLE 3
Occurrence of Virus in Some Commercial Cultivars of Garlic and Shallot Grown in the U.K.[a]

	Virus observed by electron microscopy ISEM					
Cultivar	**OYDV (potyvirus)**	**LYSV (potyvirus)**	**Unidentified potyvirus/es**	**SLV (carlavirus)**	**Unidentified [(carlavirus(es)]**	**Original source**
Garlic						
Fructidor clone A	+	+	−	−	+	France
Fructidor clone B	−	+	+	−	+	France
Blanc de la Drome	−	−	+	−	+	France
Blanc du Commun	+	−	+	+	+	France
Moulinin	+	−	+	−	−	France
Rosé du Var	+	+	+	+	+	France
Printanor	−	−	+	−	+	France
Goulurose	−	+	+	−	+	France
Ail du Nord	−	+	+	−	+	France
Kent cv.	+	+	−	+	+	Kent, UK
Isles of Scilly cv.	+	+	+	−	+	Scilly Isles
Star	−	+	+	−	+	Unknown
Shallot						
Noordhollandse Strogele	+	−	+	+	−	Holland

[a] Results based on the studies of Walkey et al.[4] ISEM decoaraiton tests were used to identify OYDV (onion yellow dwarf), LYSV (leek yellow stripe), and SLV (shallot latent). The antibodies to OYDV, LYSV, and SLV viruses did not decorate the particles of the unidentified viruses.

same cultivar (i.e., *cv.* ''Fructidor''), varied considerably in the individual viruses forming the complex. In some cultivars, such as ''Rose du Var'' and an unnamed cultivar from Kent, the situation was even more complex than earlier studies had suggested, in that these cultivars were infected by four or five different viruses. The ''decoration'' tests indicated that *cv.* ''Rose du Var'' was infected with OYDV, LYSV, and a third so far unidentified potyvirus, together with SLV and another unidentified carlavirus The Kent cultivar was infected with OYDV, LYSV, SLV, and an unidentified carlavirus or viruses. These results confirm the widespread occurrence of OYDV in garlic (6/11 cultivars tested) and showed for the first time the widespread occurrence of LYSV in garlic (7/11 cultivars tested). The same study also showed that some garlic cultivars are infected by SLV.

Much diagnostic work is required to identify the remaining poty- and carlaviruses that have been observed in garlic. It is possible that some of these may be viruses that occur in other ornamental Liliaceous bulbs, but which do not infect or are rare in other *Allium* crop species. Recent unpublished ISEM studies on some of these unidentified garlic viruses have indicated that hippeastrum mosaic virus (a potyvirus) is present in some clones of the *cv.* ''Fructidor'', and alstromeria mosaic virus (another potyvirus) in garlic from the Isles of Scilly. In the same tests, narcissus latent virus (a carlavirus) was detected in the *cvs.* ''Rose du Var'' and ''Blanc du Commun''.[94]

Considering the vegetative propagation of garlic such results are not surprising and it is quite likely that future studies will reveal a very varied range of different viruses in local garlic cultivars in different countries, and in different localities of the same country.

In addition to the viruses mentioned so far, there are a number of reports of other vegetable viruses infecting garlic (Table 1). None appear to be widespread in their occurrence in garlic and generally they appear to be of little or no economic significance to the crop. Tobacco rattle virus has been reported in garlic in East Germany without producing any symptoms[95] and tobacco mosaic virus was isolated from symptomless garlic in the Soviet Union.[96] More recently in Yugoslavia, cucumber mosaic virus (CMV) caused conspicuous yellow striping symptoms which turned necrotic and eventually caused the leaf to split in two.[86] These plants may also have been infected with other latent viruses, so the symptoms reported may not be attributable to CMV alone.

3. Control Measures

a. Introduction

Virus diseases of garlic may be controlled by the use of healthy planting material, from which the viruses have been eradicated by meristem tip culture. Certification schemes are required to ensure that the nucleus of mother stock plants are maintained under virus-free conditions and are indexed at regular intervals to ensure their freedom from infection. The certification scheme would require multiplication of the nuclear stocks under conditions that would maintain their health status, and the distribution of the certified material to the grower. Commercial crops of healthy bulbs, should be grown at sites isolated from other *Allium* crops or any other crop that might be a source of infection.

As the viruses so far identified in garlic are not known to be seed transmitted in other *Allium* species, it is unlikely that they would be seed transmitted in garlic either. Therefore, in the long-term viruses might be controlled in garlic by growing garlic crops from seed rather than from vegetatively propagated cloves. The discovery that some garlic cultivars still produce seed[71] might make this possible. If these cultivars were not commercially acceptable to the grower in their present state, they could form nevertheless the basis of a breeding program to produce agronomically acceptable cultivars.

b. The Production of Virus-Free Garlic

Meristem-tip culture has been used by a number of workers to produce virus-free garlic[4-6,97-101] (see chapter on *Allium* Tissue Culture). Walkey et al.[4] described an improved medium and procedure for garlic meristem-tip culture, which is equally satisfactory for the tip culture of shallots. Meristem-tips measuring 0.5 to 0.8 mm in diameter were aseptically dissected under a binocular microscope on a sterile air bench, and grown on 10 ml of culture media in flat-bottomed glass tubes (Figure 6A). The cultures were incubated at a constant temperature of 23°C and illuminated under daylight fluorescent tubes for 16 h/d. The medium consisted of Gamborg's B5 mineral salts[102] containing 30 g/l sucrose, 1 mg/l nicotinic acid, 1 mg/1 pyridoxin HCl, 10 mg/1 thiamine HCl, 100 mg/1 inositol, and 0.6% agar at pH 5.7. No auxins or cytokinins were required to produce rooted plantlets (Figure 6A) and the cultures were transferred to fresh medium every 2 to 3 weeks. Rooted plantlets were transferred to peat "Jiffy" pots (Figure 6B) and hardened off in a plastic covered seed-propagation tray for 2 to 3 weeks and later transplanted into larger pots containing John Innes No. 3 compost. Following virus indexing, plants negative for virus were transplanted into sterilized soil in an insect-proofed gauze-house. The plantlets did not usually form bulbs if grown in a glasshouse at 22°C, and required a cold-period treatment before they would produce normal bulbs (Figure 7). Virus-free bulbs have been found to be between 26 and 39% larger than infected bulbs depending on the cultivar.

Between 25 and 50% of all garlic and shallot plants regenerated by this method were virus-free.[4] If infected parent plants were given a high temperature treatment of 7 d at 30°C + 14 d at 36°C + 21 d at 38°C before the tips were excised for culture, the percentage of virus-free plants produced increased from 25 to 85% (Table 4). Preconditioning of the parent plant at 30°C prior to the high temperature treatment was essential if the plants were to survive. Shallots were found to be more sensitive than garlic to high temperatures and did not survive a 36°C treatment.[4]

Resurgence of viruses in garlic following meristem-tip culture has been quite frequently

A B

FIGURE 6. (A) Rooted meristem-tip culture of garlic grown on Gamborg's B5 medium (×2). (B) Regenerated garlic plantlet following transfer from tube culture to a peat ''Jiffy'' pot (×1).

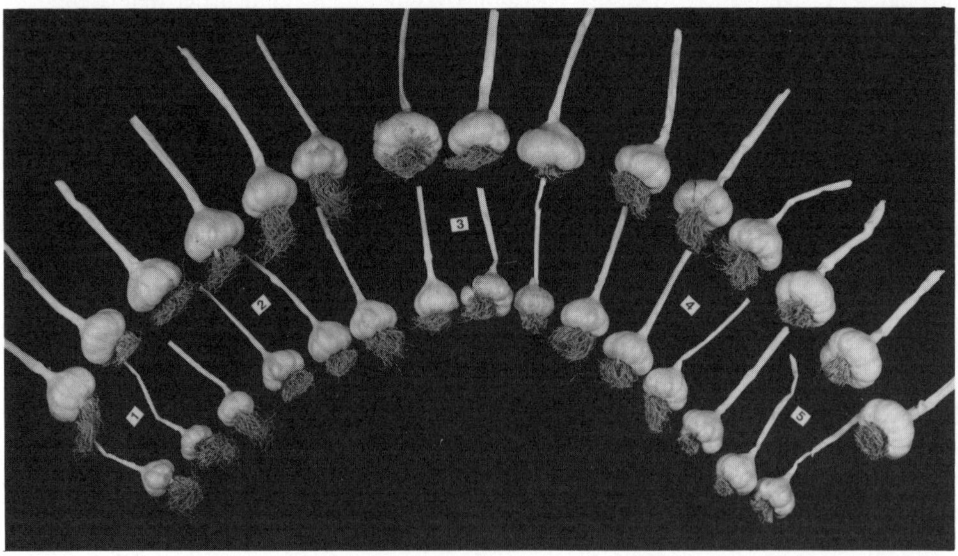

FIGURE 7. Virus-free garlic bulbs produced by meristem-tip culture compared with infected bulbs of the same cultivars. Three bulbs of each cultivar (1) ''Rosé de Var'', (2) ''Printanor'', (3) ''Blanc de la Drôme'', (4) ''Moulinin'', and (5) ''Frictidor'' (× 1/6).

TABLE 4
Effect of Thermotherapy on Virus Eradication from Garlic and Shallot[a]

Cultivar	Temperature treatment (°C)	No. tips cut	No. cultures established in soil	No. plantlets virus-free	% cultures virus-free
Garlic					
Fructidor	20	12	4	1	25
	38	24	19	16	84.2
Blanc de la Drôme	20	12	8	2	25
	38	24	13	11	84.6
Shallot					
Noordhollandse Strogele	20	42	19	5	26.3
	33	51	35	9	25.7

[a] Results based on the studies of Walkey et al.[4] Treatments were applied to sprouted bulbs grown in soil before the tips were excised. The garlic bulbs were preconditioned for 7 d at 30°C and 14 d at 36°C before treatment for 21 d at 38°C. The shallot bulbs were grown at 20°C or 33°C throughout period treatment.

observed[94] and repeated virus indexing of the plants by the sensitive ELISA[15] and ISEM[16] serology procedures is essential to confirm absolute freedom from infection. Indexing of mature bulbs over a period of at least 2 years is recommended.

III. VIRUSES OF OTHER *ALLIUM* CROPS AND SPECIES

Virus infection has been reported in a number of commercially less important *Allium* crop species and various *Allium* ornamental species (Table 2). OYDV is probably the most important virus to infect these species, as it infects the vegetatively-propagated multiplier onion,[11] tree onion,[11] and Welsh onion.[9] Although these crops are not widely grown commercially in Europe, they could be a potential reservoir of the virus for infection of adjacent onion crops. This could be particularly relevant in respect of private gardens, where these perennial onions are most likely to be grown. Similarly, the little cultivated serpent garlic (*A. scorodoprasum*) could be a reservoir of SLV[66] for infection of shallots, leeks, and garlic.

An unidentified virus 640 nm in length and referred to as garlic mosaic virus has also been isolated from wild *Allium* species in the USSR.[103] Affected species included *A. hymenorrhizum, A. longicuspis, A. oschanini, A. proliferum, A. galanthum,* and *A. rotundum.*

The other vegetable viruses listed in Table 2 are not of economic importance or widespread in *Allium* crop species, so although those *Allium* hosts might act as a disease reservoir, they are not a threat to main *Allium* crops.

IV. VIRUS-LIKE DISEASES OF *ALLIUM*

In addition to viruses, onions in particular, may be affected by mycoplasma which induce virus-like yellowing symptoms. A disease, attributed to a mycoplasma called ''aster yellows'' causes a yellowing which starts at the base of the leaf and spreads upward. The leaves flatten and develop yellow and green streaks.[30] In onions grown for seed production, it causes proliferation of the flower umbel in which the flower pedicels are abnormally long and the flower is distorted and usually sterile.[104] Stored bulbs which are infected will often sprout prematurely.[105]

The disease is transmitted by a leaf hopper, *Macrosteles fascifrons*[106] and the mycoplasma bodies may be seen in the electron microscope in ultrathin sections of the sieve elements of the phloem from diseased leaves. Aster yellows has a very wide host range, infecting more than 300 vegetables, weeds, and ornamental species in 48 families,[30] including such economically important crops as carrot, celery, lettuce, aster, onion, and shallot.

This disease of onion or one caused by a related mycoplasma has also been reported in Poland,[107] East Germany,[50] Czechoslovakia,[108] France,[109] and Rumania.[110] Leek plants are also reported to be affected in Czechoslovakia,[108] and shallots in the U.S.[11]

Although the disease is not generally widespread or of great economic significance in general onion production, it can cause losses locally, especially in crops for seed production. Control of the disease may be achieved by controlling the leaf hopper vector with insecticides and by removing weed hosts of the vector. Leaf hoppers also have a preference for hibernating on cereals and grasses, so onion crops should be planted as far as possible from these crops.

V. CONCLUSIONS

To obtain successful control of a virus disease, the virus must be correctly identified and its epidemiology understood. In recent years, the work of Bos and his colleagues on OYDV, LYSV, and SLV in the Netherlands has greatly advanced the diagnosis and handling of *Allium* viruses. Much work remains to be done, however, on the identification of the virus complexes infecting vegetatively propagated *Allium* species such as garlic. The correct identification of these viruses will assist in the establishment of measures for their control and the prevention of reinfection of healthy planting material produced by meristem-tip culture.

Good hygienic practices have enabled OYDV to be controlled in onion crops, but changing cultural practices may present new threats if the practices are not carefully monitored. The problems with such changes were highlighted in the European low countries by the outbreaks of LYSV disease in leeks in the 1970s. Here an increase in all-the-year-round cropping of leeks led to the virus epidemic, but once a leek-free break period was introduced, the disease was brought under control. In recent years, there has been an increase in the all-the-year-round cultivation of onions in many western European countries with the introduction of the overwintering Japanese cultivars. This new cultural practice could result in a build-up of OYDV, should the virus become established in the onion crop. In addition, as has happened in the U.K., growers are returning to growing a proportion of the onion crop from sets. The combination of onion sets and overlapping onion crops could present an ideal situation for the spread of OYDV. Great care must be taken therefore to ensure that the sets remain free of OYDV.

The use of host resistance to control or reduce the occurrence of *Allium* viruses has still to be exploited. Resistance could be particularly useful in controlling the nonpersistent, aphid-transmitted viruses which are the main type of virus found in alliums. The aphid-transmission of these viruses cannot usually be controlled by insecticidal sprays.

Plant breeding could also be useful in the future in controlling viruses in garlic and shallot, for as most of the viruses concerned are not thought to be seed transmitted, growing these crops from seed could avoid the problem. At present this is not possible, because garlic cultivars will not produce seed and shallot is preferably clonally propagated. If, however, the breeder developed new cultivars that could be grown from seed that were of economic value, this method of control would be feasible.

In conclusion, more work is required on the viruses of locally important clones of vegetatively-propagated alliums. These crops which may be diverse forms of *A. cepa* or other *Allium* species, are not of importance on a world-scale, but are often of considerable economic value locally. Many are only vegetatively propagated and some are shy to flower or are completely infertile. The provision of healthy planting material of these crops by meristem-tip culture procedures and the introduction of programs to maintain their health status would be of considerable benefit in many developing countries.

ACKNOWLEDGMENT

I wish to thank Mr. M. J. W. Webb, Mr. N. Paludan, and Prof. T. A. Zitter for providing photographs and the Institute of Horticultural Research, Wellesbourne, for providing various facilities in the preparation of this manuscript.

REFERENCES

1. **Delecolle, B. and Lot, H.,** Garlic viruses: detection and partial characterization with immune electron microscopy of three viruses in different garlic populations with mosaic, *Agronomie,* 1, 763, 1981.
2. **Bos, L.,** Viruses and virus diseases of Allium species, *Acta Hort.,* 127, 11, 1982.
3. **Pena-Iglesias, A. and Ayuso, P.,** Characterisation of Spanish garlic viruses and their elimination by in vitro shoot apex culture, *Acta Hort.,* 127, 183, 1982.
4. **Walkey, D. G. A., Webb, M. J. W., Bolland, C. J., and Miller, A.,** Production of virus-free garlic (*Allium sativum* L.) and shallot (*A. ascalonicum* L.) by meristem-tip culture, *J. Hort. Sci.,* 62, 211, 1987.
5. **Quiot, J., Messiaen, C., Marrou, J., and Leroux, J.,** Regeneration by meristem culture of garlic cloves chronically infected with garlic mosaic virus, *Acta 3rd Congr.,* 429, 1972.
6. **Havranek, P.,** Virus-free garlic clones obtained from meristem cultures, *Ochr. Rostl.,* 8, 291, 1972.
7. **Laroche, M. and Verhoyen, M.,** Essai de multiplication rapide in vitro de l'echalote (*Allium ascalonicum* L.), *Meded. Fac. Landbouww. Rijksuniv. Gent,* 45, 323, 1980.
8. **Fujieda, K., Ando, Y., and Fujita, Y.,** Propagation of Welsh onion through shoot tip culture, *J. Fac. Agric. Kyushu Univ.,* 22, 89, 1977.
9. **Costa, A. S., Costa, C. L., Nagai, H., and Kitajima, E. W.,** Welsh onion, a source of the onion yellow dwarf virus, *Biologico,* 37, 157, 1971.
10. **Giddings, N. J.,** Report of the Department of Pathology, *W. Virg. Agric. Exp. Stn.,* 44 pp, 1916.
11. **Brierley, P. and Smith, F. F.,** Reaction of onion varieties to yellow-dwarf virus and to three similar viruses isolated from shallot, garlic and Narcissus, *Phytopathology,* 36, 292, 1946.
12. **Bos, L., Huijberts, N., Huttinga, H. and Maat, D. Z.,** Leek yellow stripe virus and its relationship to onion yellow dwarf virus; characterization, ecology and possible control, *J. Plant Pathol.,* 84, 185, 1978.
13. **Bos, L., Huttinga, H., and Maat, D. Z.,** Shallot latent virus, a new carlavirus, *Neth. J. Plant. Pathol.,* 84, 227, 1978.
14. **Huttinga, H.,** Purification by molecular sieving of a leek virus related to onion yellow dwarf virus, *Neth. J. Plant Pathol.,* 81, 81, 1975.
15. **Clark, M. F. and Adams, A. N.,** Characteristics of the microplate method of enzyme-linked immunosorbent assay for the detection of plant viruses, *J. Gen. Virol.,* 34, 475, 1977.
16. **Derrick, K. S.,** Quantitative assay for plant viruses using serologically specific electron microscopy, *Virology,* 56, 652, 1973.
17. **Calvert, E. L. and Harrison, B. D.,** Outbreaks of tomato black ring virus in onion and leek crops in Northern Ireland, *Hort. Res.,* 2, 115, 1963.
18. **Protsenko, L. V.,** Influence of climatic conditions on resistance in onion to mosaic, *Nauch. Trudy khar'kov. Sel'-khoz. Inst.,* 43, 223, 1964.
19. **Schmelzer, K., Wolf, P., and Gippert, R.,** Gemusepflanzen, in *Die Virusen an Gemusepflanzen, Obstgewachen und Weinreben in Europa,* Schmelzer, K. and Spaar, D., Eds., Akademie Verlag. Berlin, 1, 1977.
20. **Razvyazkina, G. M.,** Das Zwiebelmosaikvirus und seine Verbreitung im Freiland, *Sitz Ber. Akad. Landwirtsch.,* 115, 69, 1971.
21. **Cheremushkina, N. P.,** Electron microscopy ultrathin sections study of the relationship between Allium mosaic virus and its vector-mite Aceria tulipae L., *Comm. Inst. Biol. Pedol.,* 21, 107, 1974.
22. **Cheremushkina, N. P.,** The role of Aceria tulipae in spreading onion mosaic virus, *Comm. Inst. Biol. Pedol.,* 28, 174, 1975.
23. **Melhus, I. E., Reddy, C. S., Henderson, W. J., and Vestral, E.,** A new virus disease epidemic on onions, *Phytopathology,* 19, 73, 1929.
24. **Henderson, W. J.,** Yellow dwarf, a virus disease of onions and its control, *Res. Bull. Iowa Agric. Exp. Stn.,* 188, 209, 1935.
25. **Chamberlain, E. E.,** Plant virus diseases in New Zealand, *N.Z.D.S.I.R. Bull.,* 108, pp. 1, 1954.
26. **Allam, E. K. and Ghalwash, F.,** Effect of onion yellow dwarf virus on the yield of onion bulbs and seeds in Egypt, *Egypt. J. Microbiol.,* 7, 21, 1972.

27. **Bremer, H.,** Uber die bisher falschlich 'Zwiebelrutz' genannte Gelbstreifigkeit and Zwiebelsamentragern, *Phytopathology, Z.,* 10, 79, 1937.
28. **Graichen, K.,** Viruskrankheiten der Zwiebel, des Porrees und des Knoblauchs und ihre Bekampfung, *NachrBl. PflSch. DDR.,* 32, 245, 1978.
29. **Fischer, H. U. and Lockhart, B. E. L.,** High incidence of onion yellow dwarf in areas of commercial onion production in Morocco, *Plant Dis. Rep.,* 58, 252, 1974.
30. **Sherf, A. F. and MacNab, A. A.,** *Vegetable Diseases and Their Control,* John Wiley, 1986,
31. **Stefanac, Z.,** Onion yellow dwarf virus in Yugoslavia, *Acta Bot. Croat.,* 36, 39, 1977.
32. **Chamberlain, E. E. and Baylis, G. T. S.,** Onion yellow dwarf, successful eradication, *N.Z. J. Sci. Technol.,* A29, 300, 1948.
33. **Drake, C. J., Tate, H. D., and Harris, H. M.,** The relationship of aphids to the transmission of yellow dwarf of onion, *J. Econ. Entom.,* 26, 841, 1933.
34. **Tate, H. D.,** Insects as vectors of yellow dwarf, a virus disease of onions, *Iowa State Coll. J. Sci.,* 9, 677, 1940.
35. **Hardtl, H.,** Untersuchungen uber Verbreitung und Schaden der Gelbstreifigkeitsvirose bei Zwiebetn in der Pflaz, *Z. Pflakrankh. PflSch.,* 69, 587, 1962.
36. **Hardtl, H.,** Die Ubertragung der Zwiebelgelbstreifigkeit durch den Samen, *Z. Pflakrankh. PflSch.,* 79, 694, 1972.
37. **Louie, R. and Lorbeer, J. W.,** Mechanical transmission of onion yellow dwarf virus to mother onion bulbs, *Plant Dis. Rep.,* 52, 581, 1968.
38. **Schmidt, H. B. and Schmelzer, K.,** Electronenmikroskopische Vermessung des Zwiebelgelbstreifen-Virus, *Phytopathol. Z.,* 50, 191, 1964.
39. **Edwardson, J. R.,** Some properties of the potato virus-Y group, *Fla. Agric. Exp. Stn. Monogr.,* 4, 398, 1974.
40. **Brunt, A. A.,** A review of problems and progress in research on viruses and virus diseases of narcissus in Britain, *Acta Hort.,* 110, 23, 1980.
41. **Bos, L.,** Onion yellow dwarf virus, *CMI/AAB Description of Plant Viruses,* No. 158, 1976.
42. **Kupke, W.,** Die Gelbstreifigkeit, eine gefahrliche Krankheit des Porres, *Rhein. Monatschr. Gem.,* 45, 173, 1957.
43. **Gibson, R. W. and Rice, A. D.,** The combined use of mineral oils and pyrethroids to control plant viruses transmitted non- and semi-persistently by Myzus persicae, *Ann. Appl. Biol.,* 109, 465, 1986.
44. **Razvyazkina, G. M., Kapkova, E. A., and Cheremushkina, N. P.,** A mite Aceria tulipae (Eriophyide) — vector of the onion mosaic virus, *Zool. Zh.,* 48, 288, 1969.
45. **Ahmed, K. M. and Benigno, D. A.,** Control of the eriophid mite Aceria tulipae Keifer, the cause of tangle-top malady and the vector of garlic mosaic virus, *Indian J. Plant Protect.,* 12, 153, 1984.
46. **Mohamed, N. A. and Young, B. R.,** Leek yellow stripe virus in New Zealand, *N.Z. Agric. Res.,* 23, 129, 1980.
47. **Paludan, N.,** Virus attack on leek. Survey, diagnosis, tolerance of varieties and winter hardiness, *Tidsskr. Planteavl.,* 84, 371, 1980.
48. **Matthiew, J. L. and Verhoyen, M.,** La striure chlorotique due poireau, epidemiologie et incidence de la maladie, *Parasitica,* 37, 59, 1981.
49. **Verhoyen, M.,** La striure chlorotique du poireau. II. Note concernant l'epidemiologie due virus, *Parasitica,* 29, 35, 1973.
50. **Novak, J. B.,** Ein beitrag zur kenntnis der Viruskrankheiten der Zwiebelgemuse in der Tschechoslowakai. Sborn. Vys. Skol. Zem. Praze, *Sborn. Vys. Skol. Zem. Praze,* 287, 1959.
51. **Paludan, N.,** Viruses in vegetables, *Plantesygdomme Denmark,* 1976, 64, 1977.
52. **Graichen, K.,** Viruskrankheiten der Zwiebel udn des Porrees vorbeugend bekampfen, *Gartnerpost,* 6, 7, 1976.
53. **Cornuet, P.,** Maladies a virus des plantes cultivees et methodes de lutte, *Inst. Nat. Rech. Agron.,* Paris, p. 440, 1959.
54. **Heinze, K.,** Virusubertragungsversuche mit Blattlausen auf Dahlien, Zwiebeln, Wasserruben und einigen anderen, *Pfl. Sch.,* 59, 3, 1952.
55. **Hardtl, H.,** Gelbstreifigkeit bei Porree, *Gesunde Pflanzen,* 17, 235, 1965.
56. **Grancini, P.,** Malattie da virus degli ortelggi il mosaico della cipolla, *Flora,* 6, pp 19, 1951.
57. **Verhoyen, M. and Horvat, F.,** La striure chlorotique du poreau. 1. Identificatin de l'agent causal, *Parasitica,* 29, 16, 1973.
58. **Fletcher, J. T.,** Leek diseases: a socio-economic problem. MAFF, ADAS, Open Conf. Advisory Plant Pathol., Malvern, Feb 6—8, 1979, PP/O/494, 1979.
59. **Horvat, F. and Verhoyen, M.,** Inclusions virales dans le cyotplasme de cellules de pioireaux et de Chenopodium quinoa infectes parle virus provoquant la striure du poireau, *Parasitica,* 31, 53, 1975.

60. **Graichen, K.,** Untersuchungen zum Wirtspflanzenkreis des Porree-gelbstreifen-Virus (leek yellow dwarf virus), *Arch. Phytopatol. Pflsch.*, 14, 1, 1978.
61. **Weidemann, H. L.,** Die Gelbstreifigkeit des Porrees, *Gemuse (Spezbl. Feld-u. Intensiygemusebau),* 15, 128, 1979.
62. **Van der Meer, Q. M.,** Resistentie tegen preigeel streepvirus in kurrat *(Allium ampeloprasum), Zaadbelangen,* 34, 173, 1980.
63. **Stefanac, Z. and Plese, N.,** Turnip mosaic virus in two Meditteranean Allium species, *Proc. 5th Congr. Mediter. Phytopathol. Union,* 37, 1980.
64. **Delecolle, B., Lot, H., and Michel, M. J.,** Application of ELISA for detecting onion yellow dwarf virus in garlic and shallot seeds and plants, *Phytoparasitica,* 13, 266, 1985.
65. **Cadilhac, B., Quiot, J. B., Marrou, J., and Leroux, J. P.,** Mise evidence au microscope electronique de deux virus differents infectant l'ail *(Allium sativum* L.) et l'echalote (*Allium cepa* L. var. *ascalonicum),* *Ann. Phytopathol.*, 8, 65, 1976.
66. **Verhoyen, M. and Horvat, F.,** Identification des virus de l'echalote (*Allium ascalonicum* L.) en Belgique, *Parasitica,* 37, 79, 1981.
67. **Quiot, J. B., Marrou, J., Leroux, J. P., Leclant, L., and Messiaen, C. M.,** Viroses des Allium, selection sanitaire des semences d'ail, *Ann. Phytopathol.*, 7, 234, 1975.
68. **Hollings, M.,** Virology, *Ann. Rept. Glassh. Crops Res. Inst. 1967,* 108, 1968.
69. **Henderson, D. M.,** Virus yellows of shallots, *Plant Pathol.*, 108, 2, 1953.
70. **Messiaen, C., Marrou, J., and Quiot, J.,** Sanitary selection schedules for garlic and shallots in the south of France, *Proc. 18th Int. Hort. Cong. Tel-Aviv Abst.*, 301, 1970.
71. **Etoh, T.,** Fertility of the garlic clones collected in Soviet Central Asia, *J. Jpn. Soc. Hort. Sci.*, 55, 312, 1986.
72. **Liu, X., Peterson, J. F., and Crete, R.,** Detection of viruses infecting garlic, *Phytoprotection,* 66, 175, 1985.
73. **Ding, X.,** Personal communication, 1986.
74. **Messiaen, C. M. and Arnoux, M.,** Une maladie de l'ail probablement due a un virus, son influence sur le rendement, *Etudes de virol. appl. INRA,* 29, 1960.
75. **Messiaen, C. M. and Marrou, J.,** Selection sanitarie de l'ail, deux solutions possible au problems de la mosaique de l'ail, plantes sensibles saines, ou plants virosees tolerantes, *C.R. J. Phytiatr. Phytopharm. Tr. Circummedit. Marseille,* 204, 1965.
76. **Lee, Y. W., Yamazaki, S., Osaki, T., and Inouye, T.,** Elongated viruses in garlic. Garlic latent virus and garlic mosaic virus, *Ann. Phytopathol. Soc. Jpn.*, 45, 727, 1979.
77. **Abiko, K., Watanabe, Y., and Nishi, Y.,** Studies on garlic mosaic, *Bull. Veg. Orn. Crops Res. Stn. Ser A.*, 139, 1980.
78. **La, Y. -J.,** Studies on garlic mosaic virus, *Kor. J. Plant Prot.*, 12, 93, 1973.
79. **Lee, C. U.,** On improvement of garlic production by inactivation of virus in garlic, *Korean J. Plant Prot.*, 20, 6, 1981.
80. **Fischer, H.,** An uncommon virus isolated from garlic in Morocco, *Abstr. 2nd Int. Conf. ISHS Veg. Virus Working Gp. Avignon,* 28, 1975.
81. **Alhubaishi, A. A., Walkey, D. G. A., and Webb, M. J. W.,** Plant virus diseases in Yemen Arab Republic, *Br. Soc. Pl. Pathol.*, 16—18 Dec, 1986, 1986.
82. **Tulegenov, T. A.,** Virus diseases of onion and garlic in the Alina-Ata region, *Biologicheskaya,* 10, 33, 1972.
83. **Walkey, D. G. A. and Smitamana, P.,** unpublished results, 1987.
84. **Chen, M. J. and Ko, N. -J.,** Etiological studies on virus diseases of garlic in Taiwan, *Pl. Prot. Bull (Taiwan),* 21, 210, 1979.
85. **Lastra, R., Ladera, P., and Debrot, E. A.,** Purification of garlic mosaic virus, *Phytopathology,* 69, 1036, 1979.
86. **Stefanac, Z.,** Cumcumber mosaic virus in garlic, *Acta Bot. Croat.*, 39, 21, 1980.
87. **Messiaen, C. M., Youcef-Benkada, M., and Beyries, A.,** Potential yield and tolerance to virus-diseases in garlic *(Allium sativum* L.), *Agronomie,* 1, 759, 1981.
88. **Havranek, P.,** The effect of virus diseases on the yield of common garlic, *Ochr. Rostl.*, 10, 251, 1974.
89. **Pena-Iglesias, A. and Ayuso, P.,** Meristem-tip culture of garlic. Elimination of viruses from infected bulbs and rapid multiplication of virus-free clones, *Proc. XIX Int. Hort. Congr. Warsaw,* Vol. 1A, 1974, 62.
90. **Pena-Iglesias, A., Ayuso, P., Rubio-Huertos, M., and Moreno, R.,** Some characteristics of the potyvirus of some cultivated liliaceous (garlic, leek, onion) detected in Spain, 3rd Int. Congr. Virol. C456 Madrid, 10—17, Sept., 1975.
91. **Shepherd, R. J. and Pena-Iglesias, A.,** Identity of viruses of garlic and globe artichoke in Spain and the USA, *Rept. NSF Grant No. INT - 20881 US-Spain Program 1978,* 1976.
92. **Liu, X., Peterson, J. F., and Crete, R.,** Differentiation of garlic viruses, *Phytopathology,* 76, Abst. 655, 1986.

93. **Mohamed, N. A. and Young, B. R.,** Garlic yellow streak virus, a potyvirus infecting garlic (and transmitted by Myzus persicae) in New Zealand, *Ann. Appl. Biol.*, 97, 65, 1981.
94. **Walkey, D. G. A. and Webb, M. J. W.,** unpublished results, 1986.
95. **Graichen, K.,** Allium — Arten als Naturliche Wirte Nematoden-uber-Tragbarer Viren, *Arch. Phytopathol. Pfl Sch. Berlin,* 11, 399, 1975.
96. **Vasil'eva, T. Y. and Mozhaeva, K. A.,** Svoistva shtamma VTM, vydelennogo iz rastenii roda Allium, *Tr. Biol-Pochven. Inst. Dal'nevost. Nauch. Tsentr. ANSSSR,* No. 46/149, 75, 1977.
97. **Mori, K., Hamaya, E., Shimomura, T., and Ikegami, Y.,** Production of virus-free plants by means of meristem culture, *J. Cent. Agric. Expt. Stn.*, No. 13, 97, 1969.
98. **Wang, P. and Huang, L.,** Studies on the shoot meristem culture of *Allium sativum* L., *J. Chinese Soc. Hort. Sci.*, 20, 79, 1974.
99. **Ayuso, P. and Pena-Iglesias, A.,** The elimination of garlic viruses by thermotherapy and/or tissue culture, *Cell Biol. Int. Repts.*, 5, 835, 1981.
100. **Bhojwani, S. S., Cohen, D., and Fry, P. R.,** Production of virus-free garlic and field performance of micropropagated plants, *Sci. Hort.*, 18, 39, 1982.
101. **Mar'yakhina, I. Y., Cheremushkina, N. P., and Polumordvinova, I. V.,** Breeding and propagating virus-free garlic and shallot plants in tissue culture, *S. Kh. Naulk. Im. Vi. Lenina,* 4, 9, 1984.
102. **Gamborg, O. L., Miller, R. A., and Ohyama, K.,** Nutrient requirements of suspension cultures of soybean root cells, *Exp. Cell Res.*, 50, 148, 1968.
103. **Brčák, J.,** Garlic mosaic virus particles and virus infections of some wild Allium sp., *Ochr. Rostl.*, 11, 237, 1975.
104. **Larsen, R. and Walker, J. C.,** Aster yellows a hazard in onion seed production, *Wisc. Agric. Exp. Stn. Bull.*, 463, 50, 1944.
105. **Hooper, G. R., Lacy, M. L., and Vest, G.,** Mycoplasma-like bodies associated with onion bulbs sprouting in storage, *Plant Dis. Rep.*, 55, 824, 1971.
106. **Kunkel, L. O.,** Studies on aster yellows in some new host plants, *Contr. Boyce Thompson Inst.*, 3, 85, 1931.
107. **Kochman, J. and Stachyra, T.,** Beitrage zur Kenntnis der pflanzlichen Viruskrankheiten und virusverdachtigen Erscheinungen in Polen, *Nachrbldt. Pflschd (Berlin),* 12, 41, 1957.
108. **Blattny, C.,** La variabilite du niveau du virus de la jaunisse chez Allium ampeloprasum L. subsp. porrum L. (Regel), *Biol. Plant. (Praha),* 3, 34, 1961.
109. **Cousin, M. T., Schweisguth, B., Faivre-Amiot, A., Kartha, K. K., Staron, T., and Moreau, J. P.,** Contribution a l'etude du cycle des mycoplasmes chez les vegetaux a l'aide d'observations effectuees sur oignon atteint d'Aster yellow, *C.R. Hebd. Acad. Sci. Ser. D.*, 272, 830, 1971.
110. **Petre, Z. and Ploaie, P. G.,** Onion proliferation and aspermy, a new disease in Romania caused by mycoplasma, *Anal. Inst. Cerc. Prot. Plant.*, 9, 13, 1973.
111. **Gessner, R.,** Zur frage des Verhaltens und der Ubertragb-arkeit des Tabak-Ratelvirus, *Dissertation, University of Giessen,* 32, 1959.
112. **Kristensen, H. R. and Engsbro, B.,** Undersogelser og forsog vedroerende jordbarne vira. I. Rattle virus, *Tiddskr. Planteavl.*, 70, 353, 1966.
113. **Szirmai, J.,** Virusveszely a makoi hagymatermesztesben, *Novenytermeles,* 7, 63, 1958.
114. **Kovachevsky, I. and Markov, M.,** Dienekrotische Bronze fleck-enkrankheit der Tomaten, *Inst. Pflsch. (Sofia),* 1, 5, 1961.
115. **Severin, H. H. P. and Frazier, N. W.,** California aster yellows on vegetable and seed crops, *Hilgardia,* 16, 573, 1945.
116. **Smith, F. F. and Brierely, P.,** Aster yellows in shallot and gladiolus, *Phytopathology,* 38, 581, 1948.
117. **Van Slogteren, D. H. M.,** Ratelvirus als oorzaak van ziekten in bloembolgewassen ende mogelijkhede de infectie door middel van grondontsmetting te bestrijden, *Tijdschr. Plziekt.*, 64, 452, 1958.

Chapter 10

INSECT PESTS

S. K. Soni and P. R. Ellis

TABLE OF CONTENTS

I. INTRODUCTION

The production of onions and allied crops is threatened by a wide range of pests and, each year, between 10 and 25% of the world total production is lost as a result of pest attack. Some insects, for example the onion thrips, are pests in whichever climatic zone these crops are grown; others like the salt marsh caterpillar are distributed more locally. Certain pests like cutworms cause damage sporadically, while others including the onion fly or onion maggot cause severe damage every year. A few of the pests such as the red spider mite attack a very wide range of crops in many plant families, not just alliaceous crops. Others, for example the onion maggot, are specific of *Allium*. The number of severe pests is quite small and the majority of insects and mites present in a crop are either harmless or beneficial. This fact should not be overlooked in the chemical treatment of a crop.

The enormous task of controlling insect pests has exercised the minds of entomologists, farmers and advisors for more than 100 years. Even with all the sophisticated equipment, techniques and armoury of chemicals available, pests continue to reduce significantly the production of onions and related crops. There are several reasons for this: insecticide-resistant strains of pests, inefficient application of pesticides, inaccurate timing of treatments, as well as enhanced degradation of insecticides in the environment. All these problems have necessitated a rethinking of the whole approach to pest management. Well over twenty years ago it was realized that, to rely almost entirely on just one method of control, insecticides had serious drawbacks; not only did the pests become resistant to many of these chemicals, but their parasites and predators were depleted dramatically by spraying in and around the crops. Indeed, in some cases, the onion maggot problem was enhanced by the application of organochlorine insecticides which decreased the numbers of the natural enemies and induced the rapid development of resistant races of the pest. Most of these chemicals have been withdrawn because of the problems of insect resistance and their effects on the environment. Legislation relating to pesticide usage is now much stricter and the numbers of new products reaching the market place are small.

Thus, over the last two decades, research workers have investigated a wide range of alternative methods of prevention and control of pests. Some of these have been developed and are a practical proposition. Certain methods now being advocated were described first more than 80 years ago but rapidly lost favor as research workers and growers became infatuated with the dramatic success of insecticides.

In this chapter we have attempted to provide up-to-date information on the range of different methods of crop protection which includes cultural methods, biological control, the use of resistant crop cultivars and insecticides; the latter remain the most important control measure and will continue to be so for many years to come. A crucial element of the current approach is the monitoring and prediction of pest activity. This will enable cropping schedules to be timed to avoid the worst damage and treatments to be timed more accurately. Most of this information and monitoring and prediction has been gathered in the last ten years.

In the following pages, we describe the major pests of onions and their allies. However, a pest that is a serious threat to growers in Egypt may not pose a problem to growers in Scotland. We have, therefore, described in detail at least one representative species of each genus, family or group and provide references to closely related species so that readers can obtain information pertinent to their local situation.

General notes on the identification of species are provided in each section.

II. PLANT BUGS, TRUE BUGS, OR PLANT LICE, ORDER: HEMIPTERA

This order includes aphids, capsid bugs (mirids), leafhoppers, mealy-bugs, psyllids (suckers), scale insects, and whiteflies. They are all sap-sucking insects characterized by having an incomplete metamorphosis. Plant bugs have egg, nymph, and adult stages, the latter two stages usually having a close resemblance to each other. All insects in this order have mouthparts which are modified to pierce plant tissues and extract the sap. Direct damage results from the piercing of tissues, disruption of cells, loss of sap and the injection into the host plant of what are, in some cases, toxic salivary juices. Indirect damage results from the entry of other organisms through the site of injury, the transmission of viruses by several groups of plant bugs, and the excretion of sap providing a medium which is colonized quickly by sooty mold fungi. The presence of these moulds and the cast skins of the insects can reduce the marketability of crops. Several plant bugs, particularly aphids, can be severe pests of alliaceous crops. These and species of some of the plant bugs which are minor pests are listed in Table 1.

A. FAMILY: APHIDIDAE APHIDS

1. *Myzus ascalonicus* Doncaster, Shallot Aphid

The genus *Myzus* includes several very important crop pests, all of which have extremely wide host ranges in many different plant families. Three species, *M. ascalonicus, M. cymbalariae* Stroyan, and *M. persicae* (Sulzer), attack onions and their allies and are widespread throughout the world. *M. ascalonicus* (see distribution map Figure 1) is described here as a typical aphid species; key references for the other two species of *Myzus* and for certain other aphid pests of alliaceous crops are given in Table 1.

a. Distribution

M. ascalonicus is distributed widely throughout western Europe, and also occurs in Australia, New Zealand, Canada, U.S., South America, India, and Japan (Figure 1). It was

TABLE 1
Species of *Hemiptera* Damaging Onions and Allied Crops

Latin name and authority	Common name	Insect family	Crop damaged or infested	Part of plant damaged	Country	Ref.
Alphis fabae Scopoli	Black bean aphid	Aphididae	Leek	*	Worldwide	4
A. gossypii Glover	Melon aphid or cotton aphid	Aphididae	Onion	*	Worldwide	4
Bactericea tremblayi (Wagn.)	Jumping plant louse	Psyllidae	Onion	Foliage	Turkey	11
Capsodes lineolatus (Brulle)		Miridae	Onion	Foliage	Italy	12
Cryptomenus bergi Froeschner	Shield bug	Cydnidae	Onion	Bulb	Colombia	12
Labopidea allii (Knight)	Onion plant bug	Miridae	Onion	Foliage	U.S.	14,15
Labopidicola allii (Knight)	Onion plant bug	Miridae	Onion	*	U.S.	15
L. ainsliei (Knight)		Miridae	Onion	*	U.S.	15
Lygus hesperus Knight		Miridae	Onion	Seedlings	U.S.	16
L. elisus (Van D)	Pale legume bug	Miridae	Onion	Seedlings	U.S.	17
Myzus ascalonicus Doncaster	Shallot aphid	Aphididae	Onion	Bulb	U.K.	18
Myzys cymbalariae Stroyan		Aphididae	Onion	Bulb	U.K.	4
Myzus persicae (Sulzer)	Peach-potato aphid	Aphididae	Alliums	Foliage	Worldwide	18
Neotoxoptera formosana (Takahaski)	Onion aphid	Aphididae	Onion, chives, leek, garlic		Australia	19
N. oliveri (Essig)		Aphididae	Chives	*	Australia	19
Rhopalosiphum maidis (Fitch.)	Corn leaf aphid	Aphididae	Leek	*	Bulgaria	20
Trioza brassicae (Vasilev)		Psyllidae	Spring onions, leek	*	Bulgaria	21
T. tremblayi (Wagn.)		Psyllidae	Onion	Leaves	Italy	22—24
Various species	Leaf hoppers	Cicadellidae	Leek, onion	Foliage	Poland	25

* Unspecified in reference.

FIGURE 1. Distribution of *Myzus ascalonicus*.

first reported infesting stored onions sets in 1940, in Wyoming[1] and subsequently in December 1941 in stored shallots in Spalding, England.[2]

b. Biology

The wingless forms vary in color and may be pale brown, greenish-brown, straw-colored, or dirty yellow. They have a shiny, strongly-convex back, black eyes, and dark sections on the legs and antennae. They vary in size from 1.1 to 2.2 mm long. The winged forms have a black head, thorax, legs, abdominal markings, and antennae with dark abdominal appendages; they measure 1.3 to 2.5 mm long. A detailed description of this insect's anatomy and morphology has been published.[3]

The wingless forms of the closely-related *Myzus cymbalariae* are oval and slightly smaller (1.2 to 2.0 mm) than *M. ascalonicus*. They vary considerably in color being yellow-green, yellow-brown, or pigmented with red. The winged forms have dark bands across the top of the abdomen.

Myzus persicae is similar in general shape and color to *Myzus ascalonicus* and can be separated using various anatomical features.[2]

Myzus ascalonicus colonizes plants in more than 20 familes, especially the Alliaceae, Caryophyllaceae, Compositae, Cruciferae, Liliaceae, and Rosaceae.[4] *Myzus cymbalariae* colonizes a similar range of hosts to *M. ascalonicus* while *M. persicae* is more polyphagous and colonizes plants of more than 40 families.

c. Life History

No sexual forms of *M. ascalonicus* have been discovered. It is therefore believed to be exclusively parthenogenetic, producing live young by asexual means; loss of a sexual phase in the life cycle is a frequent phenomenon in all major aphid groups. Winged and wingless forms of the aphid may be found on plants. Nymphs moult several times before becoming fully grown. *M. cymbalariae* also reproduces mainly parthenogenetically but males are known in this species. *Myzus persicae* has winter and summer hosts, the life cycle alternating typically between eggs laid on *Prunus* species in winter and a wide range of hosts in summer months. However, this aphid also reproduces parthenogenetically throughout the year in glasshouses, polythene tunnels or wherever the climate permits the active stages to survive unfavorable periods.

d. Damage

Usually *M. ascalonicus* is only a pest of the stored crop while *M. persicae* can cause damage to seedlings and the growing crop under protected conditions and outdoors. The aphids feed on the fleshy scale leaves and young shoots of bulbs in storage and on growing plants. Their attack leads typically to leaf curling, puckering, and distortion of leaves and stems and malformation of flower parts. Damage caused by high numbers of aphids renders bulbs unfit for planting. The numbers of aphids build up rapidly under protected conditions. The winter may be passed on bulbs in storage and, under these conditions, the onions and other crops deteriorate. Apart from the direct damage, these aphids transmit many viruses making them severe pests (see chapter on Virus Diseases). *Myzus ascalonicus* is known to transmit more than 20 viruses including Onion Yellow Dwarf and Shallot Yellows virus, while *M. persicae* is the most important aphid virus vector transmitting more than 100 virus diseases.[4,5] A comparison of the transmission of various viruses by these two *Myzus* species was made by Doncaster and Kassanis.[2]

e. Monitoring and Predicting Pest Activity

Aphids can be monitored caught using various types. For example, aphid numbers are monitored routinely at a number of sites in Great Britain, coordinated from the AFRC Institute of Arable Crops Research at Rothamsted Experimental Station. Suction traps are used for sampling these insects and weekly bulletins issued from April to November provide details of the activity of various aphids including *M. ascalonicus, M. persicae,* and *Aphis fabae.*

f. Control

i. Cultural Methods

Stored bulbs of various crops can be protected from colonizing aphids by insect-proofed stores.

ii. Biological Control

Aphids have large numbers of natural enemies including birds, other insects, spiders, bacteria, fungi, and viruses.[6]

Attempts have been made to use parasites and predators in biological control programs[7] but few are used in commercial practice. The build-up of predators and parasites is usually in phase with, but one step behind, that of the aphids themselves and eventually the predators or parasites may therefore suppress but not prevent outbreaks.

Hymenopterous parasites have failed to show the expected promise in glasshouse studies, partly because of their slow rates of increase but also because they may not parasitise an aphid fully until it has produced several young.[8] The use of the fungus *Verticilluim lecani* was commercialized[9] as Vertalec in the U.K. but is no longer marketed. This fungus is particularly effective against three species, which are pests of *Allium,* namely *Aphis gossypii, Aphis fabae,* and *Myzus persicae* but its use was restricted to glasshouse crops including chrysanthemum, cucumbers, and tomatoes because high humidities provide conditions suitable for spore germination.

iii. Chemical Control

Many insecticides are available for the control of aphids including organophosphorus compounds, carbamates, synthetic pyrethroids, and naturally occurring plant extracts such as derris, nicotine, and pyrethrum. Most of these insecticides are equally effective under protected conditions and in the field.[10] The insecticides may be contact nonpersistent, contact persistent, or systemic in their action. Systemic chemicals are particularly effective as they are translocated to parts of the plant which may not be reached by chemical sprays yet which harbor aphid colonies. They are also much safer to use where natural enemies and insect pollinators are active. Certain other insecticides have a minimal effect on natural enemies and their use is strongly recommended. For example, pirimicarb is valuable in this respect.[10]

Strains of several aphid species, e.g., *Myzus persicae,* have developed which are moderately or strongly resistant to many carbamate, and organophosphorus compounds. These strains may be particularly troublesome in glasshouses and polythene tunnels where ideal conditions exist for the continuous and rapid development of damaging populations of aphids. Quick-acting insecticides may need to be applied immediately after aphids are discovered to prevent damage and limit the spread of viruses within the crop.

III. THRIPS OR THUNDERFLIES. ORDER: THYSANOPTERA

Thrips are small insects which can build up rapidly to very large populations and cause a great deal of damage to onions and their allies. Jones and Mann[26] considered them to be the most severe pest of these crops and attacks by thrips can totally destroy young plants. In addition to the direct damage resulting from their feeding activity, thrips can transmit virus diseases to crop plants. Several species attack Alliaceae; here the onion thrips is described in detail and notes and references are provided on other species recorded as pests of onions and related crops (see Table 2).

A. FAMILY: THRIPIDAE

1. *Thrips tabaci* Lindeman, Onion Thrips

a. Distribution

The onion thrips is distributed widely throughout the world, ranging from north temperate zones to the equatorial regions. It occurs in Africa, Asia, Australia, North and South America, and Europe (see distribution map, Figure 2). It colonizes onions grown from sea level up to 2000 m.[27] Like most insects, thrips have been spread by natural means and by man's trading. Such small insects can easily escape notice in food or fiber products and establish themselves rapidly in a new location.

b. Biology

Thrips have an incomplete metamorphosis and the life stages include eggs, larvae, prepupae, pupae, and adults. The pale yellow or white eggs are cylindrical, bean-shaped, and smooth. Females lay about 80 eggs which are inserted singly in an incision made with the ovipositor (egg-laying organ) in the cotyledons, leaf, or flower tissues. One end of the egg protrudes from the slit to facilitate emergence of the larva, which take 5 to 10 d depending on temperature. The larva develops through two stages before becoming a pre-pupa and this is followed by a pupal stage. Pupae may be found on the ground or in crevices below the surface of the soil. The larvae, pupae, and adults resemble each other in appearance. The adults are up to 2 mm long, slender and tapered at both ends. They are green, pale yellow, or brown with prominent reddish-brown eyes, a pair of slender jointed-antennae, and two pairs of wings fringed with hairs. When at rest, the wings are folded along the back of the insect. The mouthparts are modified to form a proboscis or cone. Within this cone are sheathed the flexible stylets used to pierce and rasp plant tissues. Immature stages of the onion thrips are wingless, paler in color and smaller than the adults.

c. Life History

In some species of thrips, male and females are found frequently; in others males occur

TABLE 2
Species of *Thysanoptera* Damaging Onions and Allied Crops

Latin name and authority	Common name	Crop damaged or infested	Part of plant damaged	Country	Ref.
Aelothrips collaris		Onion	*	India	47
Caliothrips indicus Bagnall		Onion	*	India	47
Frankliniella occidentalis (Pergande)	Western flower thrips	Onion	Seed head	U.S.	27
Heliothrips indicus Bagnall		Onion	*	India	47
Scirtothrips dorsalis Hood	Assam or chile thrips	Onion	*	India	47
Thrips angusticeps Uzel	Cabbage thrips	Onion	All parts	Europe	18
Thrips tabaci Lindemann	Onion thrips	All crops	All parts	Worldwide	27

* Unspecified in reference.

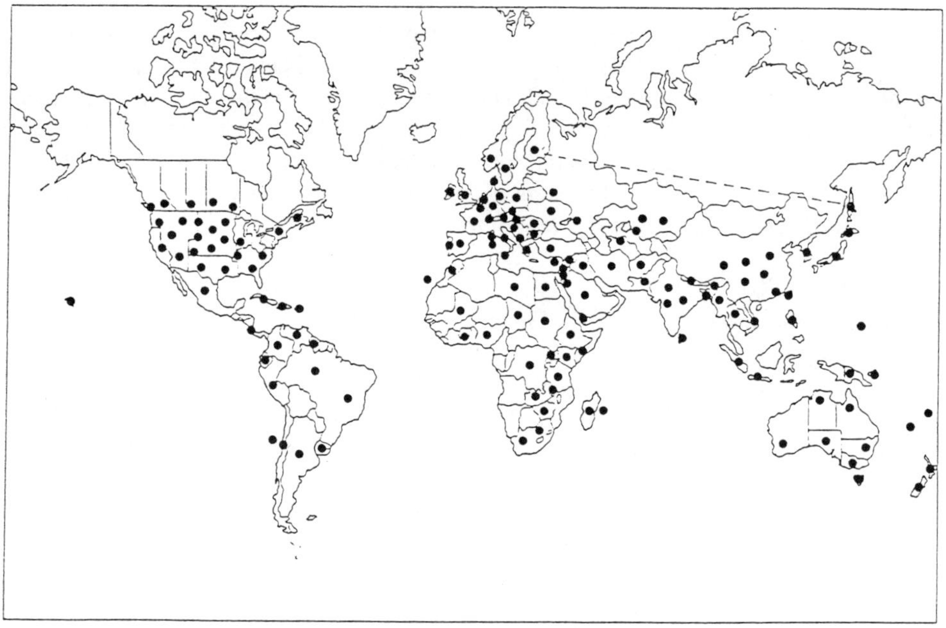

FIGURE 2. Distribution of *Thrips tabaci*.

infrequently; while in a few species males are entirely absent. In many species reproduction occurs regularly by parthenogenesis. Quite often, the sex ratio for a particular species differs from region to region or from season to season within a region. Males of *T. tabaci* are unusual in glasshouses in England but may be found locally outdoors. In other regions the sex ratio (male:female) ranges from 1:1 to 0:3000.[27]

The complete life cycle from egg to adult usually takes about 14 d but may be as short as 11 d at 30°C. The adults live for up to 20 d.[27] Prevailing conditions determine the developmental times and longevity of adults and these, in turn, determine the numbers of generations produced each year. Many species have periods of quiescience or diapause during the life cycle, either as fully grown larvae or adults. Diapausing larvae often remain in crevices in soil, leaf litter etc. and they may overwinter in this condition. In temperate climates there may be 1 or 2 generations each year. There may be several more under glass and, in regions where food is unlimited and the climate is favorable, thrips may breed continuously. *Thrips tabaci* can also overwinter in bulbs and emerge the following spring to infest newly planted crops.

d. Damage

Thrips cause highly characteristic damage consisting of silvery patches or streaks on the leaves and pseudostems which shine in the sun. This sheen results from the air which occupies the space in the damaged cells. Damage results from the piercing and rasping of plant tissues by the mouthcone. A glandular secretion which lubricates the mandibles and other mouthparts also aids digestion. The contents of the ruptured cells are ingested. Patches of damaged tissues coalesce, resulting in distortion and twisting of leaves, stems and flower parts. Plants become dessicated, are said to have a "blasted" appearance ("white blast") and they may wither and collapse. Most insects are concealed from view and feed between leaf sheaths and pseudostems or between flower parts.

Apart from the damage inflicted by feeding, the insects transmit Tomato Spotted Wilt virus. Other pathogens carried on their bodies may gain entry to plants through wounds caused by thrips feeding. The feces, small blobs of colored material surrounded by clear fluid, are colonized rapidly by sooty molds and these blemishes, together with the presence of the insects themselves, can reduce the marketability of crops.

When conditions are dry and plants are suffering from a lack of water, damage is particularly severe and entire crops can be destroyed.[28] On a worldwide basis, onion thrips is believed to cause more damage than all other pests to alliaceous crops.[26] In the U.S. the losses resulting from onion thrips alone in the absence of control measures were estimated in 1981 to be 55% of the crop.[29] In India 50% of onion crops may be lost as a result of attack by thrips.[30] Young onion plants may be destroyed by heavy infestations[31] and seed production and seed viability can be greatly reduced.[32] All alliaceous crops are damaged by *Thrips tabaci*.

e. Monitoring and Predicting Pest Activity

Thrips are caught easily on sticky traps and it is practical to monitor their activity by regular sampling. Monitoring, coupled with regular examinations of plants make it possible to identify which crops are at risk and when control treatments are required. Andaloro and Shelton[33] described the sampling procedures used for thrips in onion crops in the northeast U.S. A few plants are selected for examination at several sites chosen at random in the crop. The numbers of juvenile and adult insects are counted on each plant, paying particular attention to the area between the newest leaves where often many insects are concealed. The numbers of damaged leaves per plant and the percentage of leaf area with the typical silvery-white blotches are estimated visually. Damage ratings or reliable estimates of populations can then be derived for the crop.[34-37] In addition, it is recommended that quick checks for damage are made at the edges of onion crops, particularly where they border woods, cereal or other field crops. It thrips are found during this quick check, a more extensive examination within the crop is recommended

f. Control

A wide range of control measures is available to reduce the severity of damage to *Allium* crops.

i. Cultural Method

Turgid, well-watered plants are able to withstand thrips damage better than those which are under water stress. This knowledge, and the fact that heavy rain washes many insects off plants has been put into practice on commercial farms. For example, in Queensland, Australia onions are irrigated in a deliberate attempt to reduce damage by *Thrips tabaci;* the benefits are as great as those resulting from insecticide use.[38]

Attention to crop husbandry and good hygiene can be beneficial. The destruction and burning of severely infested plants and debris can reduce the numbers of thrips. Ploughing and other cultivation practices reduce the numbers surviving in the soil. The time of sowing and harvesting crops can also reduce the severity of injury.[27] Crop rotation can reduce thrips infestation and is important particularly where the insects persist on crop remains on or in the soil. Species such as *Thrips tabaci* which colonize large numbers of weed species can be reduced by efficient control of weeds in and around the crops.

Mixed cropping of carrots and onions also has potential for reducing damage caused by *T. tabaci* to onion crops.[39]

ii. Biological Control

Thrips have many natural enemies which help to regulate their populations in the wild. For example, Saxena[40] listed several species found in India, and Lewis[27] published a comprehensive list of parasites and predators. However, there have been few deliberate attempts to encourage or release parasites and predators into crops.[27] More recently in India, predators were released into onions as part of an integrated program of control.[41] The Commonwealth Institute of Biological Control was commissioned to seek and identify natural enemies of thrips on onion crops in Central and Southern Europe. A parasitic fungus, *Entomophthora parvispora,* was the most promising species discovered[42] although field studies showed that epizootics of the fungus depended largely on their host's density and therefore, prospects for utilizing the fungus in a biological control program were limited. Another fungus, *Verticillium lecani,* attacks thrips and a commercial preparation of this fungus "Vertalec" was used, in the U.K., for the control of *T. tabaci* in glasshouses but it is marketed no longer. Apart from these parasites, several predatory mites have potential for the control of thrips in glasshouses and these are being exploited.[42,43]

iii. Plant Resistance

Resistance in onions to *Thrips tabaci* was discovered first in the U.S. in 1934.[44] The onion "White Persian" was found to possess factors which reduced the build up of populations on its leaves and it was tolerant to the damage. The resistance was found to be related to a wide angle of divergence of the two innermost leaves and the distance apart of the leaf blades on the sheath column. In a susceptible cultivar such as "Australian Brown", better shelter amongst the leaves was offered to the thrips than with "White Persian". This resistance was found to persist under a wide range of conditions and work began in seven states to utilize it in breeding programs. Painter listed key references to this work.[45] Other resistant cultivars include: "Grano", "Sweet Spanish", "Crystal Wax", and "Yellow Bermuda",[26] "Spanish White",[46] "Bombay White",[41] and the *A. fistulosum cvs.* "Japanese Bunching" and "Nebuka".[34]

iv. Chemical Control

Prior to the second world war insecticides for control of thrips included sprays or dusts of nicotine sulfate, naphthalene, hydrated lime, and carbon bisulphide or mixtures of these compounds.[28] Organochlorine and organophosphorus insecticides were used later but several applications may be required to suppress populations. Bromophos, carbofuran, carbaryl, diazinon, endosulfan, formothion, nicotine, phorate, synthetic pyrethroids, and sulfur dust have been recommended by various workers for the control of thrips.[10]

Under glasshouse conditions, *Thrips tabaci* can be controlled by using polybutenes which act as a physical trap; mixed with the insecticide deltamethrin they are marketed in the U.K. as "Thripstick".[42]

IV. MOTHS, ORDER: LEPIDOPTERA

A. FAMILY: NOCTUIDAE

1. Cutworms

Agrotis segetum (Schiff), turnip moth; *Agrotis ipsilon* (Hufnagel), black or greasy cutworm; and *Peridroma saucia* (Hubner), variegated cutworm (*Snyonyms margaritosa*; Haworth), *porphyrea* sensu auct.

Cutworms are the larval (caterpillar) stage of species of noctuid moths and are pests of many crops. Species known to be of economic importance on alliaceous crops are listed above and in Table 3. The following species are also occasional pests of vegetable crops: *Noctua pronuba* (L.), the large yellow underwing; *Euxoa nigricans* (L.) the garden dart moth; and *Euxoa tritici* (L.) the white-line dart moth.

Mamestra brassicae (L), the cabbae moth is another noctuid species which does not feed as a typical cutworm. It attacks alliaceous crops occasionally.

a. Distribution

Cutworms are major pests of most vegetables including alliaceous crops and other crops throughout the world. In India[48] an *Agrotis* species was reported as a new pest of onions in 1967. Nikolova[49] in Bulgaria and Rivnay[50] in Israel reported cutworms damaging various vegetables. Caterpillars of one or more species of cutworms have been reported damaging vegetables in U.S.,[51] Canada,[52] and France.[53] Quiroz[54] reported *Agrotis* species as occasional pests of onions in Chile and their range extends to Brazil and as far south as New Zealand, Tasmania, and South Africa. Further details of the distribution of *Agrotis ipsilon* are given in Figure 3.

b. Biology and Life History

The caterpillars possess a strong pair of cutting mandibles or jaws, three pairs of true legs and five pairs of false abdominal legs. They are generally dull, grayish-brown with faint lines along the sides and back and, when fully grown, they may be up to 35 mm long. Sometimes the caterpillars can be confused with chafer grubs, e.g., *Melolontha melolontha,* which have longer true legs and black abdominal legs.[55] They can also be confused with leatherjackets, *Tipula* spp., which lack a distinct head and legs and are uniformly grey in color. Adult moths of various species can be distinguished by their wing markings. Details of the biology and life history of the three most important cutworm pests of alliaceous crops are given below.

2. *Agrotis segetum* (Schiff), Turnip Moth

In Europe *A. segetum* is a common and serious pest of vegetables.[6] The adult moths emerge from late May until end of June/early July. Their forewings are about 30 mm long and gray to brown with a distinct kidney-shaped mark. The hind wings are pearly white. Each female may lay up to 1000 eggs on leaves or stems of common weeds and cultivated plants or even on the soil surface or plant debris.[55] Depending on the temperature, the eggs hatch after 8 to 24 d. The caterpillars have glossy spots, each with a hair, along both sides of the body.[6] They are pale gray or grayish-brown with a dark, pale-edged line down the back. The under surface is light gray and there is a X-shaped mark on the head. The cutworms feed on plant leaves but later instars descend to the ground and adopt the typical feeding habit of cutting through soft stems close to the ground. Most of the caterpillars feed slowly and are not developed fully until late autumn, when they overwinter in the top 5 cm of soil as fully grown caterpillars or pre-pupae. Feeding then ceases and pupation occurs in early April to mid-May of the following year. The pupa is brown, 10 to 12 mm long, smooth and heavily chitinized with two spines at the posterior end.

TABLE 3
Species of *Lepidoptera* Damaging Onions and Allied Crops

Latin name and authority	Common name	Insect family	Crop damaged or infested	Part of plant damaged	Country	Ref.
Acrolepiopsis assectella (Zeller)	Leek moth	Yponomeutidae	Leek, onion, chive, garlic and shallot	Foliage and flowerheads	France	87—89
Acroleopiopsis sapprorensis (Matsumura)	Stone leek miner	Yponomeutidae	Onion	Foliage	Japan	90
Agrotis ipsilon (Hufnagel)	Black or greasy cutworm	Noctuidae	Onion	Foliage, stems	Canada	52
Agrotis segetum (Schiff)	Turnip moth	Noctuidae	Onion, leek	Foliage	U.K.	6,61
Copitarsia sp.		Noctuidae	Onion	Foliage	Chile	54
Clepsis spectrana (Treitschke)	Cyclamen tortrix	Torticidae	Onion	Foliage	U.K.	103
Cacoecimorpha pronuba (Hubner)	Carnation tortrix	Tortricidae	Leek	Foliage	France	104
Cryptoblabes gnidiella (Mill.)	Christmas berry webworm; honeydew moth	Pyralidae	Garlic	Stored bulbs	Egypt	105,106
Dyspessa ulula (Bkh.)		Cossidae	Onion	Bulb	Yugoslavia	107
Ephestia cautella (Walker)	Almond moth	Pyralidae	Onion, garlic	Stored bulbs	Egypt, India	76,77
Ephestia kuehniella (Zell.)	Mediterranean flour moth; mill moth	Pyralidae	Onion	Bulb	Egypt	76
Estigmene acrea (Drury)	Salt marsh caterpillar	Arctiidae	Onion	Whole plant	U.S.	26
Euxoa nigricans (L.)	Garden dart moth	Noctuidae	Leek	Seedlings	U.K.	55
Euxoa tritici (L.)	White-line dart moth	Noctuidae	Leek	Seedlings	U.K.	55
Faronta sp.	Army worm	Noctuidae	Onion	Foliage	Chile	54
Gymnoscelis pumilata (Hli.)		Geometridae	Onion	*	Italy	108
Heliothis armigera (Hubner)	Corn earworm; gram caterpillar	Noctuidae	Onion	Seed	India	109—111
Hydraecia mongoliensis (Urbahn)	Onion stem borer	Noctuidae	Onion	*	Japan	112
Loxostege sticticalus (L.)	Beet webworm	Pyralidae	Onion	*	USSR	113
Mamestra brassicae (L.)	Cabbage moth	Noctuidae	Onion, leek	Whole plant	U.K.	74

Noctua pronuba (L.)	Large yellow underwing	Noctuidae	Leek	Seedlings	U.K.	55
Peridroma saucia (Hubner)	Variegated cutworm	Noctuidae	Onion	*	U.S.	51

* Unspecified in reference.

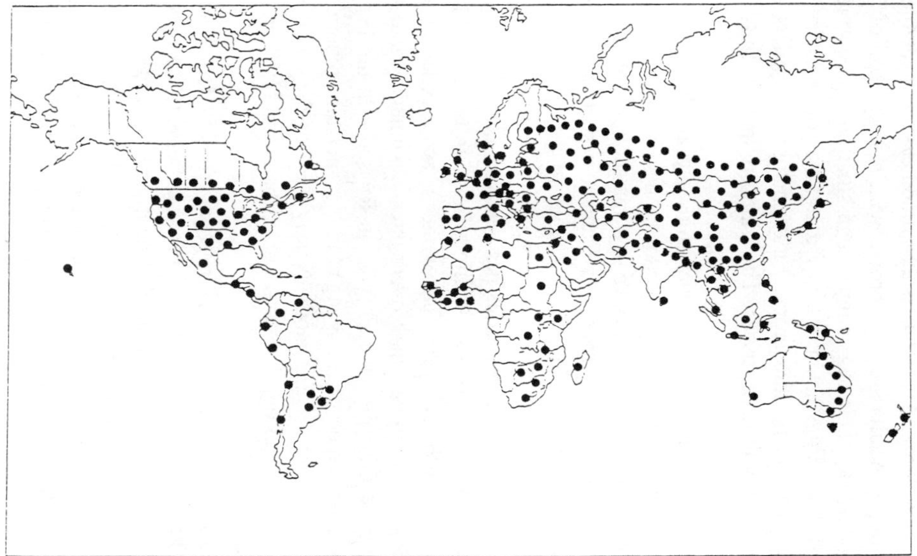

FIGURE 3. Distribution of *Agrotis ipsilon*.

A few caterpillars grow more rapidly and produce a second generation in September or October. Cutworms do not cause any appreciable damage to crops during the spring in the northern hemisphere.[55]

In Bulgaria, Genchev reported two generations of *A. segetum* each year. Most insects overwintered as fully developed sixth instar larvae which pupated in early May. Adults emerge in the second half of May and early June. In the field caterpillars occur mainly in the soil at depths of 2 to 4 cm and prefer soil humidities in the 20 to 40% RH range. On onions, the larval stage averages 34 d, pupal weight is 300 mg and each gravid female lays nearly 1000 eggs.[56]

3. *Agrotis ipsilon* (Hufnagel), Black or Greasy Cutworm

The black or greasy cutworm, *A. ipsilon,* is a common pest of vegetable crops in America and also occurs in Europe. Adults are dark brown with a wing span of 25 mm.[52] In May or early June they usually lay their eggs singly or in small groups on grasses or leaves of weeds or on the soil surface. Eggs hatch in 5 to 7 d and the young caterpillars feed initially on grass or weeds before moving into crops. Fully mature caterpillars are about 25 mm long, greasy gray to brown with faint stripes. The winter is spent as fully grown caterpillars or pupae.

In Canada the black cutworm has two generations a year, the second occurring in late August or early September. In most cases it is not thought to overwinter there but a small proportion of pupae may overwinter and survive in some sheltered areas.[53] Most infestations result from moths which migrate northwards to Canada from U.S. during May. The first generation caterpillars pupate in late July/early August to give rise to a second generation of moths in late August/early September which produce another brood of caterpillars. In Tennessee there can be up to four generations per year. Nikolova[49] in Bulgaria, and Santos and Nakano in Brazil[57] published details of its life history. In Bulgaria it has three complete generations and a partial fourth each year. The winter is spent generally in the pupal stage but some fully developed caterpillars may overwinter. Akhmedov and Abdinbekova[58] reported four generations of this species each year in Russia with the last generation overwintering as fully developed caterpillars. Wang et al.[59] from China reported on the nocturnal activities of the first generation of this moth. Nearly 75% of the adults emerged between 18:30 to 2:00 h and pairing and egg-laying occurred between 23:50 and 1:50 h and 5:00 and 5:30 h, respectively.

Rivnay[50] reported *A. ipsilon* as a serious pest of crops in Israel, causing damage in spring and autumn. There were four generations per year but temperatures over 26°C during summer months affected larval and pupal development, and reduced fecundity in subsequent generations of moths. High or low humidity also affected fecundity and 75% RH was found to be the optimum level. The author concluded that there was probably an annual imigration of the moths from North Africa to Lebanon, with many moths stopping *en route* in Israel.

4. *Peridroma saucia* (Hubner), Variegated Cutworm

Like many noctuid moths, the variegated cutworm or pearly underwing, *Peridroma saucia*, has a very wide range of host plants and has been reported as a serious pest from at least 10 countries in four continents.[51] It occurs sporadically in very large numbers, attacking many vegetables in Canada and the U.S. A bibliography[51] of its host-range, as well as a worldwide annotated bibliography of this pest has been published.[60]

It is the most important cutworm attacking vegetables in the U.S. and it can be a serious pest of onions. The adults have reddish brown forewings, 47 mm long, with whitish scales.[61]

The hindwings are translucent white with brown edges. Females lay between 300 to 2000 eggs, which are hemispherical, glossy yellowish white and strongly ribbed. The caterpillars are generally defoliators and behave as cutworms severing plants near ground level.

The body is pale grayish brown with a pale dorsal line and reddish-brown spiracular lines. The thoracic legs are yellowish-brown and the prolegs have brown lateral plates. The reddish-brown pupa has two, finely tapered, outward-curving spines. The adults migrate regularly from central and southern Europe to northern Europe and the British Isles. They fly at night and are found in temperate zones from May — November with the largest numbers in September — October. The caterpillars are active during June — July and from September — November, feeding during the night and burrowing below ground during daytime. The first generation caterpillars move below ground in August and early September to turn into earthen colored pupae which give rise to a second generation in September. This species seldom survives in northern Europe or Britian and infestations originate almost entirely from an influx of moths from southern Europe during the following year.

a. Damage

On hatching, the young cutworms feed on leaves; typical symptoms are holes made by feeding but generally no serious damage is caused.[55] Older cutworms are nocturnal, hiding during daytime under stones, leaves, in thick grass or in the soil. At night they feed sometimes on leaves but typically they attack the stem of crop plants cutting the top from the tap root near the soil surface. They can be particularly troublesome in row crops where the caterpillars tend to move from plant to plant, destroying sections of rows.[28]

Typical symptoms of crop damage are poor top growth or wilting caused by caterpillars feeding on the stem or roots near the soil surface. If plants are found completely severed from the roots, cutworms are the most likely culprits but sometimes damage of this type may be caused by the grubs or caterpillars of chafer beetles. Damage caused by slugs may also be confused with that caused by cutworms. Typically, damage caused by cutworms is most noticeable during hot dry summers in light and medium soils while slugs generally prefer moist conditions in the summer and autumn on heavier soils. Slugs also leave a slime trail whereas cutworms do not.[55] In India, damage by cutworms[48] was more serious on young rather than on old onion plants and infestations were heavier in dry rather than in irrigated fields. Severe damage by cutworms on onions and other crops was also reported in Hungary.[62]

b. Monitoring and Predicting Pest Activity

A knowledge of the pests' biology is a prerequisite in formulating a rational approach to pest management. Entomologists monitor the activity of adult cutworm moths and damage caused by caterpillars. Also, agarometeorologists use temperature and rainfall data to provide an index of egg development and caterpillar survival. In the U.S, Wilson et al.[63] found that traps baited with a sex attractant were significantly more efficient than ultraviolet light traps in the spring to monitor adult populations of *Peridroma saucia* and *A. ipsilon*. However, during the summer and autumn flights the ultraviolet light traps caught more adults. Clement et al.[64] used modified Wong traps baited with the pheromone and concluded that the optimum size for a Hercon dispenser was 12.90 cm^2 with a total exposed edge of 15.24 cm. Lofstedt et al.[65] found optimum attraction of males of turnip moth, *A. segetum,* to sex pheromones in a wind tunnel and in fields to be from 1 to 30 μg lures. Peak responses to the synthetic pheromones were, however, lower than to female glands.

In the U.K. a model was devised[66] to estimate an index of larval survival to third instar, based on temperature rate of development relationships and mortality due to daily rainfall. This survival index was highly correlated with year-to-year changes in numbers of reported attacks by *A. segetum*. Esjberg[67] published details of a program for monitoring cutworms in Denmark. A sex pheromone trap for the turnip moth provided a more efficient monitoring system than that possible with light traps. Using a synthetic preparation of the turnip moth sex pheromone and a new trap design, adult catches were improved more than tenfold compared to a single trap baited with virgin females. Trapping data were used to forecast the levels of insect activity and to determine the optimum time for insecticide treatment. A statistical model for forecasting attack-damage thresholds was developed on the basis of historical records and data from the current year's monitoring.

c. Control

i. Cultural Methods

Cutworms favor weedy land which provides cover for egg-laying females and food for newly hatched larvae.[55] Crops planted following heavy weed cover are more likely to be damaged than those on land kept free of weeds. In laboratory studies on the effects of soil moisture, Nikolov[68] found that the optimum moisture content in the soil was 45% for *A. ipsilon* and only 5 to 25% for *A. segetum.* Young turnip moth caterpillars can only survive in dry soil and more of them reach maturity in hot dry summers than in years with above average rainfall in June and July. Therefore, regular irrigation of susceptible crops should prevent severe infestations.[55]

ii. Biological Control

In Egypt[69] various preparations of *Bacillus thuringiensis* were tested in the laboratory against three noctuid moths, including *A. ipsinol.* Two of the preparations were effective against the caterpillars of this moth. In Missouri and adjacent states in the U.S., Puttler and Thewke[70] identified a nematode (*Hexamermis arvalis*), found commonly in clover fields, parasitizing *A. ipsilon* as well as caterpillars of two other noctuid species. Parasites from overseas were introduced into western Australia for the control of noctuid pests.[71] In Hungary, larvae of *A. segetum* were infested by a number of parasites. For example 25.9% parasitism by *Peleteria nigricornias* (Mg) was recorded[62] and the authors concluded that parasites may play an important role in suppressing outbreaks of cutworms.

iii. Chemical Control

Cutworms attack a wide range of host plants and they tend to wander over the soil surface in search of food. Until recently poison baits were therefore located under covers in infested areas of fields late in the day so that the bait remained moist and attractive. These baits attracted the caterpillars in large numbers and resulted in up to 80% larval mortality. However, in the U.K. poison baits have been replaced by sprays.

In New York State, Eckenrode et al.[72] reported that synthetic pyrethroid compounds were the most effective insecticides against *A. ipsilon* on onions in glasshouse and field trials. The effectiveness of these compounds was confirmed by Harris et al. in laboratory and field trials on onions and other crops in Canada.[73]

In Great Britain[55] chemical control is recommended on susceptible crops while the caterpillars are young and feeding on foliage.

5. *Mamestra brassicae* (L); the Cabbage Moth

Caterpillars of the cabbage moth, although primarily major pests of brassicas, occasionally attack onions and leeks;[74] they attack all parts of the plants. This moth's range extends from Europe to Japan and into subtropical Asia including India; it is very common in the British Isles and continental Europe. The adults, which fly at night, appear in late May and early June and have grayish brown forewings mottled with dark brown; the wingspan is 42 mm.[6,56,74] The hindwings are grayish brown and pale toward the base with a terminal white line. The globular, finely reticulated, and ribbed pale brownish eggs are laid singly or in groups of 20 to 100 on the undersides of leaves. The young caterpillars are light green but vary in color as they grow older; they may remain light green or change to a darker green, brown or even black on the dorsal surface while the under surface and legs vary from

FIGURE 4. Distribution of *Ephestia cautella*.

green to yellow. The caterpillars feed for 4 to 5 weeks and then burrow into the soil to form glossy reddish-brown pupae.

There is usually one generation a year but occasionally there may be a partial second generation, the caterpillars feeding until October before pupation. In the U.K., a range of organophosphorus, carbamate and synthetic pyrethroid compounds, as well as *Bacillus thuringiensis,*[75] can be used to control the caterpillars. Other foliage-feeding noctuids are listed in Table 3.

B. FAMILY: PYRALIDAE

1. *Ephestia cautella* (Walker), Almond Moth

Synonyms: *Cadra cautella,* dried currant moth, fig moth, tropical warehouse moth.

The almond moth is a pest of stored products, damaging dried fruits and stored vegetables. It causes severe losses in warehouses and stores. It is primarily a pest of subtropical and tropical regions but occurs indoors in other regions. An infestation of the closely-related *Ephestia kuehniella* Zell. was reported from Egypt.[76]

a. Distribution

It is a cosmopolitan species and has been transported throughout the world on dried food products. (Figure 4).

b. Biology and Life History

Carter[61] described the stages and life history of this moth in detail. Adults are 21 mm long with a grayish brown head, thorax, abdomen and forewings, and white hindwings. The head bears simple slender antennae. Adult moths fly at night and commonly are seen resting on warehouse walls during the day. The eggs are oval, highly sculptured and grayish white turning pinkish as they develop. The females scatter up to 350 eggs over stored food products. The eggs hatch in 4 to 14 d and the larvae commence feeding. The larvae are pink or yellowish white and are covered with yellowish-brown hairs; the head is reddish brown with two dark brown patches on either side. They produce webbing which shrouds the produce. When fully grown, the larvae crawl into the stored products and pupate in a silken cocoon; the pupae are yellowish-brown or reddish-brown. At 20°C the life cycle is completed in 64 d and there may be up to six generations a year when conditions are favorable.

c. *Damage*

Onions and garlic may be damaged severely in warehouses. Their cast skins and frass and the webbing which shrouds the foodstuffs and containers reduce the marketability of stored products. Up to three larvae may be found in a single onion bulb and one larva can damage several bulbs.[77]

d. *Monitoring and Predicting Pest Activity*

Light traps are used in warehouses to indicate the presence of pest Lepidoptera and to reduce their numbers. A green light was found to be more attractive than ultraviolet light or a combination of both in tests on the almond moth in Georgia.[78]

Male almond moths were attracted to crude extracts and a synthesized form of the almond moth sex pheromone but the crude extract remained attractive to the males longer than the synthetic pheromone.[79] The latter was also evaluated in field trials in Vietnam where it attracted male almond moths.[80]

e. *Control*

i. Biological Control

Moths have many natural enemies. The minute egg parasites which belong to the hymenopterous insect family, the Trichogrammatidae, are important and have been used against pest Lepidoptera in several biological control programmes. *Trichogramma* species suppressed populations of the almond moth in stimulated warehouse experiments.[81]

ii. Genetic Control

The potential of the sterile-male release technique for controlling the almond moth has been investigated in the U.S.[81-83] At Gainesville in Florida, a combination of an atmosphere permeated with a sex-pheromone and sterile males in a ratio of 2 sterile: 1 fertile reduced the reproduction of almond moths. These studies indicated that the technique may be a practicable proposition.

iii. Chemical Control

Organochlorine insecticides were used for many years to control this pest. Bulbs were dipped in solutions containing DDT or methoxychlor and these treatments prevented infestations of the moth.[84]

Organophosphorus carbamate and synthetic pyrethroid compounds have been used more recently. However, the pest has developed strains which are resistant to insecticides including malethion.[85,86]

Various fumigants including aluminium phosphide have been used to control the almond moth.[77]

C. FAMILY: YPONOMEUTIDAE

1. *Acrolepiopsis assectella* (Zeller), Leek moth

Also known as *Acrolepia assectella* and misspelling, *Acrolepiopis assectella.*

The leek moth is a severe pest of leek, particularly in western Europe, and it also attacks onion, chive, garlic and shallot. The caterpillars can feed on the bulb, leaves, and flowering stems. Several aspects of the biology and control of this insect have been studied in detail in recent years by French research workers.[87-89] The closely related species *Acrolepiopsis sapporensis* Matsumura, the stone leek miner, is an occasional pest of onions in Japan.[90]

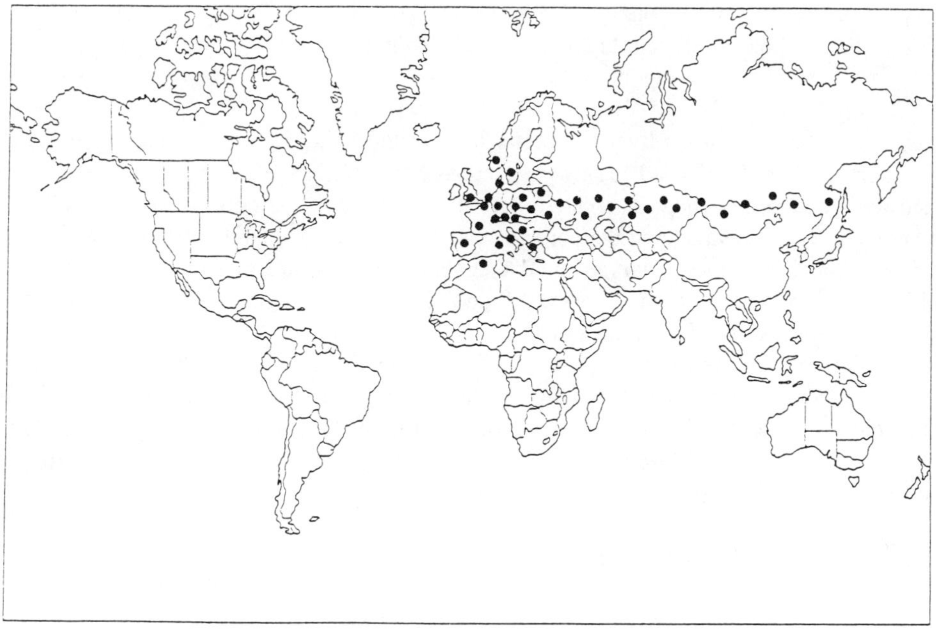

FIGURE 5. Distribution of *Acrolepiopsis assectella*.

a. Distribution

The leek moth is distributed throughout most of Europe, southern Scandinavia and parts of Russia; it has also been reported from Algeria[87] and Hawaii[61] (Figure 5). The insect was first reported in Britain in 1943,[91] confined initially to south and east England, particularly in coastal regions.[92-94] It has spread further inland in Britain over the last thirty years.[95]

b. Biology

The eggs of the leek moth are irridescent white, oval with raised reticulated sculpturing on the surface. Eggs are laid one at a time on leeks or onions at or near ground level, but each female may lay several around one plant. Each female may lay up to 200 eggs. Egg development lasts 4 to 8 d, depending on temperature. The caterpillar is grey at first but later changes to a yellowish-green. The head and abdominal legs are yellowish-brown. When fully grown, the caterpillars are 13 mm long; they spin an open network, delicate cocoon and form a reddish-brown pupa on dead vegetation near the host plant or among the stalks of seed heads.

The moth is small, dark brownish-grey and has a wing span of 16 to 18 mm. The forewings are pale brown and possess white scales on their hind edge; these scales form a distinctive triangular mark when the wings are folded. The hindwings are pale gray and darker toward the apex. The head and thorax are dark brown, and the abdomen grayish brown. Further details of the biology of this species are given by Carter.[61] Adult moths are attracted to host plants by volatile sulfur compounds released from alliaceous plants.[88]

c. Life History

This pest overwinters as a pupa or as a moth. As soon as temperatures rise to 12 to 13°C, the adults become active; this is usually during April in France. Eggs are laid on food plants (leek, garlic, onions, or shallot) at or near ground level.[91,94] Overwintered plants of leek, perennial onion, and garlic are chosen commonly for egg-laying in Britain.[94] The larval stage lasts 25 to 30 d, depending on temperature. The insect can complete up to 5 generations per year in France,[89] occasionally 6 in southern Europe and 8 in Algeria[87] but only 2 in Britain; the complete life cycle takes from 1 to 2 months depending on the season.

d. Damage

Initially the caterpillars live as miners, leaving the epidermis of the host plant intact. They bore through the folded leaves producing a "shot-hole" effect which is seen easily when the leaves are opened. The caterpillars then migrate to the core of the plant where a vertical cavity is excavated and the young leaves are destroyed. Plants rot and die as a result of this damage. In onions, larvae mine the hollow leaves and penetrate the stem base, inducing bulb rot. Several caterpilars are found commonly in a single plant. The tip of the flowering stem may become honeycombed with tunnels and the seed head disintegrates. In parts of France, damage to flower heads on the seed crops results from the third generation of caterpillars and is the most damaging.[96] Disruption of the tissues facilitates the entry of pathogens and the onion fly. As with many insect/crop relationship, the amount of economic damage depends on the balance between the speed of tissue consumption by the caterpillars and the rate of regeneration of plant tissues.[97]

e. Monitoring and Predicting Pest Activity

The monitoring and prediction of leek moth activity has been the subject of considerable research, particularly in France, and now forms an integral component of systems for the prevention and control of the pest. As with most insects, the leek moth requires a fixed number of accumulated thermal units (day degrees [D°]) to complete its development. This information can be used to predict leek moth activity in the field and forecast likely emergence dates of the adult months. Originally the warnings to farmers of the likelihood of leek moth attacks were based on observations of the development of the pest in the laboratory.[98] To provide more precise information, two methods of trapping the insect have been used to determine emergence dates.[99] One was to record dates of moth emergence in field cages placed over infested leek plots. The second, more accurate method was to use sex-phermone traps. Laboratory-reared, virgin female moths were maintained in traps fixed above the ground. Male moths emerging from previously infested crops were attracted to the pheromones released by the females and counted. The main component of the leek moth sex pheromone has been identified as (2)-11-hexadecenal[100] and traps baited with this substance were as effective as virgin females in attracting males. In field trials in France and Spain, the most effective baits were those which used polyethylene and rubber dispensers for the pheromone.[101]

f. Control

i. Cultural Methods

Cutting off all leaves from leek crops after hatching of the last larvae in the season (usually in September) and before the growth of the durable winter leaves in October can reduce the severity of the damage caused by this pest.[97] A reduction in the numbers of pupae that overwinter in fields can be achieved by deep ploughing and frequent cultivations; pupae are destroyed by the mechanical damage or are exposed to predators and unfavorable conditions.

ii. Biological Control

Bacillus thuringiensis is used in France to control the leek moth.[89] The potential of two solitary parasitic wasps *Diadromus collaris* (Grav.) and *D. pulchellus* Wesm. has also been studied.[102]

iii. Chemical Control

Goix[89] in France, provided a list of carbamate, organophosphorus, and synthetic pyrethroid compounds and their dose rates currently recommended for use against the leek moth. Applications are timed according to the pest's activity. The value of pest monitoring is well illustrated by the studies of Goix in the Creuse Department, Massif Central region France. Although the leek moth is a widespread and serious pest throughout this region, the moth population is usually low in certain districts at higher altitudes and leeks do not require protection. However, in some years, the moth can occur in these districts in devastating numbers and insecticides are then required. Pest monitoring enables the farmers to predict seasons when the crop is at risk and chemical sprays are required.

V. TWO WINGED FLIES (TRUE FLIES), ORDER: DIPTERA

Several different types of true flies are important pests of alliaceous crops; these include root flies, bulb flies, craneflies, and leaf miners. The adult flies feed largely on nectar and pollen. Whilst some have mouthparts adapted for piercing and sucking liquid, they do not damage plants. Female flies lay their eggs in or around host plants and from these emerge the larvae (maggots) which cause the damage to crops. The larvae grow rapidly, moulting several times, and pupate when fully grown. After a few weeks or following a period of rest, the adult flies emerge to complete the life cycle.

A. FAMILY: ANTHOMYIIDAE

1. *Delia antiqua* (Meigen), Onion Fly, Onion Maggot

Also known as *Chortophila antiqua, Hylemia antiqua, Hylemya antiqua, Pegomya cepetorum, Phorbia antiqua, Anthomyia ceparum* (Bouche), *Phorbia cepetorum* (Meade).

The onion fly is a major pest attacking bulbing and green onions in many parts of the world; the maggots can also cause serious damage to related crops like garlic, leeks, and chives. The most serious damage occurs in onions grown from seed but damage can also occur in older onion plants grown from transplants or sets.

a. Distribution

The onion fly is believed to have originated in the palaearctic region and to have spread to North America by the middle of the 19th centry. It is now indigenous in many parts of the northern hemisphere (Figure 6) and also occurs in Brazil[114] and Colombia.[115]

b. Biology

The adult fly, 8 to 9 mm long, closely resembles the common house fly but is slightly smaller and greyish in color. It has a dark stripe down the mid-line of the abdomen in addition to four brownish stripes on the throax.[116] The adult female can be identified by the length of the 5th pre-alar bristle on the thorax, which is nearly a third of the length of the other bristles.

The eggs are elongate, white, and about 1.2 mm long. They show a distinctive pattern on the surface.

The creamy-white maggots, about 8 mm long when fully grown, have no legs and can be distinguished into 1st, 2nd, and 3rd instar stages. They develop fully in about 3 weeks, leave the onion bulbs and burrow 5 to 8 cm into the soil to develop into ovoid, chestnut-brown pupae, the resting stage of the insect.

In many parts of the world, onion fly has several generations a year, e.g., in parts of Europe and U.S. there are three generations in the major onion growing regions (see Figure 7).

FIGURE 6. Distribution of *Delia antiqua*.

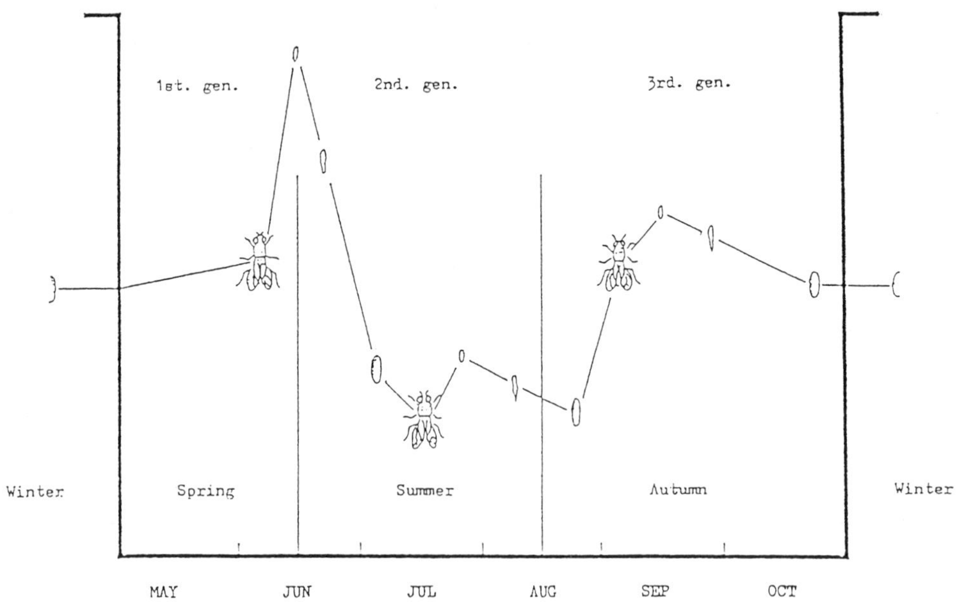

FIGURE 7. Schematic representation of the seasonal distribution of *Delia antiqua* life stages. (After Ellis and Eckenrode, 1979, with permission).[174]

c. *Life History*

Diapausing pupae tide the insect over unfavorable periods,[117] usually the winter months in North temperate regions. Occasionally during spells of high temperatures during summer, the pupae enter a resting stage termed aestivation. In the British Isles, the adults emerge in spring, generally during the third week of May, coinciding with the seedling stage of maincrop onions, which are sown normally in March. About 1 week after emergence, the females mate and soon start to lay eggs in the soil near the base of young seedlings. Under favorable conditions, the eggs hatch within 1 week and the tiny maggots burrow upward into the pseudostem of young bulbs. The maggots feed and develop over the next 3 to 4 weeks and then pupate in the soil. Depending on the summer temperatures, a second generation of adults emerges within 3 weeks during mid-July. Sometimes a partial third generation of adults emerges in late summer in southern Britain but normally the second generation of maggots develops into diapausing pupae which remain in the soil until the following spring.

At the limits of its geographical range, e.g., Scotland and Scandanavia, onion fly has only one or occasionally two generations a year,[118] while in more southerly regions including the northern United States, southern parts of Canada,[119] Egypt, and France it can have up to three complete generations. In Turkey[120] there may be up to four generations. In Israel[121] there are two peaks of adult activity during November — December and February — March. Pupae produced during the spring may estivate during the hot summer months and give rise to a generation of flies in November.[121]

d. Damage

The larvae damage plants as their hooked mouth parts bore into the slightly swollen bases of the bulbs.[122]

In north temperate regions, maggots of the first generation which cause the most severe damage, often during late spring and early summer when plants are small and vulnerable. Some damage also occurs in August as a result of the second generation of maggots.[116] The first sign of damage is wilting of the leaves, which become yellowish and flaccid and eventually the whole plant collapses and dies. Almost identical symptoms are caused in onions attacked by the seed corn maggot, *Delia platura* (Meigen).

If symptoms appear very early in the seedling loop or "flag" stage of an onion crop, they can usually be attributable to seed corn maggot. The onion fly emerges later and does not start egg laying until the crop is more advanced.[116]

When present in large numbers, maggots can destroy a high proportion of young plants. In the absence of any control measures, Harris et al.[123] reported losses of 20 to 40% of onion seedlings in Ontario, Canada. If attacked seedlings die before the maggots are fully grown, the maggots move to adjacent plants leaving a patchy crop. With onions grown from sets, the maggots enter the bulbs from the base or the side. A mature bulb may contain 20 to 30 maggots. Attach by second and third generation maggots can result in distorted growth and unmarketable crops. Primary damage by maggots can also lead to secondary infection by fungal and bacterial pathogens.[122]

Damage by onion maggot may be confused with that caused by seed corn maggot (see above) or onion stem nematode which commonly causes twisting of the leaves and rotting of the basal parts[116] (see chapter on Nematodes).

e. Monitoring and Predicting Pest Activity

Considerable effort has been made to predict onion fly emergence patterns and thus peak infestations in order that control measures can be made more effective. In Britain, there are 2 to 3 generations each year. Most insects require a fixed number of accumulated thermal units, commonly called day degrees (D°), to complete their development from one generation to the next. Day degrees are calculated from the numbers of daily accumulated thermal units above a base temperature. The relevant formula is:

$$D^\circ = T\ \mathrm{max} + T\ \mathrm{Min} - \mathrm{base}\ t^\circ$$

where t° is the base temperature (°C).[123] For onion fly this has been shown to be 5°C. A scheme using life cycle events based on day degree accumulations was used by Lui et al.[124]

TABLE 4
Species of *Diptera* Damaging Onions and Allied Crops

Latin name and authority	Common name	Insect family	Crop damaged or infested	Part of plant damaged	Country	Ref.
Ceroxys latiusculus			Onion	Bulb	Japan	205
Delia antiqua (Meigen)	Onion fly onion maggot	Anthomyiidae	Onion, garlic	Seedling bulbs	U.K., Canada, U.S., Turkey, Israel, Korea	116,117, 120—122,227
Delia platura (Meigen)	Bean seed fly seed corn maggot	Anthomyiidae	Onion, garlic, leek	Seedling bulbs	Turkey, U.K., U.S., Canada, Japan	120,183,186, 189,190
Eumerus amoenus Lw.	Onion bulb fly	Syrphidae	Onion	Bulb	Egypt	201,202,212
Eumerus strigatus (Fallen)	Small narcissus fly	Syrphidae	Onion	Bulbs	New Zealand, Rumania, Japan, USSR, Finland	200,204,205, 208,211
Eumerus tuberculatus (Rondani)	Small narcissus fly; lesser bulb fly	Syrphidae	Onion	Bulb	Japan, Russia, Finland	205,207,208, 211
Euxesta anna		Ottitidae	Onion	Bulb	Japan	205
Euxesta notata		Ottitidae	Onion	Bulb	Japan, U.S.	205,228
Hylemya alliaria (Fonseca)	Onion maggot	Anthomyiidae	Onion	Seedlings	Egypt	229—232
Liriomyza cepae (Hering)	Stone leek miner	Agromyzidae	Onion, garlic	Bulbs	Germany, Hungary, Netherlands, Taiwan, Italy	214—217,219
Liriomyza chinensis (Kato)		Agromyzidae	Onion	*	China, Japan, Malaya, France	213,218
Liriomyza huidobrensis (Blanchard)		Agromyzidae	Onion	*	U.S.	213
Liriomyza nietzkei (Spencer)		Agromyzidae	Onion, leek, garlic	*	Germany, Italy, Switzerland	213,219,220
Liriomyza trifolii (Burgess)		Agromyzidae	Onion, chives	*	Venezuela, Mauritius	213,221
Lonchea chorea (F.)		Anthomyiidae	Onion	Bulbs	Japan	205
Merodon equestris (F.)	Large narcissus fly; bulb fly	Syrphidae	Onion	Bulb	New Zealand	200

Phytomyza horticola (Goureau)	Chrysanthemum leaf miner	Agromyzidae	Onion	Leaves	Yugoslavia	233,234
Suillia lurida (Mg.)	Garlic fly	Heleomyzidae	Garlic, leek/onion	Leaves bulbs	Yugoslavia, Czechoslovakia	235,236
Suillia univittata (von Roser)		Heleomyzidae	Garlic, leek/onion	Leaves	Italy	237
Syritta pipiens (L.)		Syrphidae	Onion	Bulb	Rumania	204
Tipula paludosa (Meigen)	European crane fly; leatherjackets	Tipulidae	All onion crops	Seedlings	U.K., Canada	6,238
Tritoxa flaxa (Wied)		Ottitidae	Onion	Bulb	U.S.	239,240

* Unspecified in reference.

in Canada, to predict the peaks in field populations. With these data the numbers of chemical sprays required to control onion fly during the growing season was reduced from 7 and 12 to just 2. This method of forecasting can be combined with the use of traps to predict pest infestations in the crop accurately.

Eckenrode et al.[125] reported results of trapping trials over a four year period in New York State where there are three distinct generations of onion fly. They established that the average air thermal unit accumulations, above the base threshold of 5°C, were 378, 642, and 697 D°C respectively, for the first, second, and third generations. These data can thus be used to predict the emergence and presence of the pest in the crop.

Various methods can be employed to trap the flies and assess population trends in the field. Inverted screen cone traps on a wire frame, interception traps, and yellow sticky traps have been used commonly to monitor flies.[122] Several factors influencing the efficiency of traps for onion fly have been investigated.

Ishikawa et al.[126] in investigations on the color preferences of the onion fly reported white-2, in the high reflectance range of 400 to 600 nm, to be the most attractive in a range of 18 colors. Vernon and Borden[127] on the other hand, found maximum consistent response to gray, yellow (3 hues), white, and blue. In further investigations Vernon[128] reported that colors with peak reflective wavelengths between 400 and 700 nm, at or above 30% of reflectance were the most preferred. This group of workers suggested that attraction is determined by color hue or by saturation, and the magnitude of a response is determined by the intensity of key wavelengths, white being the most effective for use in monitoring onion fly.[129]

Plant volatiles which attract onion fly have been used to make the traps more effective. Traps baited with 2-phenylethanol[130] were successful in attracting males and females. Sulfur compounds such as dipropyl disulfide and methyl propyl disulfide attract larvae and adults. In addition tetramethylpyrazine, heptanal, and ethyl acetate[131] enhanced the attractancy of dipropyl disulfide. Other chemicals, like dimethyl-, methyl propyl-, and dipropyl-trisulfides, have also been found to be attractants and oviposition sitmulants for onion fly.[132] In laboratory investigations, polythene granules containing propyl disulfide and encased in gelatine capsules attracted more males and females than a check trap baited with untreated granules.[133] Soni and Finch[134] evaluated 27 volatile sulfur-bearing chemicals as attractants for newly hatched larvae. Of the 13 chemicals which attracted the larvae, onion oil and methyl trisulphide elicited the greatest response.

In tests in Canada, Pierce et al.[135] reported methyl propyl disulfide and *cis*- and *trans*-propenyl propyl disulfide as attractants and oviposition stimulants. A mixture of thiopropyl-containing compounds from onions was additive in the attraction of males and females.[136]

More recently Dindonis and Miller[137] suggested that a blend of chemical, rather than a single key chemical, was the more effective host finding stimulus. Furthermore, microbial activity on damaged tissues enhanced the attractancy of a blend of onion volatiles. In field trials in Japan, beet molasses[138] and mature onion pulp which contains 2-phenylethanol,[139] a potent attractant, were found to be the most attractive baits for adult onion flies. In addition, enzymatic yeast hydrolysate was highly attractive to the flies, females being more responsive than males.[140] Clearly there is a potential for utilizing a combination of color preferences, onion volatiles, and microbial activity to increase the efficiency of traps for onion fly.

f. Control

Damage caused to alliaceous crops by onion fly maggots can be reduced by cultural, biological, and chemical measures. A variety of natural enemies of eggs, maggots, pupae, and adults can also help to reduce populations of onion fly.

i. Cultural Methods

Good husbandry and crop hygiene can help to reduce or eliminate sources of maggot infestations. Attacked plants and infested debris should be removed and burned. Removal of cull and volunteer plants[122] also reduced the sources of infestations for the following crop of onions.

On a commercial scale, isolation of onion, garlic, and leek fields from one another and, where feasible, a crop rotation of 2 or more years reduce maggot populations considerably. Minimizing herbicide and cultivation damage to the crop and clearing weeds from ditch banks and field boundaries can help to reduce populations.[122] Finch and Eckenrode[141] reported that, in New York State, onions left in the field after harvest were a major source for overwintering populations. Numbers of overwintering pupae ranged from 60,000 to 67,000/ha. At least 95% of these pupae produced adults the following spring, thus providing a source of infestation for the new crop. These authors also reported that cull piles and volunteer plants were not as important as previously thought because maggots from cull piles were unable to establish successfully on volunteer plants. Eckenrode and Nyrop[142] pointed out that injury caused bulbs during lifting, prior to drying, and harvesting, increased the likelihood of maggot infestation in the field. Therefore, care should be taken to minimize mechanical damage to bulbs.

ii. Biological Control

There are many parasites and predators of all stages of the life cycle of the onion fly, which can help to reduce populations in the field. In Poland[143] *Trybliographa rapae* (Westw.) was the most important of the six parasites studied. Heavy infestations of crops by maggots were accompanied by 20 to 30% parasitism but where the infestation was slight up to 80% of the maggots and pupae were parasitized. In Rumania[144] the principal parasite of onion maggots was the braconid *Aphaereta minuta* (Nees) which comprised 95.3% of the total parasite population. The highest levels of parasitism occurred in August and September, corresponding to the second and third generations of onion fly.

Methods for mass rearing *Aleochara bilineata* (Gylh.), an important predator of several dipterous pests including onion fly, have been developed in Belgium[145] and Canada.[146] In Canada[147] three staphylinid and four hymenopterous species were found to parasitize onion fly pupae, but only the braconid *Aphaereta pallipes* (Say) and the staphylinid *Aleochara bilineata* were significant mortality agents. In laboratory studies in Canada,[148] adults of *Aphaeraeta pallipes* parasitized 70% of the pupae of onion fly. In field cage experiments the parasite was successful in parasitizing onion fly pupae.[149] An ichneumonid, *Phypadenon trichops* (Thoms.) was reported from Germany to be a potential biological control agent for *Delia antiqua*,[150] and *Beauverie bassinia* and *Metarhizium anisopliae* have been reported as fungal pathogens of *D. antiqua* in the U.S.[151] In cool and moist conditions in spring and autumn, the parasitic fungus *Entomopthora muscae* can infect large number of adults but this rarely reduces damage.[122]

Coomans and Grisse[152] in Belgium reported that third instar maggots and pupae were susceptible to the nematode *Steinernema bibionis*. Examination of the adult onion flies from France and Denmark showed that a few were parasitized by the nematode *Heterotylenchus aberrans* but attempts to transfer it from infested flies to adults, in U.S., were not successful.[153]

iii. Genetic Control

In the Netherlands and Canada, much effort has been devoted to the use of laboratory reared sterile adults in the control of wild populations of onion fly in the field. In field-cage experiments in Canada[154] semisterile mutants displayed a lower rate of increase than the wild type in each of two generations. By the third generation, the mutant population contained a significantly smaller ratio of females with high rates of hatching than the untreated population, indicating a lower reproductive potential in the presence of mutants. In the Netherlands, attempts were made to use radiation-induced mutagenesis to achieve genetic control of onion fly. In field cages, the fertility of eggs was reduced from the standard 90 to 54% in some laboratory produced strains. These effects were counteracted by changes in diapause response and this effect was not investigated further.[155]

Sterility can also be induced by chemical manipulation of the diet of the adult flies. Ticheler et al.[156] and Theunissen et al.[157-158] described appropriate techniques and also reported the results of small scale field trials in the Netherlands. Loosjes[118] has summarized this work and the technique is now practiced commercially on a small scale in parts of the Netherlands.

iv. Plant Resistance

The use of partially or wholly resistant plant cultivars to reduce the damage caused by insect pests is an attractive proposition. It would reduce or eliminate toxic and environmental hazards associated with insecticides and costly specialized insecticide application equipment would not be needed by growers. The potential of selecting for resistance is being investigated by researchers in the U.K.[159] U.S.,[160] Japan,[161] the Netherlands,[162] and USSR.[163]

Soni et al.[159] described a glasshouse method to assess resistance in *Allium cepa* to damage by onion fly. The technique enables onion cultivars and breeding lines to be screened so that only the most promising cultivars need to be tested in field scale trials requiring greater resources. In addition the ability of mature bulbs of 44 onion cultivars to support the pest was evaluated; some cultivars supported only small populations, indicating a degree of partial resistance.[164] In further trials, about 20 cultivars from a total of 375 screened demonstrated varying degrees of resistance to onion maggot attack (Soni and Ellis, unpublished data). In these trials, two of the cultivars, Pukekohe Long Keeper from New Zealand and Robusta (Rijnsburger) from the Netherlands consistently exhibited a higher level of tolerance to onion fly attack than the other cultivars.

Certain microorganisms colonizing onion seed and seedlings have been shown to influence egg-laying by onion fly around host plants. The bacterium *Klebsiella pneumoniae*[165] was identified as a major colonizer of the onions which were most attractive to the onion fly. The growth of these microorganisms on plant surfaces metabolizes chemical precursors in onions and converts them to volatile compounds which stimulate onion fly oviposition.[166] Of the 16 bacterial isolates tested, species of *Pseudomonas* (an isolated of *Pseudomonas* B) were as stimulatory as a mixture of microbes in determining oviposition.[167] In Japan[168] soil microorganisms, in conjunction with alkylthio-compounds produced by intact onions, enhanced egg laying by the onion fly.

Ellis et al.,[160] demonstrated the combined effects of onion cultivars and their microbial colonisers on resistance to attack by onion maggot. Japanese bunching onions *(Allium fistulosum)* were less damaged compared to *A. cepa* cultivars.

Ikeshoji[169] illustrated the importance of microorganisms on onions in attracting flies. Onion cultivars susceptible to onion maggot damage were associated with larger numbers of root bacteria than those which were resistant. The bacterial counts were related positively to certain volatile chemicals thought to be attractants for onion fly. Factors which influence the resistance in *Allium* species to attack by the onion fly have been reviewed[170] and include plant species, physiological, biochemical and environmental conditions.

Results of studies in Japan suggested a correlation between the chemical composition of onion cultivars and the degree of resistance to attack. Cultivars containing larger amounts of propylthio compounds, which attract larvae and egg laying by adults, were more susceptible than resistant cultivars to attack by onion maggot. F_1 hybrid cultivars were more resistant to larval attack than their parents, due probably to a lower concentration of propylthio compounds and a lower level of alliinase activity than in the parent cultivars.[161] Tsutsumi et al.[171-172] reported differences in the tolerance of some onion cultivars to damage by a complex of insect pests including *Delia antiqua. Cv.* Pukuhoke Long Keeper, Fusario 24 and Furanui[171] were found to be resistant to damage.

Ponti and Inggamer showed that some cultivars and breeding lines, including selections of *cv.* Rawska from eastern Europe, were less damaged than *cv.* Jumbo.[162] They also confirmed[162,173] that *A. fistulosum* is generally less susceptible than *A. cepa* to damage by onion fly. However, when the fly was offered no choice of host, the resistance of *A. fistulosum* was very much reduced, confirming that the resistance of these bunching onions is due largely to nonpreference.[160]

v. Chemical Control

Growers have relied on insecticides for the control of onion fly for more than 50 years and, in most countries, chemicals are likely to remain the principle control measure for the foreseeable future. From the late 1940s to 1960 organochlorine chemicals were used extensively to control this pest. However, the pest developed resistance to these compounds in North America and Europe and most growers now rely on organophosphorus and carbamate insecticides applied at drilling for its control on onions. Furrow treatments with currently recommended chemicals sometimes lack the persistence necessary to control mid- and late-season attack.[174] Therefore, in North America, sprays of organophosphorus compounds have been applied at 3 to 5 d intervals to control adult flies. Recent reports however indicate that in North America[175] and Europe[176] the onion fly is developing resistance to some of these organophosphorus and carbamate compounds.

More recently techniques[177] have been developed for film-coating seeds with pesticides. Onion seed treated in this way has been used in field experiments at different sites to control onion fly. Toms and Blacket[178] outlined the concepts and advantages of using pesticide-coated seeds for control of this pest as well as some seedling diseases. This technique offers potential for reducing the amounts of insecticide required to control onion fly.

vi. Integrated Control

Of the 10 soil insecticides developed over the past 25 years in Canada to control onion fly damage, only two remained effective by 1979.[149] To overcome this shortage of insecticides, in 1981 in the onion growing areas of the Holland Marsh in Ontario, Canada, growers used an integrated approach to control onion fly.[179] Treatments are now based on a knowledge of previous damage and cropping history, together with day-degree models and monitoring peak adult activity. Growers participating in the program reduced their insecticide treatments by 30 to 100%.

Whitfield et al.[180] used day-degree accumulations to correlate adult flight activity with the time of damage caused to onions by the first and second generation larvae. In their field trials, foliar sprays of parathion and carbaryl did not provide additional protection where granular insecticides had been applied at planting. Finch et al.[181] appraised the current strategies for onion maggot control in New York State and concluded that integrated control programs should be based on cultivars possessing resistance to the pest, insecticide treatments at sowing and others applied mid-season.

2. *Delia platura* (Meigen), Bean Seed Fly, Seed Corn Maggot

This fly has also been referred to as *Hylemya platura, H. cilicrura* (Rond), *Anthomyia platura,* and by Ormerod[182] as the leek or shallot fly. A comprehensive bibliography of seed maggots *H. platura* and the closely related *H. florilega* was published by Thorne.[183]

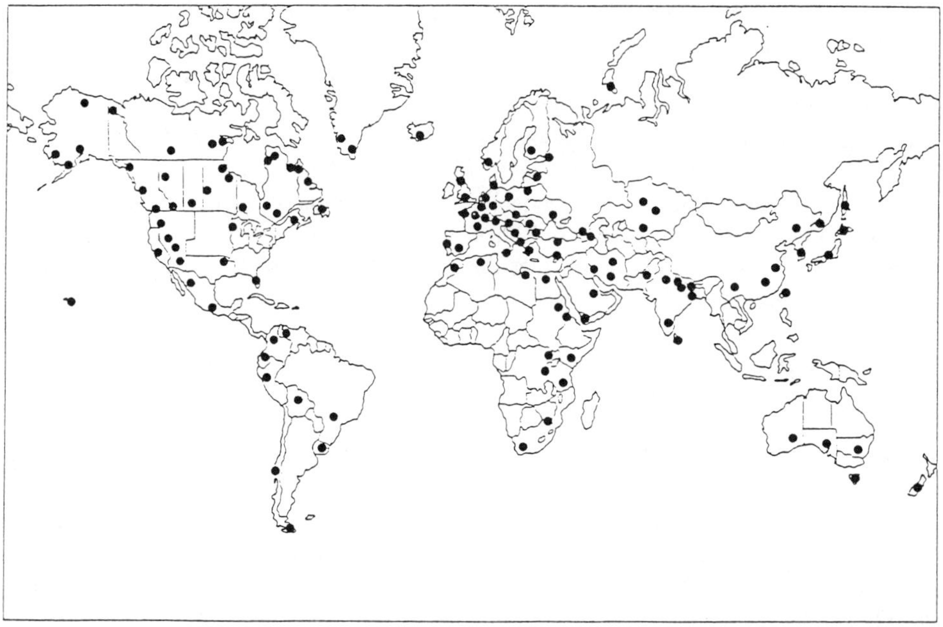

FIGURE 8. Distribution of *Delia platura*.

a. Distribution

The map (Figure 8) shows the wide distribution of this pest in almost all parts of the world. It is truly polyphagous in its host-plant range, damaging a large number of vegetable crops including several species of cultivated *Allium*. It was introduced from Europe to North America and first reported in New York in 1856.[28] Keyder and Atak[120] recorded the seed corn maggot as a pest of onions, garlic, and leeks as well as other vegetables in Turkey. From Japan, Tsutsumi and Mitsui[184] have reported it as a pest of onions.

b. Biology

The grayish-brown adult, slightly smaller than the onion fly, is about 5 mm long. The eggs are brilliant white, elongate and about 0.3×1.0 mm long. The pattern on the surface of the egg appears like irregular, rounded hexagons. The fully grown, white legless maggots are 6 to 8 mm long; they have curved mouth hooks, used to burrow into germinating seeds and young seedlings. Important distinguishing features include the presence of 5 to 8 fan-like processes on the anterior spiracles and the three-lobed posterior spiracles. The three larvel instars are spent feeding within plant tissues after which the final instar larvae enter the soil and develop into reddish-brown pupae 3 to 4 mm long.

c. Life History

In temperate regions, the pupal stage of this fly overwinters in a diapause condition. In the U.K., the first generation adults emerge during late April and up to four overlapping generations may occur each year.[185] Thorne and Eckenrode[186] reported 2 to 3 overlapping generations in New York State with fly activity from May—October. In Japan, Kuwayama et al.[187] reported 3 generations during the year with damage reaching 100% when the egg density exceeded 80 eggs per 100 cm^2. In early May in the U.K., females prefer to lay their eggs in freshly turned soil. A high content of decaying organic plant matter and manure in the soil increases the probability of infestation.[6,188] The eggs hatch within 2 to 4 d and the small maggots burrow down into the soil seeking germinating seeds and young seedlings.

After feeding for 1 to 3 weeks, the maggots are fully developed and leave the damaged plants to pupate nearby in the soil. The pupal stage may last 2 to 3 weeks before the next generation of flies emerge. Depending on soil temperatures, weather, and food availability, the life cycle from egg to adult is completed within 3 to 5 weeks. Under favorable conditions, first generation females can continue to lay eggs for up to 6 weeks.

Normally, the first generation of maggots causes economic damage in the Northern Hemisphere when germinating seeds and young seedlings are most vulnerable in early spring. Maggots of later generations feed either on decaying vegetable matter or on other host-crops.[185]

d. Damage

The flies are general scavengers in organic matter but the young maggots bore into the germinating seeds and young seedlings, especially in spring when the soil is cool and wet.[185,188] The symptoms of attack in infested plants are similar to those caused by the onion maggot. Without expert examination of the maggots, damage by this pest is often confused with damage by onion maggot. In the U.K., first generation adults of this pest usually emerge nearly 3 to 4 weeks before the onion fly and very early damage in onion crops can be attributed reliably to seed-corn maggot activity. If heavily infested, the young leaves appear yellowish, become flaccid and the plant eventually collapses and dies. Initial infestation by the maggot may facilitate secondary infections by disease organisms and hasten the collapse of the plants, unless adequate control measures are taken in time. However, once the seedlings emerge above the soil, they are less likely to be seriously damaged.[188]

e. Monitoring and Predicting Pest Activity

The emergence patterns of this pest have been studied by Thorne and Eckenrode using traps located in and near to crops.[186] The data gathered from monitoring the insect can be used to facilitate the timing and application of precautionary treatments. As with onion fly, investigations have been carried out on a range of factors which may enhance trap efficiency. Vernon and Borden,[189] in Canada, studied the color preferences of various adult dipterous pests including *Delia platura*. Using cardboards of 13 chromatic or achromatic lines, they found that adults showed a marked preference for grey and a significant response to green, although yellow was the most effective color for monitoring a wide range of insects in vegetable crops. Other workers have investigated the response of flys to plant products. Dindonis and Miller[190] reported that females showed a preference for rotting onion plants, rotten seedlings, and healthy seedlings. Yeast products were more attractive than specific food-plant volatiles, although this may have been influenced by the reproductive stage in the development of the female flies.[140] In general, females were more responsive than males to enzymatic yeast hydrolysate. In a field trial in Japan,[191] mature onion pulp was more attractive in traps than fresh pulp and dipropyl-disulphide attracted fewest flies. In later investigations, 1-phenylethanol detected in mature onion pulp was a potent attractant for *Delia platura* and *D. antiqua*.[139] Attempts have been made to utilize volatiles in traps and the catches used to estimate field population. In a simple-release and repeated-recapture experiment in the field, 2-phenylethanol was more attractive to males than females.[130] In more recent studies, valeric acid, one of a number of acids fractionated from extracts of decomposing onions, combined with 2-phenylethanol enhanced the attractiveness of traps by a factor of 3.7 to 5.7 times when compared with decomposing onion pulp.[192]

f. Control

i. Cultural Methods

Early reports suggested that earthing-up soil alongside rows of drilled onions gave some protection from this pest, particularly with leeks.[182] Ten years later Ormerod[193] recommended that "the best course is to raise all injured onions with the maggots within, and to destroy them at once". Burial of decayed organic debris and early preparation of seed beds may also prevent severe attack by seed corn maggot.[186] Rapid germination of seeds is encouraged by early ploughing, adequate moisture, and firm friable seedbeds free from compaction. This ensures establishment of seedlings before the female flies start egg-laying.[185]

ii. Biological Control

Many parasites infest adult been seed flies. Batko and Weiser[194] reported that the fungus *Strongwellsea castrans* infected adults of *D. platura* in the field in Wisconsin. Barisford and Tsao[195] reported that in the field at Athens, Georgia, large numbers of adults were infected with *Entomophthora muscae* and Matanmi et al.[196] found adult flies infected with two phycomycete fungi, *Entomophthora virulenta* and *Conidiobolus coronatus*. However, these flies were collected from flower stalks of radish. A parasitic nematode infecting adults of this pest in onion fields was found in France and Belgium.[153] Kuwayama et al.[187] reported that the cynipid *Eucoila tanebae* Ishii and the staphylinid *Aleochara bipustulata* (L) attacked *D. platura,* the former more commonly than the latter. At present, none of these natural enemies is used as an economic control measure.

iii. Plant Resistance

In Japan, Tsutsumi et al.[171] identified four onion cultivars "Pukekohe", "Long Keeper", "Fusario 24", and "Furanui" as tolerant to damage by *D. platura* and *D. antiqua.*

iv. Chemical Control

During the last 50 years various seed dressings[116] have been used effectively to protect young seedlings of many crops from maggot attack. Organochlorine insecticides, e.g., aldrin and dieldrin, were used successfully for almost 20 years to control this pest. However, the insect developed widespread resistance to these compounds in certain vegetable growing areas in Europe[197] and North America.[198] Currently, growers rely on organophosphorus, and carbamate and synthetic pyrethroid compounds for the control of this pest.[185,188,199]

B. FAMILY: *SYRPHIDAE,* HOVER FLIES

1. *Eumerus* spp., Bulb Flies

Eumerus amoenus Loew, onion bulb fly, *Eumerus strigatus* (Fallen) and *Eumerus tuberculatus* (Rondani), the last two being known as small narcissus flies or lesser bulb flies.

Certain species of bulb flies are important pests of alliaceous crops and may inflict losses exceeding 50% in the field or in stored bulbs.

a. Distribution

Eumerus species are distributed widely and have been reported as pests of *Allium* spp. in Africa, North America, Asia, Europe, and New Zealand.[200] *Eumerus amoenus* is a serious pest on onions in Egypt.[201] It is probably restricted to warmer regions since eggs failed to hatch and larvae and pupae of this species died when subjected to controlled temperatures of 10°C.[202] The insect develops most rapidly at temperatures between 25 and 35°C. *Eumerus strigatus* and *E. tuberculatus* are recorded more commonly as pests in temperate zones. *E. strigatus* has occurred in England since 1901, although it was detected in Italy in 1857,[203] and it was first reported in Rumania in 1973.[204] Bulb flies have spread as a result of man's trading activities. For example, in Japan, these insects have been discovered in onion bulbs imported from nine different countries.[205]

b. Biology

Adult hoverflies are attractive insects and familiar to most people as they are seen commonly wherever hedgerow or garden flowers are in bloom. They hover above the flowers and then alight to feed on nectar and pollen. The adults are important pollinators of many crops and the larvae feed in many different ways. They may be phytophagous, saprophagous or predators of other insects and mites. A few species are important biological control agents of pests. Like all true flies (Diptera), they have a complete metamorphosis and the life cycle includes eggs, larvae, pupae and adult flies.

Eumerus adults are similar to small solitary bees with a short, thick-set body 6 mm long. The thorax is black with a green or metallic sheen and a white line down the sides. The abdomen has three pairs of white, crescent-shaped marks. Adults are probably attracted for egg-laying by volatiles from the host plant and from damaged or decaying plant tissue. The small eggs, 0.8 mm long, are oval, white and shiny. The grayish white larvae develop through several stages and moult before becoming fully grown at 8 mm long. The larvae have several projections at the hind end, one of them being bright chestnut red. The arrangement of these projections can be used to identify the species.[203] The pupae are grayish-brown, tough and pear-shaped, about 6 mm long and are found in plants or in the soil nearby.

Cultivated and wild species of several different plant families are attacked by these flies, the most common hosts being hyacinth, iris, lily, narcissus, and onion, but the host range also includes carrot, parsnip, and potato.[206] In Russia garlic and onion are attacked.[207]

c. *Life History*

In Britain, adult flies emerge in late April or early May. Soon afterwards they lay eggs on or around host plants in batches of 5 to 40. The larvae hatch out in about 3 d and usually penetrate bulbous plants in the neck region. They may also gain entry to bulbs, swollen tap roots or tubers through a site damaged by some other pest, or decaying as a result of attack by a plant pathogen. Most larvae are developed fully by late June or early July and pupate in or near the plant. In July the second generation adults emerge and they produce larvae which infest more plants; most larvae remain in or near the bulb over winter and give rise to the first generation of flies in the following spring. A small portion of the larvae of the second generation develop rapidly and produce adults in September and October. These rarely lay eggs but are seen commonly in stores of bulbs.

d. *Damage*

After entering the plant, the larvae tunnel inside the bulb and reduce the tissues to a pulpy, rotting mass. The whole plant becomes soft or spongy and may collapse and rot. Bulbs which have been lifted for drying off and left on the surface of the ground remain attractive to the flies for egg-laying. Infested stored bulbs serve as a nucleus for the spread of diseases and the migration of larvae from bulb to bulb.

These pests can be a serious threat to onion production. *Eumerus strigatus* caused 20 to 80% and 40 to 50% loss of seed onions at one farm in two successive years in the Leningrad region,[208] while in Rumania these pests caused 10 to 15% losses of late-maturing onions.[204]

e. *Monitoring and Predicting Pest Activity*

There are few reports on this aspect of activity. However, traps baited with onion pulp have been used in Japan to monitor the seasonal activity of *Eumerus* spp.[184] and in New Zealand female *E. strigatus* were attracted to decomposing oatmeal.[209]

f. Control

i. Cultural Methods

Bulb flies are difficult to control and, when the maggots have entered the plant control measures are ineffective. Infested bulbs should be collected and burned.[201]

ii. Biological Control

Bulb flies have many natural enemies including birds, insect predators, parasites, and microbial pathogens.

Enumerus species are attacked by various insect parasites including braconids and pteromalids and in Rumania these two parasites are believed to be the principal factors limiting the populations of bulb flies in the field.[210] No attempts have been made to use these parasites in biological control programs.

iii. Chemical Control

For many years organochlorine insecticides have been recommended for the control of bulb flies.[211] Plants are sprayed with DDT and BHC and bulbs dusted with DDT before storage.[201] In countries where organochlorine insecticides have been withdrawn growers use carbamate and organophosphorus compounds applied as dusts or sprays.[212]

C. FAMILY: AGROMYZIDAE, LEAF-MINERS

1. *Liriomyza spp.*

Several species of leaf-miners colonize alliaceous crops. In his comprehensive book on Agromyzidae of economic importance Spencer[213] lists the following six species which attack these crops:

Allium cepa L. (onion)	*Liriomyza cepae* (Hering) *Liriomyza chinensis* (Kato) *Liriomyza huidobrensis* (Blanchard) *Liriomyza nietzkei* (Spencer) *Liriomyza trifolii* (Burgess) *Phytomyza horticola* (Gour)
Allium porrum L. (leek)	*Liriomyza nietzkei* (Spencer)
Allium sativum L. (garlic)	*Liriomyza cepae* (Hering) *Liriomyza nietzkei* (Spencer)
Allium schoenoprasum L. (chives)	*Liriomyza trifolii* (Burgess)

A few species of leaf-miners are polyphagous, for example *L. trifolii* has been reported as colonizing 47 plant genera in 10 families. Other species are highly specific, for example *L. nietzkei* is confined to *Allium.* Leaf miners are pests of outdoor and protected crops throughout the world. The two species which probably cause most damage to onions and garlic are *L. cepae* and *L. nietzkei. Liriomyza cepae* is of particular importance in eastern Europe and the Orient while *L. nietzkei* has caused serious losses to these crops in western Europe. In this chapter, we have selected *L. cepae* for a detailed description as it is probably the most severe leaf-miner pest of *Allium.* However, reference is made to the other species where appropriate.

Liriomyza cepae (Hering) synonyms *Phytobia cepae, Dizygomyza cepae, Allium* leaf miner, stone leek leaf miner.

a. Distribution

In Europe *L. cepae* occurs as a pest in Germany,[214] Hungary,[215] the Netherlands.[216] It has also been reported in Taiwan.[217] *L. nietzkei* is confined to Western Europe and has been reported from Germany,[213] France,[218] Italy,[218] and Switzerland.[220] *L. chinensis* is a pest in China, Japan and Malaya[213] and in France.[218] *L. trifolii* occurs on onions and chives in Venezuela,[213] and on onions in Mauritius,[221] but it is a polyphagous species found on weeds, ornamentals and other vegetables in many countries.[222]

b. Biology

As with all true flies, the life cycle includes eggs, larvae, pupae and adults, i.e., a complete metamorphosis. Spencer[213] has described in detail the species of leaf-miners which infest alliaceous crops. *Liriomyza cepae* adults have a blackish-gray thorax with yellow markings, a black abdomen and yellow antenna, and legs which are predominantly yellow with brown markings. The wings are hyaline becoming milky-white below with dark veins and range from 2.0 mm long in the male to 2.5 mm in the female. Eggs average 0.31 mm long and are yellowish white. The larvae are the same color as the eggs when fully grown, they may be 2 mm long; they develop through three stages. The reddish-brown pupae are 3.0 to 3.5 mm long.

c. Life History

Adult flies begin to emerge in the spring and their numbers build up to a peak over a few weeks, each fly living for 1 to 2 weeks. Females pierce or cut plant tissues with their ovipositor and feed on the cell contents which are released. Most eggs are inserted into the tissues towards the base of the leaves. After hatching, the larvae mine within the leaf and move down toward the base; the larval development takes about 20 d. When fully grown, the larvae emerge from the plant and fall to the ground to pupate. In Taiwan there may be up to ten generations of *L. cepae* a year[223] while in European countries there may be one only. *L. nietzkei* has two generations a year in Switzerland.[220]

d. Damage

The feeding punctures can lead to loss of cell contents, invasion by fungi and bacteria and, in some species, transmission of viruses.[222] However, the most serious damage results from the larvae mining in the tissues. This disruption of tissues can reduce the photosynthetic capability of the plant and, when infestations are very severe, leaves dessicate and collapse and seedlings may be killed. Damage by the second generation of larvae in Europe is usually less severe than the first because larger plants can withstand the attack.

Leaf miner attacks are a serious threat in Hungary where an average of 60 to 70% of onion crops may be invested by *L. cepae* with 100% infestations locally.[215] Similar levels of damage also occur in Italy.[219] This pest has spread westward in Europe over the last few decades and in the Netherlands damage may be heavy on onions.[216]

e. Control

i. Biological Control

Leaf-miners have many natural enemies. Minkenberg and Van Lenterren[222] listed the parasites and predators of *L. trifolii* and discussed the effectiveness of the various natural enemies as biological control agents. The different types of parasites of leaf miners and their behavior were described by Wardlow[224-225] who also described a scheme for monitoring the infestation of the pest and the levels of parasitism.

ii. Chemical Control

A range of organophosphorus compounds, carbamates, and synthetic pyrethroids, as well as gamma-HCH and nicotine, are used to control leaf miners in the U.K. and are applied as granules, sprays, and fumigants.[10] In Venezuela leaf-miner damage on chives is prevented by using organophosphorus compounds.[226] Steem distillates of neem *(Azadirachta indica)* leaves have given satisfactory control of leaf-miners on onion crops in Mauritius.[221]

VI. BEETLES, ORDER: COLEOPTERA

Beetles constitute the largest single order of insects in terms of numbers of species, more than 280,000 different species having been described. Several beetles are important pests of alliaceous crops including flea beetles, chafer beetles, leaf beetles, weevils, and click beetles (Table 5). Damage to crops may occur in the field, under protected cultivation or in stores. These insects are familiar to most advisers and growers and the adults of practically all species are characterized by their two pairs of wings, the forewings being modified to form tough wing cases under which the membranous flight wings (hind wings) are folded. Usually, the first segment of the thorax is large and mobile. These pests have biting mouthparts and adults and larvae may damage plants. The life cycle consists of egg, larva, pupa, and adult. Larvae usually have three pairs of legs while the pupae have free appendages. Although some species are important pests of alliaceous crops, many others are predators of considerable value as natural enemies of crop pests.

The most important beetle pests of onions and their allies are listed in Table 5. We describe below the lily beetle as an example of one of the most economically important group of beetles, the leaf beetles.

A. FAMILY CHRYSOMELIDAE: LEAF BEETLES

1. *Lilioceris* spp., Lily Beetles

There are two species of lily beetle which attack members of the family Liliaceae: *Lilioceris merdigera* (F.) Synonym *Crioceris merdigera,* and *Lilioceris lilii* Scop. lily beetles, lily criocerans.

These brightly colored beetles are important pests of ornamental lilies and sometimes *L. merdigera* damages onion crops seriously.[241]

a. Distribution

Lily beetles are distributed widely throughout Europe, extending as far east as Asia and south into North Africa. They are particularly important pests in parts of western Europe. For example, they cause severe damage to crops in France.[242] In Britain these pests are confined largely to southern England,[243] although they occur occasionally in other parts of the country.

b. Biology

A full description of the biology and life history of these beetles is given by Balachowsky and Mesnil.[242] Adult beetles are 8 mm long and the upper parts of their bodies are almost entirely bright vermilion. The head, antennae, and lower parts of the body are black. *Lilioceris merdigera* has reddish legs while those of *L. lillii* are black, the difference permitting the two species to be distinguished easily. The frons (face) is flat and streaked, the eyes are small but prominent. The pronotum is more than half the size of the elytra (wing cases) and strongly indented near the middle of each side. Eggs are about 1 mm long,

TABLE 5
Species of *Coleoptera* Damaging Onions and Allied Crops

Latin name and authority	Common name	Insect family	Crop damaged or infested	Part of plant damaged	Country	Ref.
Agriotes spp.	Wireworm	Elateridae	Onion	Bulb	U.K.	244
A. lineatus (L.)	Wireworm; common click beetle	Elateridae	All crops	Seedlings	U.K.	6
Agrypnus variabilis (Cand.)			Onion	*	New Zealand	245
Araecerus fasiculatus (Deg.)	Coffee bean weevil	Anthribidae	Garlic	Bulb	Japan	246
Carpophilus dimidiatus (Fab.)	Corn sap beetle	Nitulidae	Garlic, onion	Bulb	Egypt	76
					U.S.	247
C. hemipterus (L.)	Dried fruit beetle	Nitulidae	Onion	Bulb	Egypt	76
C. obsoletus Erichs.		Nitulidae	Garlic, onion	Stored bulbs	Egypt	105
					USSR	248,249
Ceutorhynchus jakovlevi (Scultze)	Onion weevil	Curculionidae	Onion	Foliage	USSR	250,251
C. suturalis Gyll.		Curculiuonidae	Onion	Foliage	Hungary	252
					Netherlands	253
Colaphellus alpinus (Gelb.)		Chrysomelidae	Onion	Foliage	USSR	207
Ctenicera strangulata (White)			Onion, garlic	Root	New Zealand	245
Cryptolestes spartii (Curt.)	Grain beetle	Cryptophagidae	Garlic	Stored bulbs	Egypt	105
Galeruca daurica Joann.	Leaf beetle	Chrysomelidae	Wild onion, onion	Foliage	Mongolia	254
G. tanaceti (L.)		Chrysomelidae	Chives, garlic, leek, onion, shallot	Foliage	France	255
Gronopus inaequalis Boh.		Curulionidae	Onion	*	USSR	207
Lasioderma serricorne (F.)	Cigarette beetle	Anobiidae	Onion, garlic, onion	Stored bulb	Egypt	76
				Stored bulb		
				Dehydrated bulbs	Egypt	256

TABLE 5 (continued)
Species of *Coleoptera* Damaging Onions and Allied Crops

Latin name and authority	Common name	Insect family	Crop damaged or infested	Part of plant damaged	Country	Ref.
Lilioceris lilii (Scop.)	Lily beetle	Chrysomelidae	Onion, garlic	Foliage	France	242
Lilioceris merdigera (L.)	Lily beetle	Chrysomelidae	Onion, garlic	Foliage	France	242
Oryzaephilus surinamensis (L.)	Saw-toothed grain beetle	Cucujidae	Onion	Bulb	Egypt	76
Tenebroidas mauritanucus (L.)	Cadelle	Trogositidae	Dried onions	Stored bulb	Egypt	76
Tribolium castaneum (Hbst.)	Red flour beetle	Tenebrionidae	Dried onions	*	Egypt	76
T. confusum (Dnv.)	Confused flour beetle	Tenebrionidae	Dried onions	*	Egypt	76
Typhaea stercorea (L.)	Hairy fungus beetle	Mycetophagidae	Onion, garlic	Stored bulb	USSR	257

* Unspecified in reference.

shining and dirty yellow. The larvae are a similar color to the eggs, have a slightly curved, thick, short body and six black-brown legs. The head is also black-brown. When fully grown, the larvae are 8 to 10 mm long. The segments of the body possess small, regularly arranged black plates which bear two hairs. The excrement of the larva accumulates in a mass above these plates probably as a form of camouflage but also possibly to protect the larvae from the heat of the sun and to retain body moisture. The larvae pupate in the soil in a polished, oval cell lined with varnish-like coating.

c. *Life History*

Adults emerge in the spring, often toward the end of March in parts of France. They aggravate on plants and may nibble the leaves or flowers. Shortly after emergence, the adults mate and females soon begin to lay their eggs. Females lay up to 300 eggs during the season and, if they survive to the following year, they lay more eggs in that spring. The eggs are glued to the border of leaves of host plants either singly or in small irregular groups. Incubation lasts 7 to 10 d and the young larvae immediately attack the parenchyma of leaves or flower buds. Growth is rapid and the larvae can be fully developed in 16 d. Having completed their development, the larvae descend to the soil and bury themselves in the surface layer where they hollow out a cavity which is lined with saliva. This lining produces the varnish-like pupal cell. The larvae pupate and remain in this state for 20 to 22 d after which the adults emerge.

In France, the first eggs laid in the spring give rise to adults in mid-May and the latter produce a second generation in June—July while a third generation may follow in the autumn. The generations tend to overlap and eggs, larvae and adults may be found on plants during most of the growing season. Many insects overwinter as adults in leaf litter and organic debris near damaged plants.

d. *Damage*

At first the larvae feed gregariously upon the parenchyma of the leaves progressing in regular rows. As they increase in size, they separate. Onions and leeks may sometimes be seriously damaged by larvel attack.[241]

e. *Control*

These insects are controlled most easily when they are feeding on the flowers or foliage of crops. Therefore, it is important to inspect plants for infestations early in the season. In the U.K. these insects are controlled with dusts containing gamma HCH, derris, or nicotine or with sprays of synthetic pyrethroids.

VII. MITES, ORDER: ACARINA

Mites are arthropods in the class Arachnida (which also includes harvestmen, scorpion, spiders, and ticks). They are widespread and common animals which can rival aphids in their rate of reproduction. Certain species are important pests of alliaceous crops. Mites, like insects have a tough exoskeleton and jointed limbs, but they differ from insects in their lack of wings and antennae and in the fact that their head is not separated distinctly from the abdomen. The body is usually sac-like or globular with little or no segmentation and, in most species, the adults have eight legs. Mites are small, few of them being discernible with the naked eye. The grower needs a hand-lens to see them in detail and identification of species is very difficult. Species which damage alliaceous crops feed in a similar way to

FIGURE 9. Distribution of *Tetranychus cinnabarinus*.

insect bugs by piercing plant tissues and sucking the sap released from damaged cells. Some species produce a characteristic webbing.

There are four main groups of mites: (1) Red spider mites; (2) Tarsonemid mites; (3) Acarid mites; and (4) Gall mites. Pests of alliaceous crops occur in all except the Tarsonemid group.

A. FAMILY TETRANYCHIDAE: RED SPIDER MITES, BRYOBIA MITES

1. Red Spider Mites

Tetranychus cinnabarinus Boisduval (Synonym *T. telarius* (L.)), Carmine spider mite; *T. urticae* Koch, red spider mite, two-spotted spider mite, glasshouse red spider mite, hop red spider mite; *T. cucurbitacearum* Sayed.

a. Distribution

Tetranychid mites are distributed throughout the world and, although they originated in the warmer zones, they have become widespread in temperate regions in glasshouses, plastic-covered structures and in sheltered areas outside. The distribution of *T. cinnabarinus* is illustrated in Figure 9; *Tetranychus urticae* is present in almost every country in the world. These mites are particularly severe pests of onions in the U.S.[28] but they are also severe locally in many other parts of the world.

b. Biology

These mites have five stages in their life cycle: egg, larva, protonymph, deutonymph, and adult; the nymphal stages have a feeding and a resting stage. Adult females of *T. urticae* are 0.6 to 0.85 mm long. They are green to straw colored but turn bright red when starved or ready to hibernate. The adults have a large black spot on each side of the body. Males and juveniles are straw yellow or pale green; the larvae have three pairs of legs while the nymphs and adults have four pairs. At the deutonymph stage, sexual characters are evident. Individuals with elongated bodies become males and those with rounded bodies become females. Eggs are spherical, pale, white-tinged with pink or red, and 0.14 in diameter. A very wide range of crops from many different plant families, including all the alliaceous crops is attacked by these mites.

c. Life History

In warm countries or in protected structures where conditions are favourable, the mites breed continuously and, as the complete life cycle takes only 3.5 to 30 d depending on temperature,[258] numerous generations may be completed during one crop. In cooler areas, activity increases with a rise in temperature.

In temperate zones, towards the end of autumn, mite reproduction slows down. Only adult females survive the winter. They become bright red and leave the plants to hibernate in crevices. Here they remain in diapause until temperatures increase in the following spring.

d. Damage

Larvae, nymphs, and adults feed on the chloroplasts of cells, producing characteristic yellowish, speckled feeding marks. As populations build up, these marks coalesce and leaves or parts of flowers shrivel and die. Mites migrate to the top of plants and produce silken webs; ropes consisting of many threads infested with mites are carried from plant to plant and spread the infestation throughout the crop. Mites are moved within the crop by other agents also, for example insects and wind.

The severity of an attack is estimated best by examining the damage. Thresholds have been determined for some nonalliaceous crops.[258]

e. Control

i. Cultural Methods

Severely infested plants should, if possible, be removed from crops and destroyed. In temperate regions, infested crops growing outdoors or under protection should be removed before mites hibernate, thus reducing the numbers of adult females which can enter diapause and initiate an infestation in the following season.

ii. Biological Control

Red spider mites have many natural enemies including parasitic and predatory insects and mites. Broad spectrum insecticides which reduce the numbers of parasites and predators should be avoided wherever possible.

The predatory mite *Phytoseiulus persimilis* Athias-Henriot is now used widely on protected crops as a biological control agent for red spider mite.[258] Another predatory mite *Agistemus exsertus* Gonz (Acarina: Stigmaaeidae) is a valuable predator of tetranychid mites in Egypt and methods have been devised for mass rearing it.[259]

iii. Chemical Control

Over several decades, many strains of red spider mites have developed resistance to certain acaricides, mainly because the mites breed rapidly and the same chemicals have been used continuously over a period of years.

Prior to the second world war, a miscellany of chemicals were used to control mites. These chemicals were replaced by several different groups of acaricides including organophosphorus compounds, dicofol quinomethionate, tetradifon, carbamates, derris (Rotenone), and petroleum oils.[260,261] Some of these are systemic and others are contact in action. Growers

FIGURE 10. Distribution of *Petrobia latens*.

should consider using several different acaricides if a program of repeated spraying is planned. Flucythrinate is effective in Egypt.[262]

2. Bryobia Mites

Petrobia latens Muller, Brown wheat mite, stone mite. This species feeds largely on grasses and cereals but may also infest a wide range of other crops including apple, carrot, cherry, lettuce, and onions and their allies. In some parts of the world it is the most abundant and serious mite pest of onions and garlic.[263,264]

a. Distribution

Petrobia latens has been reported as a pest in Australia,[265] south/western U.S.,[26] Egypt,[263] Spain,[264] and Britain[266] (see Figure 10).

b. Biology and Life History

These mites breed parthenogenetically and males are unknown. Adult females are about 0.7 mm long, flattened, oval, and have short bristles on the body. They are reddish-brown in color and have four pairs of legs, the first pair being twice as long as the other three pairs. They have mouthparts modified to pierce plant cells and suck the contents. The females lay eggs on stones (hence their name), leaves, fragments of wood, and in the soil.

c. Damage

Feeding marks are typical of tetranychid mites. As the yellow, speckled marks coalesce, large patches of dehydrated tissue develop. The leaves shrivel and die as a result of this damage. Centers of infection in fields become yellow.

d. Control

Several predatory insects and mites have been shown to exert some control of *Petrobia latens* in the field[264] but none has been reared for release in onion or garlic crops.

Usually infestations are controlled by acaricides; those found to be effective include several organophosphorus compounds.[263,265,266]

B. FAMILY ACARIDAE: BULB MITES

1. *Rhizoglyphus callae* Oudemans, *Rhizoglyphus robini* Claparede

Synonym: *Rhizoglyphus echinopus* (Fumouze and Robin). These mites, which attack bulbs, corms, and tubers of several different species of plants, are occasionally serious pests on stored onion and garlic bulbs.[26]

a. Distribution

These mites are distributed widely in Europe, America, and Israel where they are pests of bulbous or tuberous flower crops.

b. Biology and Life History

The adults are glistening and yellow white in color, with a globular body up to 0.9 mm long. Males are usually smaller than females but both sexes have four pairs of stout legs. The female lays up to 100 eggs, 0.2 mm in diameter, singly on corms, bulbs, or tubers. The young have three pairs of legs but in other respects resemble adults. They feed on plant tissues and moult three times before becoming adult. The complete life cycle takes less than 30 d when conditions are favorable. A specialized immature stage, the hypopus, develops in adverse conditions and these minute shiny brown mites attach themselves to insects and other animals by which they are dispersed.

c. Damage

These mites feed on bulbs which usually have been damaged by some other pest or disease, or by mechanical means. Established colonies build up rapidly and bulbs in store may be reduced to a pulpy, rotting mass.

d. Control

Strict hygiene and disposal of damaged bulbs reduce the risks of infestation. Acaricides applied to bulbs in store probably reduce the number of bulb mites.

C. FAMILY ERIOPHYIDAE: GALL MITES

1. *Aceria tulipae* (Keifer)

Synonym: *Eriophyes tulipae:* wheat curl mite, garlic mite.

This mite is a serious pest of garlic but also occasionally damages onions and shallots. It is a vector of Wheat Streak, Garlic Mosaic, and Onion Mosaic Viruses. The symptoms of attack are known as tangle-top malady of garlic in India.[267]

a. Distribution

This pest is distributed widely in Europe, North and South Africa, North and South America, Europe and parts of Asia (India), and New Zealand. Recently, this mite has been reported on garlic in South Africa[268] and Egypt.[269]

b. Biology and Life History

These gall mites are minute, between 0.1 and 0.25 mm long, and are only visible under a hand lens or microscope. They are worm-like in shape and have two pairs of legs at the anterior end of the body. The females lay round, colorless eggs, 0.6 mm in diameter, on bulbs. There are two nymphal stages similar in appearance to the adults. The mites enter diapause under short- and long-day conditions, thereby avoiding unfavorable periods in winter and summer.[270] They overwinter on wild cereals and grasses and on garlic bulbs.

c. *Damage*

The pest is found most often on garlic bulbs but it also infests onion and shallots; it was recorded by Chamberlain and Evans[271] on stored onion sets in northern England. The mite causes heavy losses in stored garlic bulbs and also damages shallots in New Zealand.[272]

Symptoms of damage include yellow streaks and spiral or twisted growth of the leaves, a slowing down in plant development and shrivelling of bulbs in store. Secondary rots develop in stored bulbs.[273,274] The symptoms on foliage are confused sometimes with those caused by viruses.

d. *Control*

When the mites enter the folds in the leaves of growing plants, they are difficult to control. Foliar acaricide sprays using organophosphorus compounds are effective in the field. Alternatively, bulbs are dipped in solutions of these compounds prior to planting out or prior to storage.[274,275] Fumigation of bulbs with methyl bromide has also been recommended.[272] In India a granular formulation of aldicarb applied to the garlic crop and followed by sprays of cypermethrin was effective in controlling mites and preventing the virus diseases transmitted by them.[267]

VIII. SLUGS AND SNAILS, ORDER: GASTROPODA

A. FAMILY: ARINIDAE (SLUGS); HELICIDAE (SNAILS)

1. *Deroceras reticulatum* (Mull), *Arion hortensis* Fer. and *Helix aspersa* Mull

Slugs and snails are familiar to most people and can be troublesome pests of cultivated plants in gardens, market gardens, small holdings and on farms.[6] The commonest species in the U.K. are the field slug, *Deroceras reticulatum* (Mull), the garden slug, *Arion hortensis* Fer. and the garden snail, *Helix aspersa* Mull. Godan[291] lists the species of economic importance in other countries.

a. *Distribution*

Slugs and snails are distributed throughout the world.[291]

b. *Biology*

Slugs and snails have soft, unsegmented bodies with a well developed head which bears two pairs of retractable tentacles. The anterior pair is sensitive to smell and possibly taste, whereas the posterior pair, which bear the eyes at their tips, are responsive to light and smell.[6,292] Slugs and snails move along the soil or on plants on a slimy muscular foot and prefer a moist humid environment. They are more abundant on heavy clay soils with a calcareous content than on light sandy loams.

Slugs are streamlined, lack a spiral shell and can move through crevices in the soil. They can tolerate lower amounts of calcium in their environment than snails. The garden slug, *Arion hortensis,* is black with a yellow foot. Immature field slugs *Deroceras reticulatum* are fawn but adults can vary in color from light brown to fawn with a chalky-white slime.[18] Slugs were introduced from Europe into U.S. but, apart from one indigenous species, European slugs are not reported to cause any damage there.[6]

The garden snail, *Helix aspersa,* is widespread, and is sometimes a serious pest of agricultural crops. It is 3.5 to 5 cm long with a gray-brown shell with five whorls.[6,18] If disturbed, the head and foot are retracted within the well developed, hardened shell. In

TABLE 6
Species of *Acarina, Diplopoda,* and *Symphyla* Damaging Onions and Allied Crops

Latin name and authority	Common name	Family	Crop damaged or infested	Part of plant damaged	Country	Ref.
Acarina						
Aceria tulipae (Keifer)	Wheat curl mite garlic mite	Acaridae	Garlic, onion	Foliage	India, Bangladesh, South Africa, Egypt, U.K. Venezuela, Bulgaria, Cuba	267 268 269,271 274,275 276,277 278,279
Caloglyphus krameri (Berl.)	Bulb mite	Acaridae	Garlic	Bulb	Argentina	280,281
Petrobia latens (Muller)	Brown wheat mite, stone mite	Tetranychidae	Onion, garlic	Foliage	Spain, Australia	264 265
Rhizoglyphus callae Oudemans		Acaridae	Onion, garlic	Bulb	Egypt	26
Rhizoglyphus echinopus (Fum. & Rob.)	Bulb mite	Acaridae	Garlic, onion	Bulb	Spain, Argentina, USSR, India, China	264 281 282—286 287
Rhizoglyphus robini (Claparede)		Acaribae	Onion, garlic	Bulb	Israel	288
Tetranychus cinnabarinus (Biois.)	Carmine spider mite; red spider mite	Tetranychidae	Onion	Bulb	U.S. Egypt	28 262,263
T. cucurbitacearum Sayed		Tetranychidae	Onions	*	Egypt	263
T. urticae (Koch)	Two-spotted spider mite	Tetranychidae	Garlic, leek	*	Spain, Greece	264 289

TABLE 6 (continued)
Species of *Acarina*, *Diplopoda*, and *Symphyla* Damaging Onions and Allied Crops

Latin name and authority	Common name	Family	Crop damaged or infested	Part of plant damaged	Country	Ref.
Diplopoda						
Blaniulus guttulatus (Bosc)	Spotted millipede	—	All	Bulb	Widespread	18,290
Symphyla						
Scutigerella immaculata (Newport)	Glasshouse symphlid	—	All	Roots	Widespread	18

* Unspecified in reference.

temperate regions, during winter, snails tend to hibernate, often in groups, under leaves, or stones. The open end of the shell is covered with an operculum with a small hole for breathing.[6]

c. Life History

Although slugs and snails are hermaphrodite, they generally pair and cross fertilize. Under favorable circumstances, some species can breed throughout the year. The eggs, which are spherical, pale and translucent, are laid in hollows in the soil. Eggs can hatch within a few days, except for those laid in winter which may not hatch until the following spring.[291] In temperate regions, field slugs take 12 to 15 months to complete their life cycle and gardenslugs take 6 to 12 months. Snails become adult in 12 months and may live up to 5 years.[18]

d. Damage

Slugs and snails feed mainly at night, browsing on decaying vegetable matter as well as on buds, flowers, and all other parts of a wide variety of plants.[291] They attack onions at all stages and occasionally cause enough damage to render small areas of crops patchy.

e. Control

Slugs and snails can be controlled in various ways. Baits, consisting of molluscicides such as methiocarb or metaldehyde are used widely for their control but they may be hazardous to livestock. Usually, the molluscicides are mixed in bran and sold as pellets to be scattered around plants. Thrushes and blackbirds readily eat snails and thus reduce their numbers.[18] Slugs and snails are attracted to weed species for food and shelter. Removal of weeds in and around crops can reduce the numbers of these pests.

REFERENCES

1. **Fronk, W. D.,** An economically important aphid new to the United States, *Pan-Pac. Entomol.*, 31, 190, 1955.
2. **Doncaster, J. P. and Kassanis, B.,** The shallot aphis, *Myzus ascalonicus* Doncaster, and its behaviour as a vector of viruses, *Ann. Appl. Biol.*, 33, 66, 1946.
3. **Doncaster, J. P.,** The shallot aphis, *Myzus ascalonicus* sp. n, *Proc. R. Entomol. Soc. London. B*, 15, 1946.
4. **Blackman, R. L. and Eastop, V. F.,** *Aphids on the World's Crops: An Identification and Information Guide*, Chichester: John Wiley & Sons, 1984, 466.
5. **Kennedy, J. S., Day, M. F., and Eastop, V. F.,** A conspectus of Aphids as vectors of plant viruses, *London: Commonwealth Institute of Entomology*, 1962, 114.
6. **Jones, F. G. W. and Jones, M. G.,** *Pests of Field Crops*, Publ. Edward Arnold, London, 1984, 392.
7. **Huffaker, C. B.,** *Biological Control*, Plenum Press, New York, 1971, 511.
8. **Wyatt, I. J.,** Aphid control by parasites, in *Biological Pest Control. The Glasshouse Experience*, Hussey, N. W. and Scopes, N. E. A., Eds., Poole, Dorset, The Blandford Press, 1985, 134.
9. **Hall, R. A.,** Aphid control by fungi, in *Biological Pest Control. The Glasshouse Experience*, Hussy, N. W. and Scopes, N. E. A., Eds., Poole, Dorset: The Blandford Press, 1985, 138.
10. **Scopes, N. E. A. and Ledieu, M.,** *Pest and Disease Control Handbook*, Croydon: BCPC Publications, 1983, 693.
11. **Klimaszewski, S. M. and Lodos, N.,** Further data about jumping plant lice of Turkey (Homoptera, Psylloidea), *Turkiye Bitki Koruma Dergisi*, 3, 3, 1979.
12. **Quaglia, F.,** Notes on the ecology and behavior of *Capsodes* (Horistus) *lineolatus* (Brulle) (Rhynchota, Miridae), injurious to onion in Tuscany, *Frustula Entomol.*, 1, 247, 1978.
13. **Garcia, G. C. A. and Bellotti, A. C.,** Preliminary study of the biology and morphology of *Cryptomenus bergi* Froeschner, a new pest of cassava, *Rev. Colomb. Entomol.*, 6, 55, 1980.

14. **Kelton, L. A.,** Labopidea Uhler in North America, with descriptions of a species and a new genus (Heteroptera: Miridae), *Can. Entomol.*, 111, 753, 1979.
15. **Henry, T. J.,** The onion plant bug genus *Labopidicola* (Hemiptera: Miridae): economic implications, taxonomic review, and description of a new species, *Proc. Entomol. Soc. Wash.*, 84, 1, 1902.
16. **Fye, R. E.,** Damage to vegetable and forage seedlings by overwintering *Lygus hesperus* (Heteroptera: Miridae) adults, *J. Econ. Entomol.*, 77, 1141, 1984.
17. **Fye, R. E.,** Damage to vegetable and forage seedlings by the pale legume bug (Hemiptera: Miridae), *J. Econ. Entomol.*, 75, 994, 1982.
18. **Edwards, C. A. and Heath, G. W.,** *The Principals of Agricultural Entomology*, Chapman and Hall, London, 1964, 418.
19. **Carver, M.,** *Neotoxoptera* Theobald and *Pterocallis* Passerini (Homoptera: Aphididae) in Australia, *J. Aust. Entomol. Soc.*, 19, 139, 1980.
20. **Khristova, E.,** The specific composition of aphids from vegetable crops in the Plovdiv district, *Gradinar. Lozar. Nauka*, 1, 55, 1964.
21. **Konov, Y.,** A new pest of onions in Bulgaria, *Rastit. Zash*, 9, 41, 1961.
22. **Wagner, W.,** *Trioza tremblayi*, eine neue Blattfloh-Art aus Italien (Homoptera-Psyllina), *Boll. Lab. Entomol. Agr. Portici B*, 19, 263, 1961.
23. **Tremblay, E.,** Risultati di prove di lotta contro la psilla della cipolla *(Trioza tremblayi* Wagner), *Ann. Fac. Sci. Agrar. Univ. Napoli, Portici*, 30, 15, 1965.
24. **Domenichini, G.,** I Tetrastichini (Hymenoptera Eulophidae) paleartici ed i loro ospiti, *Boll. Zool. Agr. Bachicdtin*, 6, 61, 1965.
25. **Nowacka, W. and Zoltanska, E.,** Leafhoppers (Cicadodea, Homoptera) on umbelliferous and alliaceous vegetables, *Rocz. Nauk Roln*, 4, 33, 1974.
26. **Jones, H. A. and Mann, L. K.,** *Onions and their Allies, Botany, Cultivation and Utilization.* Leonard Hill Interscience, London, 1963, 286.
27. **Lewis, T.,** *Thrips, Their Biology, Ecology and Economic Importance*, Academic Press, London, 1973, 349.
28. **Metcalf, C. L. and Flint, W. P.,** *Destructive and Useful Insects: their Habits and Control*, McGraw Hill, New York, 1939, 981.
29. **Schwartz, P. H. and Klassen, W.,** Estimate of losses caused by insects and mites to agricultural crops, in *Handbook of Pest Management in Agriculture*, Pimentel, D., Ed., Vol. 1, CRC Press, Boca Raton, FL.
30. **Mote, U. M.,** Control of onion thrips *(Thrips tabaci* Lind). *Pesticides*, 12, 42, 1978.
31. **Schmutterer, H.,** *Pests of Crops in North East and Central Africa*, Gustav Fischer Verlag, Stuttgart, 1969, 296.
32. **Lall, B. S. and Singh, L. M.,** Biology and control of the onion thrips in India, *Indian J. Econ. Entomol.*, 61, 676, 1968.
33. **Andaloro, J. T. and Shelton, A. M.,** Insects of onions and cabbage. Onion thrips, *N.Y. State Agric. Exp. Stn., Geneva Publ.*, 750.75, 2, 1983.
34. **Coudriet, D. L., Kishaba, A. M., McCreight, J. D., and Bohn, G. W.,** Varietal resistance in onions to thrips, *J. Econ. Entomol.*, 72, 614-615, 1979.
35. **Edelson, J. V.,** A sampling method for estimating absolute numbers of thrips on onions, *Southwest. Entomol.*, 10, 103, 1985.
36. **Suman, C. L. and Wahi, S. D.,** Sequential sampling plan for the onion thrips, *Thrips tabaci* (Lind.), *Entomology*, 6, 265, 1981.
37. **Freuler, J. and Fischer, S.,** Methodes de denombrement des populations du thrips de l'oignon, *Thrips tabaci* Lind, *Mitteilungen der Schweizerischen Entomolgischen Gesellschaft*, 57, 163, 1984.
38. **Passlow, T.,** Control of *Thrips tabaci* Lind. in onion crops in the Lockyer Valley, *Queensl. J. Agric. Sci.*, 14, 53-72, 1957.
39. **Uvah, I. I. I. and Coaker, T. H.,** Effect of mixed cropping on some insect pests of carrots and onions, *Entomol. Exp. Appl.*, 36, 159, 1984.
40. **Saxena, R. C.,** Observations on some predators and parasites of *Thrips tabaci* Lind, *Bull. Entomol.*, 22, 97-100, 1981.
41. **Saxena, R. C.,** Integrated approach for the control of *Thrips tabaci* Lind, *Indian J. Agric. Sci.*, 45, 434-436, 1977.
42. **Hussey, N. W.,** Thrips and their natural enemies, in *Biological Pest Control: The Glasshouse Experience*, Hussey, N. W. and Scopes, N. E. A., Eds., Blandford Press, Poole, 1985, 53.
43. **Ramakers, P. M. J.,** Biological control of *Thrips tabaci* with *Amblyseius* spp., *Bulletin of the IOBC/WPRS Working Group on Integrated Control in Glasshouses, Vantaa*, 3(3), 203, 1980.
44. **Jones, H. A., Bailey, S. F., and Emsweller, S. L.,** Thrips resistance in the onion, *Hilgardia*, 8, 215, 1934.
45. **Painter, R. H.,** *Insect Resistance in Crop Plants*, Macmillan, New York, 1951, 520.

46. **Lall, B. S. and Verma, S. K.,** *Sci. Cult.*, 24, 575, 1959.
47. **Shinde, N. N. and Sontakke, M. B.,** Bulb crops, Onions, in *Vegetable Crops in India*, Bose, T. K. and Som, M. G., Eds., Naya Prokash, Calcutta, 1986, 542.
48. **Srivastava, R. P. and Pandey, S. N.,** A new cutworm pest of onion in India, *Labdev J. Sci. Technol.*, 6-B, 54, 1968.
49. **Nikolova, V.,** Noctuids — dangerous pests of plants. The silver Y moth, *Rastit. Zash.*, 19, 13, 1971.
50. **Rivnay, E.,** A contribution to the biology and phenology of *Agrotis ypsilon Rott. Z. Angew Entomol.*, 53, 295, 1964.
51. **Rings, R. W., Johnson, B. A., and Arnold, F. J.,** Host range of the variegated cutworm on vegetables: a bibliography, *Bull. Entomol. Soc. Am.*, 22, 409, 1976.
52. **Anon,** Black cutworm *Agrotis ipsilon* (Hufnagel), *Insect Identification Sheet No. 34, Agriculture Canada*, 1979, 2.
53. **Poitout, S. and Bues, R.,** The principal injurious noctuids, *Phytoma*, 337, 39, 1982.
54. **Quiroz, E. C.,** Pests of onion, *Invest. Prog. Agric.*, 9, 43, 1977. **Ministry of Agriculture, Fisheries and Food,** Cutworms, *Leaflet 225, MAFF, 1984.*
55. **Ministry of Agriculture, Fisheries and Food,** Cutworms, *Leaflet 225, MAFF, 1984.*
56. **Genchev, N.,** An ecological study of *Agrotis segetum* Schiff. (Lepidoptera: Noctuidae), *Rast Vud. Naukia*, 5, 127, 1968.
57. **Santos, H. R. and Nakano, O.,** Biological information on the cutworm *Agrotis ipsilon* (Hufnagel, 1776) (Lepidoptera: Noctuidae), *An. Soc. Entomol. Braz.*, 11, 33, 1982.
58. **Akhmedov, R. M. and Abdinbekova, A. A.,** The phenology of Tortricids *Agrotis ypsilon* and *Barathra brassicae* in Azerbaidzhan, *Zool. Zh.*, 52, 451, 1973.
59. **Wang, Q., Huang, G. H., and Yang, Y. G.,** Preliminary studies on nocturnal activities of adult black cutworms, *Insect Knowledge*, 20, 106, 1983.
60. **Rings, R. W., Musick, G. J., Keaster, A. J., and Luckmann, W. H.,** Geographical distribution and economic importance of the black, glassy, and bronzed cutworms, *Coop. Plant Pest Rep.*, 1, 173, 1976.
61. **Carter, D. J.,** Pest *Lepidoptera* of Europe with special reference to the British Isles, Publ. Dr W. Junk, Holland, 1984, 431.
62. **Homonnay, F. and Csehi, E.,** The damage caused by *Agrotis segetum* and the role of parasites in suppressing an outbreak of it, Ed. Szalay-Marzso, L, *XVIIth Sci. Conf. Plant Prot.*, 2, 351, 1967.
63. **Willson, H. R., Semel, M., Tebcherany, M., Prostak, D. J., and Hill, A. S.,** Evaluation of sex attractant and blacklight traps for monitoring black cutworm and variegated cutworm, *J. Econ. Entomol.*, 74, 517, 1981.
64. **Clement, S. L., Hill, A. S., Levine, E., and Roelofs, W. L.,** Trap catches of male *Agrotis ipsilon* with synthetic sex pheromone emitted from different dispensers, *Environ. Entomol.*, 10, 521, 1981.
65. **Lofstedt, C., Linn, C. E., Jr., and Lofovist, J.,** Behavioural responses of male turnip moths, *Agrotis segetum,* to sex pheromone in a flight tunnel and in the field, *J. Chem. Ecol.*, 11, 1209, 1985.
66. **Bowden, J., Cochrane, J., Emmett, B. J., Minall, T. E., and Sherlock, L. L.,** A survey of cutworm attacks in England and Wales, and a descriptive population model for *Agrotis segetum* (Lepidoptera: Noctuidaea), *Ann. Appl. Biol.*, 102, 29, 1983.
67. **Esbjerg, P.,** Fangst af agerugler *(Agrotis segetum)* og nedborsmaling som baggrund for knopormevarsling, (Forecasting of cutworm attacks based on trapping of turnip moths) *(Agrotis segetum)* and measuring of precipitation), *Tidsskr for Planteavl.*, 87, 371-377, 1983.
68. **Nikolov, N. K.,** The role of soil moisture in the development of some species of the genus *Agrotis* (Lepidoptera: Noctuidae), *Rastenievud. Nauki*, 17, 102, 1980.
69. **Afify, A. M., Hafez, M., and Merdan, A. I.,** Preliminary investigations on the virulance of 13 *Bacillus* preparations against 3 Egyptian Noctuids, *Anz. Schaedlingskd. Pflanzenschutz*, 42, 54, 1969.
70. **Puttler, B. and Thewke, S. E.,** Field and laboratory observations *Hexamermis arvalis* (Nematoda: Mermithidae) a parasite of cutworms, *Ann. Entomol. Soc. Am.*, 64, 1102, 1971.
71. **Michael, P. J., Woods, W., Lawrence, P. J., and Fisher, W.,** Introduced parasites for the control of Australian noctuid pests, in *Proceedings of the Fourth Australian Applied Entomological Research Conference, Adelaide, Pest control: Recent Advance and Future Prospects*, 1984.
72. **Eckenrode, C. S., Robbins, P. S., and Webb, D. R.,** Control of seedcorn maggot, cabbage maggot, and black cutworm (1975 insecticide research report), *Plant Sci. Entomol.*, 13, 5, 1976.
73. **Harris, C. R., Svec, H. J., and Chapman, R. A.,** Potential of pyrethroid insecticides for cutworm control, *J. Econ. Entomol.*, 71, 692, 1978.
74. **Ministry of Agriculture, Fisheries and Food,** Cabbage caterpillars, *Leaflet 69, MAFF, 1982.*
75. **Ministry of Agriculture, Fisheries and Food,** Control of pests and diseases of field vegetables, *Booklet 2382*, 1982, 195.
76. **Ibrahim, M. M., Koura, A., and El-Halfawy, M.,** Ecological and biological studies on some insects infesting dried onions, in U.A.R., *Agric. Res. Rev.*, 48, 59, 1970.

77. **Lal, S. S. and Varma, B. K.,** Attack of *Cadra cautella* (WCK) in stored onions at Hyderabad, A. P., *Bull. Grain Technol.,* 12, 238, 1975.
78. **Kirkpatrick, R. L., Yancey, D. L., and Marzke, F. O.,** Effectiveness of green and ultraviolet light in attracting stored-product pests in traps, *J. Econ. Entomol.,* 63, 1853-1855, 1970.
79. **Takahashi, F., Masui, A., Kuwahara, Y., Ishii, S., and Fukami, H.,** [Studies on sex pheromone of Pyralidae IV, The male response to the female sex pheronome of the almond moth *Cadra cautella* Walker (Phycitinae).] *Botyu-Kagaku,* 37, 56, 1972.
80. **Zeleny, J., Spitzer, K., and Ondracek, J.,** Tropical moths (Lepidoptera) attracted by synthetic sex pheromones in Vietnamese agro-ecosystems, *Acta Entomol Bohemoslov,* 79, 254, 1982.
81. **Brower, J. H.,** *Trichogramma:* potential new biological control method for stored-product Lepidoptera, in, *Proceedings of the Third International Working Conference on Stored-Product Entomology,* 1984, 454.
82. **Brower, J. H. and Tiltone, E. W.,** Potential for control of *Cadra cautella* (Walker) by release of fully or partially sterile males, *Int. J. Appl. Radiat. Isot,* 26, 720, 1975.
83. **Hagstrum, D. W., McLaughlin, J. R., Smittle, B. J., and Coffelt, J. A.,** Sterile males in a sex pheromone permeated atmosphere to reduce reproduction of *Ephestia cautella, Environ. Entomol.,* 7, 759, 1978.
84. **Banaag, A. F., Bandong, P. P., and Luis, R. A.,** Two concentrations of DDT and methoxychlor in the control of insects affecting garlic in storage, *Philipp. Agric.,* 45, 36, 1961.
85. **Shukla, R. M. and Srivastava, A. S.,** Studies on the resistance to insecticides in *Cadra cautella* Walker. I. Selection of a strain resistance to malathion, *Indian J. Entomol.,* 46, 117, 1984.
86. **Shukla, R. M. and Srivastava, A. S.,** Resistance to insecticides in *Caudra cautella* Walker. II. Cross-resistance characteristics of the malathion-resistant strain, *Indian J. Entomol.,* 46, 201, 1984.
87. **Lecomte, C.,** First observations on the biology and damage caused by the leek teneid *Acrolepiopsis* (Acrolepia) *assectella* Zeller (Microlepidoptera Plutellidae) on the coast of Algeria, *Bull. Soc. Hist. Nat. Afr. Nord.,* 67, 49, 1976.
88. **Lecomte, C. and Pouzat, J.,** Electroantennogram study of the reception of olfactory stimuli of plant origin in *Acrolepiopsis assectella, Entomol. Exp. Appl.,* 40, 13, 1986.
89. **Goix, J.,** The leek tineid, *Phytoma,* 379, 37, 1986.
90. **Yamada, H. and Koshihara, T.,** Interspecific attractiveness of female sex pheromone components of the diamondback moth, *Plutella xylostella* (L.) and their analogue to the stone leek miner, *Acrolepiopsis Sapporensis Matsumara Kontyu,* 48, 104, 1980.
91. **Jary, S. G. and Rolfe, S. W.,** The leek moth, *J. Ministry Agric.,* p. 35, 1945.
92. **Simmonds, S. P.,** *Acrolepiopsis assectella* (Zell.) (Lep., Yponomeutidae) on onions in Devon, *Entomol. Mon. Mag.,* 110, 1316, 1974.
93. **Simmonds, S. P.,** Onions attacked by *Acrolepia assectella* (Zell.), *Plant Pathol.,* 23, 49, 1974.
94. **Jary, S. G., Rolfe, S. H. W., and Carpenter, K.,** The leek moth, *J. Ministry Agric.,* 12, 563, 1948.
95. **Becker, P.,** Leek Moth occuring inland, *Plant Patho.,* 10, 42, 1961.
96. **Rahn, R.,** La teigne du poireau *Acrolepia assectella* Zeller, elements de biolgie et mise au point d'avertissements agricoles fondes sur le piegeage sexuel des males, *C. R. Seanc. Acad. Agric. France,* 52, 997, 1966.
97. **Boscher, J.,** Remarks on the counting of "attacks" with reference to an example: the leek moth *Acrolepiopsis assectella* Zell., *C. R. Seanc. Acad. Agric. France,* 65, 924, 1979.
98. **Rahn, R.,** The leek moth, *Acrolepiopsis assectella* Z., possibility of issuing general agricultural warnings on the basis of sexual trapping using synthetic products, *Ann. Zool. Ecol. Anim.,* 11, 611, 1979.
99. **Rahn, R.,** Methodes d'elevage et de piegeage sexuel utilisables pour la mise au point d'avertissements agricoles concernant la teigne du poireau, *Phytoma,* 20, 21, 1968.
100. **Renou, M., Descoins, C., Priesner, E., Gallois, M., and Lettere, M.,** A study of the sex pheromone of the leek moth, *Acrolepiopsis assectella, Entomol. Exp. Appl.,* 29, 198, 1981.
101. **Rahn, R.,** Sex-pheromone trapping of the leek moth, *Acrolepiopsis assectella* Z. (Lepid. Yponomeutidae - Acrolepiinae) using Z 11 H DAL. Results of the 1981 programme, *Agronomie,* 2, 957, 1982.
102. **Kalmes, R., Hedderwick, M. P., Rojas-Rousse, D., and Chessel, D.,** Interactions between two parasitoids *Diadromus collaris* and *D. pulchellus* (Ichneumonidae): the survival of male parasitoids in pupae of *Acrolepiopsis assectella* with or without intra- and interspecific competition, *Can. J. Zool.,* 61, 627, 1983.
103. **Alford, D. V.,** The host range of *Clepsis spectrana* (Treits) (Lepidoptera: Tortricidae), *Plant Pathol.,* 25, 166, 1976.
104. **Grill, D.,** Les maladies et ravageuirs du poireau, *Phytoma,* 371, 39, 1985.
105. **Abul-Nasr, S. E., Assem, M. A., and El-Sherif, A. R. A.,** Rates of infestation of the main insects attacking stored garlic bulbs, *Bull. Soc. Entomol. Egypt,* 58, 31, 1974.
106. **Swailem, S. M. and Ismail, I. I.,** On the biology of the honey dew moth *Cryptoblabes gnidiella,* Milliere, *Bull. Soc. Entomol. Egypt,* 56, 127, 1972.
107. **Aleksov, S.,** *Dyspessa ulula* Bkh. (Lepidoptera, Cossidae), a dangerous pest of onion in Macedonia, *Zast. Bilja,* 20, 331, 1969.

108. **Tremblay, E.,** Osservazioni su un insetto dannoso ai fiori. La *Gymnoscelis pumilata* Hb. (Lepidoptera-Geometridae) (syn. *Eupithecia pumilata* Hb., *Tephroclystia pumilata* Hb.), *Boll. Zool. Agrana Bachicdtin,* 19, 199, 1961.
109. **Sinha, S. N. and Chakrabarti, A. K.,** Outbreak of *Heliothis armigera* (Hubner) on various crops at Karnal, *Indian J. Entomol.,* 45, 475, 1983.
110. **Kashyap, R. K. and Pandita, M. L.,** Nature and extent of damage of *Heliothis armigera* (Hubner) in seed crop of onion, *Haryana Agric. Univ. J. Res.,* 15, 102, 1985.
111. **Darshan Singh., Deol, B, S., Gill, S. P. S., and Labh Singh.,** Field screening on onion cultivars (seed crop) against gram caterpillar, *Punjab Hortic. J.,* 23, 242-243, 1983.
112. **Saito, O.,** Life history of onion stem borer, *Hydraecia mongoliensis* Urbahn in Hokkaido, *Annu. Rep. Soc. Plant Prot., N. Jpn.,* 35, 136, 1984.
113. **Petrukha, O. I. and Tribel, S. A.,** *Loxostege sticticalis* in the Ukraine, *Zashch. Rast.,* 4, 40, 1974.
114. **d'Araujo e Silva, A. G., Goncalves, C. R., Monteiro Galvao, D., Lobo Goncalves, A. J., Gomes, J., Nascimento Silva, M., and de Simoni, L.,** Fourth catalogue of the insects that live on plants in Brazil, their parasites and predators. Insects, food-plants and natural enemies, *Rio de Janeiro, Minist. Agric., Lab. Cent. Pat. Veg.,* 1, 622 pp. 1968.
115. **Anon.,** Onion cultivation, *Rev. Agric., (Bogota),* 3, 739-742, 1914. **Ministry of Agriculture, Fisheries and Food,** Onion fly, *Leaflet 163, MAFF* 1981.
116. **Ministry of Agriculture, Fisheries and Food,** Onion fly, *Leaflet 163, MAFF* 1981.
117. **Dustan, A. G.,** Onion maggot, *Dep. Agric. Ottawa, Can.,* No. 89, 1948.
118. **Loosjes, M.,** Ecology and the Genetic Control of the Onion Fly, *Delia antiqua* (Meigen), Ph.D. thesis, State University of Leiden, Netherlands, 179, 1976.
119. **Perron, J. P. and LeRoux, E. J.,** Sampling method for the study of natural populations of the onion fly, *H. antiqua,* in Quebec, *Ann. Entomol. Soc. Quebec,* 7, 95, 1962.
120. **Keyder, S. and Atak, U.,** Studies on vegetable flies (*Hylemya* spp.) *Erenkoy Bolge Zirai Mucadele Arastirma Enstitusu, Turkey,* 10, 136, 1972.
121. **Yathom, S.,** The life history of the onion fly *Hylemyia antiqua* Meigen in Israel, *Spec. Bull. Israel Minist. Agric. Res. Stn.,* No. 25, 1960.
122. **Andaloro, J. T. and Eckenrode, C. J.,** Insects of onion, onion maggot, Publ. *N. Y. State Agric. Stn., Geneva Publ.,* 2, 1983.
123. **Finch, S. and Collier, R. H.,** Monitoring and forecasting dipterous insect attacks in vegetable crops, *Proceedings of the International Symposium on Diptera Pests in Vegetable Crops,* 1987, 10.
124. **Liu, H. J., McEwen, F. L., and Ritcey, G.,** Forecasting events in the life cycle of the onion maggot, *Hylemya antiqua* (Diptera: Anthomyiidae): application to control schemes, *Environ. Entomol.,* 11, 751, 1982.
125. **Eckenrode, C. J., Vea, E. V., and Stone, K. W.,** Population trends of onion maggots correlated with air thermal unit accumulations, *Environ. Entomol.,* 4, 785, 1975.
126. **Ishikawa, Y., Tanaka, S., Matsumoto, Y., Yamashita, K., Yoshida, M., and Shirai, E.,** Color preference of the onion fly, *Hylemya antiqua* Meigen (Diptera: Anthomyiidae) with reference to ultraviolet reflection, *Appl. Entomol. Zool.,* 20, 20, 1985.
127. **Vernon, R. S. and Borden, J. H.,** Spectral specific discrimination by *Hylemya antiqua* (Meigen) (Diptera: Anthomyiidae) and other vegetable infesting species, *Environ. Entomol.,* 12, 650, 1983.
128. **Vernon, R. S.,** Aspectral zone of colour preferences for the onion fly, *Delia antiqua* (Diptera: Anthomyiidae), with reference to the reflective intensity of traps, *Can. Entomol.,* 118, 849, 1986.
129. **Vernon, R. S. and Bartel, D. L.,** Effect of hue, saturation, and intensity on color selection by the onion fly, *Delia antiqua* (Meigen) (Diptera: Anthomyiidae) in the field, *Environ. Entomol.,* 14, 210, 1985.
130. **Ishikawa, Y., Matmumoto, Y., Tsutsumi, M., and Mitsui, Y.,** Experimental field trapping of onion and seed-corn flies, *Hylemya antiqua* and *H. platura* (Diptera: Anthomyiidae), using 2-phenylethanol for population estimation, *Appl. Entomol. Zool.,* 19, 75, 1984.
131. **Ikeshoji, T., Ishikawa, Y., and Matsumoto, Y.,** Attractants against the onion maggots and flies, *Hylemia antiqua,* in onions inoculated with bacteria, *J. Pestic. Sci.,* 5, 343, 1980.
132. **Ishikawa, Y., Ikeshoji, T., and Matsumoto, Y.,** A propylthio moity essential to the oviposition attractant and stimulant of the onion fly, *Hylemia antiqua* Meigen, *Appl. Entomol. Zool.,* 13, 115, 1978.
133. **Soni, S. K. and Finch, S.,** Onion fly: evaluation of attractants for the adult, *29th Ann. Rep. Natl. Veg. Res. Stn. 1978,* 44, 1978.
134. **Soni, S. K. and Finch, S.,** Laboratory evaluation of sulphur-bearing chemicals as attractants for larvae of the onion fly, *Delia antiqua* (Meigen) (Diptera: Anthomyiidae), *Bull. Entomol. Res.,* 69, 291, 1979.
135. **Pierce, H. D., Jr., Vernon, R. S., Borden, J. H., and Oehlschlager, A. C.,** Host selection by *Hylemya antiqua* (Meigen): identification of three new attractants and ovistimulants, *J. Chem. Ecol.,* 4, 65, 1978.
136. **Vernon, R. S., Judd, G. J. R., Borden, J. H., Pierce, H. D., Jr., and Oehlschlager, A. C.,** Attraction of *Hylemya antiqua* (Meigen) (Diptera: Anthomyiidae) in the field to host-produced oviposition stimulants and their non host analogues, *Can. J. Zool.,* 59, 872, 1981.

137. **Dindonis, L. L. and Miller, J. R.,** Onion fly and little house fly host finding selectively mediated by decomposing onion and microbial volatiles, *J. Chem. Ecol.*, 7, 419, 1981.
138. **Nishizawa, S., Asahi, A., Nakamura, H., Tahara, S., Mizutani, J., and Obata, Y.,** Attraction of flies to some baits in an onion fly, *Mem. Fac. Agric. Hokkaido Univ.*, 8, 102, 1972.
139. **Ishikawa, Y., Ikeshoji, T., Matsumoto, Y., Tsutsumi, M., and Mitsui, Y.,** 2-Phenylethanol: an attractant for the onion and seed-corn flies, *Hylemya antiqua* and *H. platura* (Diptera: Anthomyiidae), *Appl. Entomol. Zool.*, 18, 270, 1983.
140. **Miller, J. R. and Haarer, B. K.,** Yeast and corn hydrolysates and other nutritious materials as attractants for onion and seed flies, *J. Chem. Ecol.*, 7, 555, 1981.
141. **Finch, S. and Eckenrode, C. J.,** Influence of unharvested, cull-pile, and volunteer onions on populations of onion maggot (Diptera: Anthomyiidae), *J. Econ. Entomol.*, 78, 542, 1985.
142. **Eckenrode, C. J. and Nyrop, J. P.,** Impact of physical injury and commercial lifting on damage to onion bulbs by larvae of onion maggot (Diptera: Anthomyiidae), *J. Econ. Entomol.*, 79, 1606, 1986.
143. **Anasiewicz, A. and Sandner, H.,** The role of parasites as a factor limiting the numbers of the onion fly-*Phorbia antiqua* Meig. (Diptera: Anthomyiidae), *Pol. Pismo Entomol.*, 46, 367, 1976.
144. **Gherasim, V. and Lacatusu, M.,** *Aphaereta minuta* (Nees) (Hymenoptera: Braconidae), principal factor limiting the populations of Diptera injurious to onion crops in the Socialist Republic of Rumania, *Am. Inst. Cercet. Prot. Plant,* 12, 229, 1977.
145. **Hertveldt, L., Keymeulen, M. van, and Pelerents, G.,** Large scale rearing of the entomophagous rove beetle *Aleochara bilineata* (Coleoptera: Staphylindiae), *Mitt. Biol. Bundesanst. Land Forstwirtsch, Berlin-Dahlem,* 218, 70, 1984.
146. **Whistelcraft, J. W., Harris, C. R., Tolman, J. H., and Tomlin, A. D.,** Mass-rearing technique for *Aleochara bilineata* (Coleoptera: Staphylinidae), *J. Econ. Entomol.*, 78, 995, 1985.
147. **Tomlin, A. D., Miller, J. J., Harris, C. R., and Tolman, J. H.,** Arthropod parasitoids and predators of the onion maggot (Diptera: Anthomyiidae) in Southwestern Ontario, *J. Econ. Entomol.*, 78, 975, 1985.
148. **Whistlecraft, J. W., Harris, C. R., Tomlin, A. D., and Tolman, J. H.,** Mass rearing technique for a braconid parasite, *Aphaereta pallipes* (Say) (Hymenoptera: Brachonidae), *J. Econ. Entomol.*, 77, 814, 1984.
149. **Harris, C. R., Svec, H. J., Tolman, J. H., and Tomlin, A. D.,** A rational integration of methods to control onion maggot in southwestern Ontario, *Proceedings British Crop Protection Conference — Pests and Diseases,* 1982.
150. **Plattner, H. C. and Naton, E.,** Testing the action of pesticides on beneficial arthropods. The state of preparatory work in Bavaria, *Bayer. Landwirtsch. Jahrb.*, 52, 143, 1975.
151. **Rizzo, D. C.,** Age of three dipteran hosts as a factor governing the pathogenicity of *Beauveria bassiana* and *Metaarhizium anisopliae, J. Invertebr. Pathol.*, 30, 127, 1977.
152. **Coomans, P. and Grisse, A. de,** Susceptibility of *Delia antiqua* (Meigen) to *Steinernema* (Neoaplectana) *bibionis* (Bovien), *Meded Fac. Landbouwwet. Rijksuniv. Gent.*, 50, 155, 1985.
153. **Stoffolano, Jr., J. G.,** Nematode parasites of the face fly and the onion maggot in France and Denmark, *J. Econ. Entomol.*, 62, 792, 1969.
154. **Reid, J. A. and McEwen, F. L.,** Semi-sterile mutants in the onion maggot, *Hylemya antiqua* (Diptera: Anthomyiidae), population effects in field cages, *Can. Entomol.*, 111, 749, 1979.
155. **Robinson, A. S., Herfst, M., and Vosselman, L.,** Genetic control of *Delia antiqua* (Meigen) (Diptera: Anthomyiidae). Sensitivity to diapause interfering with a field cage experiment using a homozygous chromosomal translocation, *Bull. Entomol. Res.*, 70, 103, 1980.
156. **Ticheler, J., Loosjes, M., Noordink, J. P. W., Noorlander, J., and Theunissen, J.,** Field experiments with the release of sterilised onion flies, *Hylemya antiqua* (Meig.). The sterile-insect technique and its field applications, *Proceedings of the Joint FAO/IAEA Division of Atomic Energy in Food and Agriculture, Vienna,* 1974, 137.
157. **Theunissen, J., Loosjes, N., Noordink, J. P. W., Noorlander, J., and Ticheler, J.,** Genetic control of the onion-fly, *Hylemya antiqua,* in the Netherlands, *Proceedings of the Seventh British Insecticide and Fungicide Conference, Brighton,* Vol. 3, British Crop Protection Council, London, 1974, 981.
158. **Theunissen, J., Loosjes, M., Noordink, J. P. W., Noorlander, J., and Ticheler, J.,** Small-scale field experiments on sterile-insect control of onion fly, *Hylemia antiqua* (Meigen), *Proceedings of the Joint FAO/IAEA Division of Atomic Energy in Food and Agriculture, Vienna,* 1975, 175.
159. **Soni, S. K., Freeman, G. H., and Ellis, P. R.,** A glasshouse screening method to assess resistance in onions to damage by onion fly *(Delia antiqua)* larvae, *Ann. Appl. Biol.*, 105, 565, 1984.
160. **Ellis, P. R., Eckenrode, C. J., and Harman, G. E.,** Influence of onion cultivars and their microbial colonisers on resistance to onion maggot, *J. Econ. Entomol.*, 72, 512, 1979.
161. **Ikeshoji, T., Matsumoto, Y., Nagai, M., Komochi, S., Tsutsumi, M., and Mitsui, Y.,** Correlation between the susceptibility of onion cultivars to the onion maggot, *Hylemya antiqua* Meigen (Diptera: Anthomyiidae), and the biochemical characteristics of onion plant, *Appl. Entomol. Zool.*, 17, 507, 1982.

162. **Ponti, O. M. B. de and Inggamer, H.,** Resistance to the onion fly in *Allium cepa* and *Allium fistulosum, 3rd Eucarpia Allium Symposium, Wageningen,* 21, 1984.
163. **Kireev, V. K., Shapiro, I. D., and Asyakin, B. P.,** Onion resistance to onion fly *Delia anyiqua* Mg., *Annu. Plant Resist. Insects Newsl.,* 7, 74, 1981.
164. **Soni, S. K., Baldwin, R. E., and Ellis, P. R.,** Evaluation of onion cultivars for rearing onion fly in the laboratory, *Ann. Appl. Biol., (Tests of Agrochemicals and Cultivars, Supplement)* No. 7, 108, 172, 186.
165. **Miller, J. R., Harris, M. O., and Breznak, J. A.,** Search for potent attractants of onion flies, *J. Chem. Ecol.,* 10, 1477, 1984.
166. **Hough, J. A., Harman, G. E., and Eckenrode, C. J.,** Microbial stimulation of onion maggot oviposition, *Environ. Entomol.,* 10, 206, 1981.
167. **Hough, J. A., Eckenrode, C. J., and Harman, G. E.,** Nonpathogenic bacteria affecting oviposition behaviour in the onion fly, *Environ. Entomol.,* 11, 585, 1982.
168. **Ikeshoji, T., Ishikawa, Y., and Matsumoto, Y.,** Ecological and chemical interactions between onion and onion maggot, *Hylemya antiqua* (Diptera: Anthomyiidae), *Rev. Plant Prot. Res.,* 14, 141, 1981.
169. **Ikeshoji, T.,** S-Propenylcysteine sulfoxide in exudates of onion roots and its possible decompartmentalization in root cells by bacteria into attractant of the onion maggot *Hylemya antiqua* (Diptera: Anthomyiidae), *Appl. Entomol. Zool.,* 19, 159, 1984.
170. **Ellis, P. R. and Eckenrode, C. J.,** Factors influencing resistance in *Allium* sp. to onion maggot, *Bull. Entomol. Soc. Am.,* 25, 151, 1979.
171. **Tsutsumi, M., Mitsui, Y., Nagai, M., and Komochi, S.,** Differences in tolerance of some onion varieties and seasonal damage fluctuation caused by *Diptera* spp., *Res. Bull. Hokkaido Natl. Agric. Exp. Stn.,* No. 136, 31, 1983.
172. **Tsutsumi, M., Mitsui, Y., Uragami, A., and Yoshikawa, H.,** Differences in tolerance of onion varieties to feeding by the onion maggot, *Hylemia antiqua* Meigen, *Annu. Rep. Soc. Plant Prot. N. Jpn.,* 20, 465, 1985.
173. **Anon.,** Breeding research. Vegetables. Resistance to insects and mites, Annual Report, IVT, Wageningen, Netherlands, 1985.
174. **Ellis, P. R. and Eckenrode, C. J.,** The onion maggot and its control in *N.Y. Food Life Sci. Bull.,* No. 79, 1979.
175. **Harris, C. R., Tolman, J. H., and Svec, H. J.,** Onion maggot (Diptera: Anthomyiidae) resistance to some insecticides following selection with parathion or carbofuran, *Can. Entomol.,* 114, 681, 1982.
176. **Markkula, M. and Kurppa, S.,** Resistance of insects and mites to pesticides in Finland, *Ann. Agric. Fenn.,* 24, 161, 1985.
177. **Thompson, A. R., Percivall, A. L., and Edmonds, G. H.,** Onion fly: seed film-coating, *35th Ann. Rep. Natl. Veg. Res. Stn. 1984,* 32, 1985.
178. **Toms, A. M. and Blackett, J. A.,** The use and potential of seed coatings, *10th International Congress of Plant Protection, Brighton,* 523, 1983.
179. **Madder, D. J.,** Pest management in muck crops in the Holland Marshes, *Proceedings of the 28th Annual Meeting, Canadian Pest Management Society,* 1981.
180. **Whitfield, G. H., Carruthers, R. I., and Haynes, D. L.,** Phenology and control of the onion maggot (Diptera: Anthomyiidae) in Michigan onion production, *Agric. Ecosystems, Environ.,* 12, 189, 1985.
181. **Finch, S., Cadoux, M. E., Eckenrode, C. J., and Spittler, T. D.,** Appraisal of current strategies for controlling onion maggot (Diptera: Anthomyiidae) in New York State, *J. Econ. Entomol.,* 79, 736, 1986.
182. **Ormerod, E. A.,** Report of observations of injurious insects during the year 1882, Simpkin, Marshall Co., London, 1983.
183. **Thorne, J. E.,** Bibliography of the seed maggot *Hylemya platura* and *H. florilega* (Diptera: Anthomyiidae), *N.Y. State Agric. Exp. Stn., Special Rep.,* 37, 18, 1980.
184. **Tsutsumi, M. and Mitsui, Y.,** The seasonal prevalence of three species of Diptera on onion in Sapporo District, *Annu. Rep. Soc. Plant Prot. N. Jpn.,* 33, 119, 1982.
185. **Ministry of Agriculture, Fisheries and Food, United Kingdom,** Bean seed flies, *Advisory Leaflet, MAFF,* 760, 1982.
186. **Thorne, J. E. and Eckenrode, C. J.,** Emergence patterns of the seedcorn maggot, *Delia platura* (Diptera: Anthomyiidae), *Environ. Entomol.,* 14, 182, 1985.
187. **Kuwayama, S., Hori, M., Takizawa, M., Endo, K., Kato, S., Sakurai, K., and Tsutsumi, M.,** Studies on the biology and control of the seed-corn maggot in Hokkaido, *Hokkaido Natl. Agric. Exp. Stn.,* 96, 1970.
188. **Eckenrode, C. S., Robbins, P. S., and Webb, D. R.,** Control of seedcorn maggot, cabbage maggot, and black cutworm (1975 insecticide research report). *N.Y. Food Life Sci. Bull.,* No. 63, 1976.
189. **Vernon, R. S. and Borden, J. H.,** Spectral specific discrimination by *Hylemya antiqua* (Meigen) (Diptera: Anthomyiidae) and other vegetable-infesting species, *Environ. Entomol.,* 12, 650, 1983.
190. **Dindonis, L. L. and Miller, J. R.,** Host-finding responses of onion and seedcorn flies to healthy and decomposing onions and several synthetic constituents of onion, *Environ. Entomol.,* 9, 467, 1980.

191. **Ishikawa, Y., Ikeshoji, T., Matsumoto, Y., Tsutsumi, M., and Mitsui, Y.,** Field trapping of onion and seed-corn flies with baits of fresh and aged onion pulp, *Appl. Entomol. Zool.*, 16, 490, 1981.
192. **Ishikawa, Y., Matsumoto, Y., Tsutsumi, M., and Mitsui, Y.,** Mixture of 2-phenylethanol and n-valeric acid, a new attractant for the onion and seed-corn flies, *Hylemia antiqua* and *H. platura* (Diptera: Anthomyiidae), *Appl. Entomol. Zool.*, 19, 448, 1984.
193. **Ormerod, E. A.,** Report of observations of Injurious Insects and Common Farm Pests, during the year 1892, 16th Report, Simpkin, Marshall, Hamilton, Kent & Co., Ltd., 1893.
194. **Batko, A. and Weiser, J.,** On the taxonomic position of the fungus discovered by Strong, Wells and Apple: *Strongwellsea castrans* gen. et sp. nov. (Phycomycetes: Entomophthoraceae), *J. Invert. Pathol.*, 7, 455, 1965.
195. **Berisford, Y. C. and Tsao, C. H.,** Field and laboratory observations of an entomogenous infection of the adult seedcorn maggot, *Hylemya platura* (Diptera: Anthomyiidae), *J. Ga. Entomol. Soc.*, 9, 104, 1974.
196. **Matanmi, B. A., Libby, J. L., and Maxwell, D. P.,** Two phycomycetes infecting root maggot adults in Wisconsin, *Environ. Entomol.*, 3, 1030, 1974.
197. **Gostick, K. G., Powell, D. G., and Slough, C.,** Dieldrin resistance onion fly *(Delia antiqua* (Meig.)) in England, *Plant Pathol.*, 20, 63, 1971.
198. **Harris, C. R., Hitchon, J. L., and Manson, G. F.,** Distribution of cyclodiene-insecticide resistance in seed maggot complex in relation to cropping practices in southwestern Ontario, *J. Econ. Entomol.*, 59, 1483, 1966.
199. **Harris, C. R. and Turnbull, S. A.,** Toxicity of some insecticides to insecticide-susceptible strains of the onion, cabbage, and seedcorn maggots (Diptera: Anthomyiidae) and the darksided cutworm (Lepidoptera: Noctuidae), *Can. Entomol.*, 112, 1029, 1980.
200. **Helson, G. A. H.,** Insect pests, Narcissus bulb fly, *N. Z. J. Agric.*, 127, 79, 1973.
201. **Assem, M. A. and Nasr, E. S. A.,** A syrphid fly *Eumerus amoenus* Loew injurious to onion in the U.A.R. (Diptera: Syrphidae), *Agric. Res. Rev. Cairo*, 45, 27, 1967.
202. **Farag, S. S. and Doss, S. A.,** Biological studies on the onion bulb fly, *Eumerus amoenus* Loew (Diptera: Syrphidae), *Agric. Res. Rev. Cairo*, 59, 79, 1981.
203. **Woodville, H. C.,** Narcissus flies, in *Narcissus Pests*, HMSO, London, 1970, 40.
204. **Gherasim, V.,** Musca bulbilor - *Eumerus stigatus* Fall. (Diptera: Syrphidae) un daunator nou al culturilor de ceapa din R.S. Romania, *Am. Inst. Cercet. Prot. Plant*, 11, 141, 1976.
205. **Aihara, E., Matsumoto, N., and Muramatsu, T.,** Dipterous insects detected on imported onion bulbs with a tentative pictorial key to larvae, *Res. Bull. Plant Prot. Serv. Jpn.*, 21, 75, 1985.
206. **Becker, P.,** *Pests of Ornamental Plants*, HMSO, London, 1974, 175.
207. **Preobrazhenskii, V. V.,** Pests of onion in Buryatia, *Zashch Rast.*, 15, 39, 1970.
208. **Zabirov, Sh. M.,** On the ecology of the onion-bulb fly *Eumerus strigatus* Fall (Diptera: Syrphidae), *Entomol. Obozr.*, 42, 736, 1963.
209. **Doane, J. F.,** Attraction of the lesser bulb fly *Eumerus strigatus* (Diptera: Syrphidae) to decomposing oatmeal, *N. Z. Entomol.*, 7, 419, 1983.
210. **Gherasim, V. and Lacatusu, M.,** *Aphaereta minusa* (Nees) Hymenoptera: Braconidae) principalul factor limitativ al populatiilor de diptere daunatoare culturilor de ceapa din R.S. Romania, *Am. Inst. Cercet. Port. Plant*, 12, 229, 1977.
211. **Kanervo, V. and Tiittanen, K.,** Control of larvae of the onion fly *(Hylemyia antiqua* Merg.) and the small narcissus flies *(Eumerus strigatus* Fall. and *E. tuberculatus* Rond, *Valt. Maatalouskoetoiminnan Julk.*, 196, 14, 1962.
212. **Assem, M. A., Salam, A. M. A., and Yousef, K. E. H.,** Efficiency of insecticidal dusts in controlling the bulb fly, *Eumerus amoenus* Loew on stored onion bulbs (Diptera: Syrphidae), *Bull. Soc. Entolol. Egypte, Econ. Ser.*, 6, 217, 1972.
213. **Spencer, K. A.,** *Agromyzidae (Diptera) of Economic Importance*, Dr. Junk, B. V., The Hague, 1973, 418.
214. **Hering, M.,** Eine neue gellenerzeugende Agromyzide (Dipt.), *Z. Wiss. Insekt Biol.*, 22, 319, 1927b.
215. **Shanab, L. M. A. and Bognar, S.,** Adatok a voroshagyma hazankban elofordulo kartevoirol. XVII. Novenyvedelmi tudomanyos ertekezlet (1967 februar 20-24), *Vologtott anyaga, II Kotet*, Budapest, Hungary, 637, 1967.
216. **Frankenhuyzen, A. van,** The onion leaf-miner, a little-known pest of onions and leeks in the Netherlands, *Gewasbescherming*, 8, 59, 1977.
217. **Yau-I, Chu. and Hong-Chin., F. Chou.,** Studies on the *Allium* leaf miner (Phytobia cepae Hering, Agromyzidae). I. Identification and morphology, *Plant Prot. Bull., Taiwan*, 12, 166, 1970.
218. **Martinez, M.,** Contribution a l'etude des Agromyzidae de France (Dipt.) (2e note), Liriomyza nietzkei Spencer et Liriomyza chinensis (Kato) deux especes d'importance economiques presentes en France, *Bull. Soc. Entomol. Fr.*, 87, 302, 1982.
219. **Suss, L.,** Liriomyza nietzkei Spencer (Diptera, Agromyzidae) dannosa alle cultivazioni di cipolla *(Allium cepa)* in Italia, *Bull. Zool. Agr. Bachicoltura*, 12, 205, 1975.

220. **Freuler, J., Antonin, P., Raymond, J. C., Raboud, G., and Jean, P.,** La mouche mineuse de l'oignon, Liriomyza Nietzkei Spencer (Dipt., Agromyzidae): biologie, repartition, geographique et importance economique, *Rev. Suisse Vitic. Arboric Hortic.*, 10, 111, 1978.
221. **Fagoonee, I. and Toory, V.,** Contribution to the study of the biology and ecology of the leaf-miner *Liriomyza trifolii* and its control by neem, *Insect Sci. Appl.*, 5, 23, 1984.
222. **Minkenberg, O. P. J. M. and Lenteren, J. C. van,** The leafminers *Liriomyza bryoniae* and *L. trifolii* (Diptera: Agromyzidae), their parasites and hostplants: a review, *Agricultural University, Wageningen Papers,* 86, 50, 1986.
223. **Hong-Chin., F. Chou., and Yau-I, Chu.,** Studies on the *Allium* leaf miner (Phytobia cepae Hering, Agromyzidae). II. Life history and habits, *Plant. Prot. Bull. Taiwan,* 12, 177, 1970.
224. **Wardlow, L. R.,** Leaf-miners and their parasites, in *Biological Pest Control, the Glasshouse Experience,* Hussey, N. W. and Scopes, N. E. A., Eds., Blandford Press, Poole, Dorset, 1985, 62.
225. **Wardlow, L. R.,** Control of leaf-miners on chrysanthemums and tomatoes by parasites, in *Biological Pest Control, the Glasshouse Experience,* Hussey, N. W. and Scopes, N. E. A., Eds., Poole, Blandford Press, Poole, Dorset, 1985, 129.
226. **Labrador, J. R.,** Resultados preliminares obtenidos de un ensayo en siembras de cebolla en rama para el combatede minadores (Agromyza spp), Departamento Fitosanitario, Facultad de Agronomia, *Universidad del Zulia, Venezuela,* No. 69, 10.
227. **Choi, S. Y., Lee, S. W., and Chung, B. K.,** Effects of side-furrow-treated carbofuran (3G) on the control of onion maggots *(Hylemia antiqua* Meigen: Anthomyiidae) and on the growth response of garlics, *Korean J. Plant Prot.*, 22, 271, 1983.
228. **Everts, K. L., Schwartz, H. F., Epsky, N. D., and Capinera, J. L.,** Effects of maggots and wounding on occurrence of *Fusarium* basal rot of onions in Colorado, *Plant Dis.*, 69, 878, 1985.
229. **d'Assis Fonseca, E. C. M.,** A new species of *Delia* (Dipt., Anthomyiidae) attacking onions in Egypt and Iraq, *Entomol. Mon. Mag.*, 101, 198, 1965.
230. **Abul-Nasr, S. E., Assem, M. A., and El-Sherif, A. R. A.,** Effect of seeding date on the onion maggot, *Delia alliaria* Fonseca infestation (Diptera: Anthomyiidae). *Bull. Soc. Entomol. Egypt,* 58, 93, 1974.
231. **Abul-Nasr, S. E., Assem, M. A., and El-Sherif, A. R. A.,** Susceptibility of different local and American onion varieties to the onion maggot *Delia alliaria* Fonseca infestation in seedbeds (Diptera: Anthomyiidae), *Bull Soc. Entomol. Egypt,* 58, 163, 1974.
232. **Abul-Nasr, S. E. and El-Sherif, A. R. A.,** Biological studies on the onion maggot *Delia alliaria* Fonseca (Diptera: Anthomyiidae), *Bull. Soc. Entomol. Egypt,* 58, 375, 1974.
233. **Ancev, E. and Postlovski, M.,** Contribution to knowledge of *Phytomyza horticola* Goureau (Dipt. Agromyzidae), *Zast. Bilja,* 29, 273, 1978.
234. **Pagliarini, N. and Spasic, R.,** Some observations on the occurence and harmfulness of the leaf-miner *Phytomyza horticola* Goureau, *Zast. Bilja,* 35, 17, 1984.
235. **Pejcic, P.,** Muva beloga luka - *Suillia lurida* Meig., *Zast. Bilja,* 15, 577, 1964.
236. **Dusek, J.,** *Suillia lurida* (Meigen) (Dipt. Helomyzidae) ein neuer knoblauchschadling in der CCSR, *Folia Zool.*, 12, 189, 1963.
237. **Ciampolini, M., and Suss, L.,** New findings on the garlic fly *Suillia univittata* (von Roser) (Diptera, Heleomyzidae), *Boll. Zool. Agrar. Bachic.*, 17, 19, 1983.
238. **Downing, W.,** New occurrence and host for *Tachinaephagus zealandicus* Ashmead (Hymenoptera: Encyrtidae), *Ohio J. Sci.*, 75, 62, 1975.
239. **Anon.,** Research Station, Vancouver, British Columbia, Research Branch Report, *Agric. Can.*, 12, 1977.
240. **Downing, W.,** Sympatry of *Tritoxa incurva* (Wied.) (Diptera: Otitidae), *Proc. Entomol. Soc. Wash.*, 79, 250, 1977.
241. **Karde, K. W., Hammond, P. M., and Severa, F.,** *A Field Guide in Colour to Beetles,* Octopus Books, 45.
242. **Balachowsky, A. and Mesnil, L.,** *Les Insectes Nuisibles aux Plantes Cultivees, Leurs Moeurs, Leur Destruction.* Volume 2, Lechevalier, Paris, 1936, 1475.
243. **Becker, P.,** *Pests of Ornamental Plants,* HMSO, London, 1974, 91.
244. **Griffiths, D. C.,** Susceptibility of plants to attack by wireworms *(Agriotes* spp.), *Ann. Appl. Biol.*, 78, 7, 1974.
245. **Wightman, J. A. and Morrison, G.,** Wireworms (Coleoptera: Elateridae) associated with damaged onions and garlic crops near Blenheim, *N. Z. Entomol.*, 6, 438-441, 1978.
246. **Nagano, M.,** Coffee-bean weevil injurious to garlic, *J. Econ. Entomol.*, 25, 92, 1979.
247. **Connell, W. A.,** Hosts of *Carpophilus dimidiatus, J. Econ. Entomol.*, 68, 279, 1975.
248. **Popova, L. G. and Skopinskii, V. F.,** An important part of the work, *Zashch. Rast.*, 7, p38, 1977.
249. **Bel'skaya, N. M. and Popova, L. G.,** Injurious insects in cargoes from India, *Zashch. Rast.*, 2, 42, 1978.
250. **Ter-Simonyan, L. G. and Blinova, Z. P.,** Pests and diseases of onion *Zashch. Rast.*, 5, 49, 1981.
251. **Arkhipov, G. E.,** Pests of onion, *Zashch. Rast.*, 10, 53, 1984.

252. **Hajdu, J. and Nagyimre, E.,** Control of onion foliage pests, *Zoldsegtermesztesi Kutato Intezet Bulletinje,* 17, 53, 1984.
253. **Franssen, C. J. H. and Koert, J. L.,** Ceutorrhynchus suturalis, een voor uien schadelijke snuitkever, *Tijdschr. Plantenziekten.,* 68, 235, 1962.
254. **Shurovenkov, B. G.,** The leaf beetle *Galeruca daurica* Joann in the steppes of Mongolia, *Zool. Zh.,* 44, 775-776, 1965.
255. **Gennatas, J.,** On the subject of an unusual infestation by galerucids in Alsace, *Phytoma,* 29, p9, 1977.
256. **Dimetry, N. Z., Shoukry, A., and Aboul-Zahab, A.,** On the biology of the cigarette beetle *Lasioderma serricorne* (Fabricius) attacking dehydrated onion in Egypt, *Proceedings of the Fourth Conference of Pest Control,* 1978, 132.
257. **Popova, L. G.,** *Typhaea stercorea* - a pest of maize grain, *Zashch. Rast.,* 14, 45, 1971.
258. **Scopes, N. E. A.,** Red spider mite and the predator *Phytoseiulus persimilis,* in *Biological Pest Control, the Glasshouse Experience,* Hussey, N. W. and Scopes, N. E. A., Eds., Blandford Press, Poole, 1985, 43.
259. **Rasmy, A. H.,** Eine methode zur massenzucht der raubmilbe *Agistemus exsertus* Gonz. (Acarina, Stigmaeidae), *Anz. Schadlingskd, Pflanze Umweltschutz,* 48, 55, 1975.
260. **Anon.,** Red spider mite on glasshouse crops, *Ministry of Agriculture Fisheries and Food,* Advisory Leaflet No. 224, HMSO, London, 1976, 10.
261. **Anon.,** Red spider mite on outdoor crops, *Ministry Agric. Fish.,* Advisory Leaflet No. 226, HMSO, London, 1976, 5.
262. **Abdel Fattah, S. A. S., Sharaf, I. M. F., and El Sabae, A.,** Performance of *flucythrinate* (Cybolt) against a wide spectrum of agricultural pests in Egypt, *Arab J. Plant Prot.,* 1, 74, 1983.
263. **Abo El-Ghar, M. R. and Osman, A. A.,** Ecological and control studies on mites associated with onions in Egypt, *Z. Angew. Entomol.,* 73, 439, 1973.
264. **Estal, p. del, Arroyo, M., Vinuela, E., and Budia, F.,** Acaros que atacon el cultivo del ajo en Espana, *Am. Inst. Nac. Invest. Agrar. Agricola,* 28, 131, 1985.
265. **Passlow, T. and Turner, J. W.,** Control trials against the brown wheat mite, *Queensl. J. Agric. Anim. Sci.,* 22, 103, 1965.
266. **Anon.,** *Bryobia mites, Ministry Agric. Fish. Food,* Advisory Leaflet No. 305, HMSO, London, 1974, 6.
267. **Ahmed, K. M. and Beingno, D. A.,** Control of the eriophyid mite *Aceria tulipae* Keifer, the cause of tangle-top malady and the vector of garlic mosiac virus, *Indian J. Plant Prot.,* 12, 153, 1985.
268. **Anon.,** Annual report of the secretary for Agricultural Technical Services for the period 1 July 1978 to 30 June 1979, *Rep. Dep. Agric. Tech. Ser., South Africa,* 114, 1980.
269. **Zaher, M. A. and Abou-Awad, B. A.,** A new species and new record of some eriophyid mites in Egypt (Eriophyoidea: Eriophyidae), *Acarologia,* 21, 61, 1980.
270. **Sukhareva, S. I. and Sapozhnikova, F. D.,** Seasonal adaptations of some species of tropical mites (Acarina: Eriophyoidea), *Vestn. Leningr. Univ. Biol.,* No. 9, 47, 1981.
271. **Chamberlain, J. A. and Evans, P. E.,** *Aceria tulipae* (Keifer) (Acarina: Eriophyidae) on *Lolium* spp in Wales, *Plant Pathol.,* 29, 99, 1980.
272. **Manson, D. C. M.,** Wheat curl mite on garlic, *N. Z. J. Agric.,* 121, 61, 1970.
273. **Anon.,** *Pest Control in Tropical Onions,* Tropical Development and Research Institute, London, 1986, 109.
274. **Dorestes, S. E.,** Avances obtenidos en el control del acaro del ajo (*Aceria tulipae* (K.)) nueve plaga de este cultivo en Venezuela, *Ing. Agron.,* 11, 13, 1963.
275. **Dorestes, S. E.,** Control del acaro del ajo *Aceria tulipae* (K.) mediante tratamiento a la "semilla", *Agron. Trop.,* 16, 199, 1966.
276. **Nackev, P.,** A study of Eriophyid mites in Bulgaria, *Rastit. Zasht.,* 16, 11, 1968.
277. **Nachev, P.,** Study on eriophyid mites in Bulgaria. X. Eriophyid mites on cereal plants, *Rastitelnozashchitna Nauka,* 4, 92, 1976.
278. **Almaguel, L., Sierra, S., and Martinez, H.,** Record of *Aceria tulipae* in Cuba, *Ciencia y Tecnica en la Agricultura, Prot. Plant.,* 8, 45, 1985.
279. **Almaguer Rojas, L., Sierra Landa, S., and Martinez, H.,** Report of the mite *Aceria tulipae* Keifer in Cuba, *Centro Agricola,* 11, 114, 1984.
280. **Rossi de Simons, N. H.,** Cuatro especies de acaros para la Argentina, *Idia,* 204, 47, 1964.
281. **Rossi de Simons, N. H.,** New information on mites in Argentina, *Idia,* 278, 22, 1971.
282. **Lazurina, M. V.,** The onion bulb mite, *Zashch. Rast.,* 8, 61, 1976.
283. **Shevchenko, V. G., De-Millo, A. P., Razvyazkina, G. M., and Kapkova, E. A.,** The taxonomic delimitation of the closely related species of Eriophyid mites *Aceria tulipae* Keif. and *A. tritici* sp. n. (Acarina, Eriophyoidea), the vectors of onion and wheat viruses, *Zool. Zh.,* 49, 224, 1970.
284. **Sukhareva, S. I. and Sapozhnikova, F. D.,** Seasonal adaptations of some species of Eriophyid mites (Acarina, Eriophyoidea) *Vesm. Leningr. Univ. Biol.,* 9, 47, 1975.
285. **Sandhu, G. S.,** New record of bulb mite *Rhizoglyphus echinopus* (Fumouze and Robin) from stored onions in India, *Sci. Cult.,* 42, 221, 1976.

286. **Rangarajan, A. V., Mahadevan, N. R., and Amalraj, C.,** Occurrence of the bulb mite *Rhizoglyphus* sp. on onion in storage. *Indian J. Entomol.,* 33, 355, 1971.
287. **Lee, H. S. and Wen, H. C.,** Field investigation of the acarid bulb mites on onion and their control, *J. Agric. Res. China,* 29, 211, 1980.
288. **Gerson, U., Yathom, S., Capua, S., and Thorens, D.,** *Rhizoglyphus robini* Claparede (Acari, Astignata: Acaridae) as a soil mite, *Acarologia,* 26, 371-380, 1985.
289. **Stathopoulos, D. G., Mentzelos, I. A., and Savvidis, S. D.,** Survey of insects and other pests on crops in Macedonia and Thrace. II, *Rep. Plant Prot. Agric. Res. Stn. Thessaloniki,* 3, 102, 1965.
290. **Soni, S. K. and Ellis, P. R.,** unpublished.
291. **Godan, D.,** *Pest Slugs and Snails: Biology and Control,* Springer-Verlag, Berlin, 1983, 445.
292. **Ellis, P. R.,** The biology and control of pests of alpine plants, *Q. Bull. Alpine Garden Soc.,* 55, 151, 1987.

Chapter 11

STORAGE DISEASES OF ONIONS

R. B. Maude

TABLE OF CONTENTS

I. INTRODUCTION

Bulb onions are a widely distributed and important food vegetable consumed in most areas of the world. On a global scale about 25 million t were produced in 1985.[1] Over 25% of bulb onions were produced in the tropics in the 1970s.[2] The onion is used mainly as a vegetable, but has had other roles in religion and medicine[3,4] in a history that goes back thousands of years. Onions from temperate regions have been distributed worldwide by the migrations of European man. Inevitably, however, with the changes in climate and cultivation systems occurring in different areas, there was an increase in the incidence of some diseases. For example, the major decline in bulb onion production in Egypt in recent years can be related to the building of the Aswam dam. The dam prevented the yearly deposits of alluvial soils on which onions were grown. Therefore the use of irrigation on unreplenished field soils became necessary. Consequently, the white rot pathogen *(Sclerotium cepivorum)* developed rapidly and caused increasing losses which led to the decline in export of bulb onions from 160,000 t in 1962 to 16,000 t in 1982.[5] Similarly, improvements in cultivars, crop husbandry, and storage methods led to an expansion in the bulb onion crop in the U.K.

in the early 1970s. This was accompanied by an increase in post-harvest diseases, particularly onion neck rot *(Botrytis allii)*.[6]

Thus, crop husbandry practices, crop protection treatments, cultivars, etc., all have a modifying effect on the incidence of plant diseases. The range of organisms, however, responsible for post-harvest diseases of onions are largely determined by the temperatures at which the crop is grown and/or stored.

Neck rot is the main disease of stored bulb onions grown in temperate areas of countries including U.S.,[7-12] Australia,[13,14] Britain,[15-18] Canada,[19-20] Czechoslovakia,[21] Denmark,[22] Germany,[23,24] Holland,[25] Israel,[26] Japan,[27] Lithuania,[28] New Zealand,[29,30] Norway,[31] Poland,[32] and Yugoslavia.[33] However, where higher temperatures predominate, for example in Texas[34] or Egypt[35] *Penicillium* and *Aspergillus* spp. are dominant with *A. niger* the common post-harvest pathogen. The application of high temperature (approx. 30°C and high humidities [>80% RH]) for long periods during post-harvest drying of temperate onions can promote the development of this range of high temperature fungi which blemish and rot dry bulb onions.[6] Bacterial rots occur less frequently in onions grown in temperate climates, particularly if they have been dried and stored at ambient temperatures, or dried at high temperatures and subsequently stored under ambient conditions. In the past the presence of bacterial organisms has been masked by the presence of primary fungal pathogens such as *B. allii* (neck rot). *Pseudomonas, Erwinia,* and occasionally *Lactobacillus* spp. can occur and may either singly, or in conjunction with fungi, cause rotting of onion bulbs.

Other fungi which cause post-harvest rots under certain circumstances include pink root *(Pyrenochaeta terrestris)*, bulb basal rot (*Fusarium oxysporum* f.sp. *cepae*), smudge *(Colletotrichum circinans)*, and white rot *(Sclerotium cepivorum.)* Economically, however, onion neck rot *(B. allii)* is the major pathogen and hence most emphasis will be given to this problem in the chapter.

Onion storage diseases have been reviewed[6] and there are several publications on these and other onion pathogens.[4,36-38]

II. FUNGAL STORAGE ROTS OF ONIONS

A. NECK ROT (*Botrytis allii* and *B. byssoidea*)

1. History of the Pathogens

Botrytis allii n.sp. was the name given in 1917 by Munn to the fungus which caused neck rot disease of onions in Michigan and New York State.[5] Based on observations of the size and shape of conidia he concluded that the fungus was distinct from other *Botrytis* species, such as *B. aclada* Fres., *B. cinerea* Pers., *B. cana* Kze. et Schon., *B. fulva* Link., *B. parasitica* Cav., and *B. vulgaris* Fr. var. *interrupta* Fr.. Walker[8,37] confirmed Munn's description of *B. allii* and described it as a gray mold neck rot, to distinguish it from another *Botrytis* species causing neck rot which he named *B. byssoidea* n.sp., mycelial neck rot. This species produced mainly mycelium in culture and few sclerotia and conidiophores. At one time these fungi were thought to be conspecific but this was not proved.[38] Recently the application of biometrics to this taxonomic relationship has upheld the distinction between *B. allii* and *B. byssoidea* at the species level.[39]

Hennebert[40,41] following examination of herbarium material and the use of adjusted spore measurements concluded that the fungus named *B. aclada* by Fresenius[42] in 1850 was the same as *B. allii*. While acknowledging that *B. aclada* is the older name, *B. allii* is retained in this account as the preferred name, as all basic and applied research during the last 70 years refers to that name.

Although *B. byssoidea* (mycelial neck rot) has been the main pathogen in certain regions of the U.S. in the past,[8,9] universally *B. allii* appears to be the major cause of gray mold neck rot of dry bulb onions. In Holland[43] *B. allii* was isolated from 69% and *B. byssoidea* from 2% of bulbs affected by neck rot; in the U.K.[44] in bulb samples taken between 1943—1946 72% were affected by *B. allii* and 26% by *B. byssoidea*. More recently in the U.K.[45] *B. byssoidea* was found to be more prevalent than *B. allii* in some localities where onions or onions and leeks had been grown sequentially.

2. The Pathogens

a. Botrytis allii Munn (Gray Mold Neck Rot)

Descriptions of *B. allii* have been given by various authors.[7,8,37,46-49] Presly[45] specified the growth characteristics and spore dimensions of a number of *Botrytis* spp. following their growth on onion and leek tissues and on agar (Table 1). Conidiophores of *B. allii* are compact, dense, and single stemmed and usually terminate as a head of branched phialides bearing single celled ellipsoidal conidia. There is no sexual state but resting sclerotia of the fungus are produced on the neck tissues of onions (hence neck rot). These germinate to produce the conidiophore state. Sclerotia are not usually produced in culture.

b. Botrytis byssoidea Walker (Mycelial Neck Rot)

Walker's[8] original description requires little addition. The spores are slightly larger and less ellipsoidal than *B. allii* conidia (Table 1) but normally conidiophores do not develop in agar culture. Sporulation can be induced by growing in culture under ultraviolet light (NUV - 300 to 400 nm) at 15°C or on onion bulb tissues under NUV at 20°C.[50] Sclerotia are produced on onion bulbs and are of similar dimensions and function to those of *B. allii*.

c. Botrytis squamosa Walker (Leaf Rot, Blast)

B. squamosa, the onion leaf rot (blast) fungus though mainly a foliar pathogen also causes small sclerotial neck rot. This is confined to white skinned onion types. The fungus characteristically produced larger conidia (Table 1) and smaller sclerotia than either *B. allii* or *B. byssoidea*.

3. Disease Symptoms

Symptoms of common onion neck rot (*B. allii*) associated with stored bulbs have been described by several authors.[6-9,49] There are no visible symptoms when infected bulbs are placed in store but 8 to 10 weeks later, necks of affected bulbs soften and beneath the onion skin a mass of black sclerotia (1 to 5 mm diameter) develop. Conidiophores of the fungus produce a gray mold on the sides of the bulbs below the sclerotia. When diseased bulbs are cut, the infected tissues are brown. In some cases the sides of the bulbs and basal plates and occasionally the roots may be affected independently of the necks.[7] Where *B. byssoidea* is the causal fungus, affected bulbs have more surface mycelium but show little sporulation. Both pathogens will attack all onion types including those with brown, white and red skin colors.

4. Disease Biology

The identification of infection sources of onion neck rot has been a major constraint to progress in the understanding of the biology of the pathogen. Wood[51] noted that the disease appeared on stored onions but there was no consistent link with field symptoms making it extremely difficult to detect the source or sources of the disease. The subsequent discovery of seed transmission of the pathogen[17,18] to onion crops, proved highly relevant in relation to previous information on sources of neck rot (Figure 1).

a. Transmission of Neck Rot by Infected Seeds

It has been reported that *B. allii* caused blasting of onion inflorescences in USA, thus causing low or zero seed yield.[7,52,53] In Israel, the fungus was isolated from onion umbels

TABLE 1
Growth Characteristics and Spore Sizes of *Botrytis* Species of *Allium* Species

Species	Source	Mycelial growth on agar[a]	Sclerotia formed on agar[a]	Spores produced on plant tissue: Spore dimensions[b] (μm)	Spores produced on plant tissue: Length/breadth ratio
B. allii	Bulb onion	Flat with many conidiophores	No	9.5—10.5 × 5.0—5.5 (10.2 × 5.25)	1.94
B. byssoidea	Salad onion	White fluffy, sterile	No	12.0—13.0 × 6.5—7.0 (12.4 × 6.8)	1.8
	Bulb onion	White fluffy, sterile	No	12.5—14.0 × 7.0—7.5 (13.0 × 7.5)	1.72
	Leek	White fluffy, sterile	No	13.0—15.0 × 7.5—8.5 (13.8 × 8.3)	1.65
B. porri	Leek	White ropy, later grey, sterile	Large (1 cm) cerebriform	14.5—16.5 × 8.5—11.0 (15.7 × 9.6)	1.64
B. squamosa	Salad onion	White fluffy, later flat on the agar surface, sterile	Small (1—2 mm diameter)	21.0—22.5 × 16.5—17.0 (21.8 × 16.7)	1.31

[a] Incubated at 20°C in darkness.
[b] Range of means produced by the isolates of each species. Average for the species given in brackets.

From, Presly, A. H., *Plant Pathol.*, 34, 422, 1985. With permission.

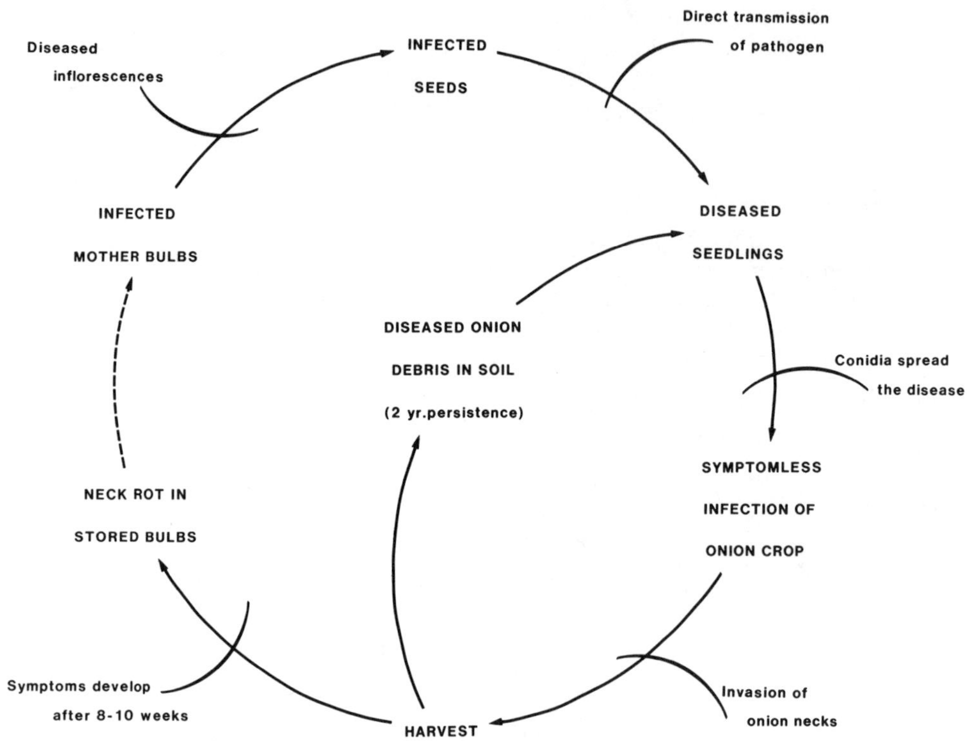

FIGURE 1. Disease cycle of *Botrytis allii* (neck rot).

in seed production fields.[26] It was possible to wash *B. allii* spores from onion seeds[7] and the fungus was isolated from a single sample of seeds in the U.K. in the 1940s.[44] Later the fungus was identified in Dutch seed samples[54] and when infected seeds were treated with benomyl, there was less infection in stored bulbs in Holland. In the early 1970s, it was demonstrated in the U.K. that many commercial seed samples bore the fungus *(B. allii)* and that these seeds transmitted the disease to emerged seedlings. Seedling infection occurs as a result of the mycelium of the fungus on or in the seed coat penetrating the tip of the cotyledon leaf.[17] The fungus then grows basipetally in the cotyledon and occasionally invades the base of the actively growing first true leaf.[17] Generally, however, *B. allii* grows downwards in leaves as they senesce. Conidiophores are produced only on moribund tissues, and the subsequently released conidia infect the tips of the next leaf and so the process is repeated. Later in the growing season, tips of leaves, the sheaths of which form the storage scales, are invaded and the fungus grows downward penetrating the well developed necks of bulbs to produce typical neck rot lesions. No field symptoms of the disease occur. Although the fungus invades and grows in green leaf tissues, it does not readily invade the dry bulb skins of or the external tissue of the necks.[7,17,37] However, there is evidence that the fungus may grow in the living tissues of leaves[17] and also that it may exist in a quiescent state in the living tissues of the onion leaf, becoming active as the leaf senesces.[55,56] These examples of symptomless infection suggest that the fungus may be capable of biotrophic growth. However, once tissues (leaf and bulb segments) are detached from onions and inoculated with *B. allii*, only necrotrophic invasion occurs.[57] In the field because no leaf symptoms are produced, healthy and infected onion crops are indistinguishable and disease expression occurs only in store (Figure 1).

b. Factors Modifying Spread of Infection and Disease Relationships

Only onions grown from infected seeds developed neck rot in U.K. stores[17,18] with the incidence of the disease directly related to the percentage of infected seeds sown[6,17,18] and to the weather conditions during the growth of the crop.[6,18,58]

The most consistent correlative parameter in this relationship was the log % neck rot and relative humidity toward the end of June recorded over a period of 7 years between 1973—1980.[59] Relative humidities in excess of 85% at this time were correlated ($R = 0.97$, $p < 0.05$) with a high incidence of neck rot in the stored crop. Thus under wet early summer conditions seeds carrying similar levels of infection gave higher numbers of neck rot infected bulbs in store than those grown in drier summers.[6,58] Contrary to the belief of many commercial onion growers, there was virtually no spread of the disease in store in crops harvested with intact foliage. There as a direct linear relationship between neck rot levels in stored bulbs and the incidence of the disease in field onion plants,[17,18] and also the incidence of seed-borne infection.[6,58] Under dry conditions this was a one to one relationship but under humid, wet conditions storage disease levels were proportionally much greater.[6,58]

Seed transmission of onion neck rot has now been reported in other European countries; Czechoslovakia,[60] East Germany,[61] and Yugoslavia.[33] In U.S., *B. allii*, *B. squamosa,* and *B. cinerea* have been isolated from diseased onion flowers.[62] However, although it was agreed that *B. allii* could be seed transmitted this was not thought to be a major source of the disease in New York State.[63] Recently (1985) transmission of the disease by a severely infected seed sample has been confirmed.[64]

The realization that the seed-borne source contributes inoculum from which the disease becomes progressive in growing crops (Figure 1) questions the validity of older views[7,65] that *B. allii* was a saprophyte of dying and dead leaves becoming parasitic only on entry of the internal neck and storage tissues of onions. The newer information indicates that there may be a nonaggressive but parasitic phase within the leaves of onions during the growth of the crop.

c. Other Sources of the Disease

i. Onion Sets

Sets command a small but increasing area (approx. 10% of that of the main crop) of dry bulb onion production in the U.K. Onion sets are susceptible to infection by *B. allii* and in 1979 they were responsible for serious outbreaks of the disease resulting in considerable losses through bulb rotting in store.[66] Some of the commonly grown cultivars for example "Sturon" and "Stuttegarter", have an inherent dormancy which allows their storage at ambient temperatures over the winter. Such onion sets if infected with *B. allii* can function as field infection sources when planted in the spring. Other cultivars, for example "Rijnsburger" types, necessarily are stored for several months at about 30°C[67,68] to prevent subsequent bolting. Neck rot is less likely to develop in these cultivars because even a relatively short exposure to 30°C will drastically reduce the growth of, or even kill *B. allii*[69] (Figure 3).

ii. Soil-Borne Inoculum

Infected onion debris[37,65] with or without resting sclerotia[37] was identified as a potential overwintering source of *B. allii* and *B. byssoidea* from which infection could be transferred to onion crops in the following spring. The fungus *(B. allii)* has been isolated from soil which bore onion crops in U.S.[70] and Israel.[71] Such sources may be important where onions follow onions on the same land or where minimal rotations are practiced as is reported to happen on certain organic soils in New York State.

B. allii can survive for two years on onion debris in sandy loam in the U.K. and the sclerotia, although they can be recovered during this time do not germinate after 6 months in the soil.[72] Thus the 3 to 4 year rotation practiced in the U.K.[73] will control this phase of the disease. In drier climates, where surface debris may not decay as rapidly, this source of infection may persist for longer periods.

iii. Onion Dumps and Overwintering Crops

Overwintering dumps or cull piles of bulb onions have been implicated in America[7,37] and Holland[25] as sources of *B. allii* for infection of spring-sown crops. Sporulating colonies of *B. allii* were found on the upper layers of onions in cull piles in Holland in June,[55] a time when the March-sown crop is young and vulnerable.[59] Onion crops may also act as infection bridges in that it has been shown experimentally in the U.K. that infected overwintered bulb onions provided a link between one spring sown crop and the next.[74] Likewise, it was found that cull piles and seed production fields in New York State, were sources of the related *Botrytis* leaf pathogen *B. squamosa* from which infection of spring sown bulb onions occured.[63] By implication, these could also be infection bridges for *B. allii*.[63]

It is possible that field sources may be important where there is little rotation of crops or where green bridging of onions occurs by the introduction of an overwintered bulb or onion seed production crop.

d. Neck Rot in Stored Bulbs

Before the realization that seed-borne *B. allii* was the main source of neck rot in stored onions in the U.K., losses of 15 to 50% occurred in some years.[75] Growers believed that the fungus infected plants close to harvest in the field and subsequently spread by bulb to bulb contact in store.[18] Current evidence however, has shown that this is not the case particularly where onions are lifted, field dried, and removed to store with their foliage intact.[6,7,18] In these cirumstances the apparently greater expression of neck rot with time is due to the progressive rotting of bulbs already symptomlessly infected when harvested.[4]

Bruising during lifting or mechanical damage during harvesting may increase neck rot in stored bulbs, presumably because of infection of damaged bulbs in the field or possibly by contact infection of exposed storage tissues in store.[11,60]

In the U.K., harvesting practices for bulb onions have changed since the mid-1970s in that the foliage of onions is now removed mechanically (topped) at harvest, the bulbs are then undercut and the crop taken directly to store. In these circumstances "topped" onions are vulnerable and must be thoroughly dried to seal the cut necks against infection by *B. allii* (see Section II.5.c.ii).

5. Disease Control

Because symptom expression occurred in the final stages of the disease in store, realistic field sources of the pathogen were difficult to identify, and hence it was almost impossible to identify effective disease control strategies. Fungicide sprays applied to onions could not be timed to relate to particular stages in pathogen attack or development and as a result were largely unsuccessful. However, with the identification of infected seeds as a major source of the disease in the U.K. and parts of Europe, seed treatments have been developed and these have been highly effective in eliminating the pathogen. In addition, broadly based strategies involving field and post-harvest drying procedures, in that they have rendered the harvested bulb less vulnerable to infection, have contributed over the years to further reductions in disease incidence.

a. Seed and Set Treatments

The disease in the U.K. was virtually eliminated by seed treatment with benomyl (1 g a.i./kg seed) or with benomyl + thiram (1 g a.i. of each/kg seed)[18] which confirmed an earlier Dutch report of reduction of neck rot in store by seed treatment.[76] The efficiency of

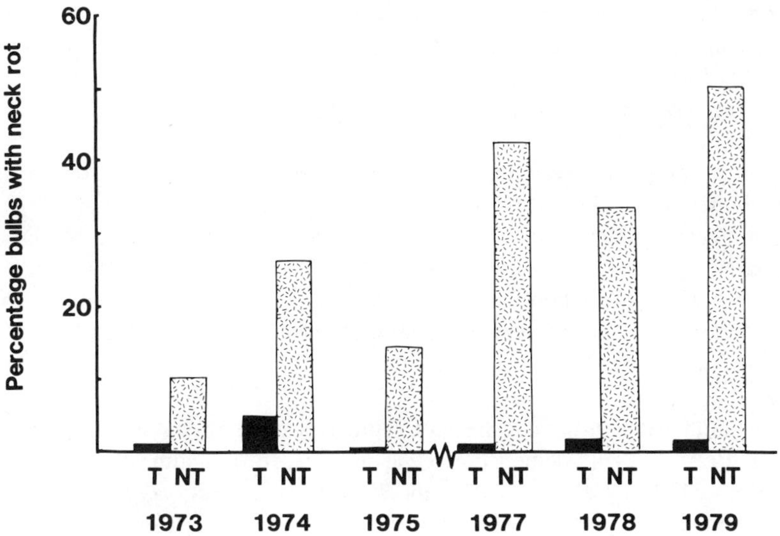

FIGURE 2. Control of neck rot in stored bulbs by seed treatment. T = seeds treated with benomyl or benomyl + thiram. NT = not treated. (Maude, R. B., *Seed Sci. Technol.*, 11, 829, 1983. With permission).

this treatment for the years 1973 to 1979 (excluding 1976)[58] is shown in Figure 2. In Czechoslovakia in 1979 seed treatment with benomyl and thiram reduced neck rot in stored bulbs from 57.5 to 3.5% and increased yields by 114%.[77] Similar results were reported from East Germany but there was a further improvement in the control of neck rot when a single fungicide spray of benomyl was applied just before the onion foliage died.[61]

The benzimidazole-based systemic fungicide benomyl is highly toxic to *Botrytis* spp. including *B. allii*. In the mixed formulation the dithiocarbamate nonsystemic fungicide thiram functions as a broad spectrum chemical protecting germinating seeds from attack by soil-borne damping-off fungi such as *Pythium* and *Phytophthora* spp. Although the benomyl/thiram formulation is no longer produced commercially in the U.K. seed merchants apply both fungicides singly to onion seeds to produce the double treatment. Tests in the U.K. and in Czechoslovakia indicate that benomyl/thiram is superior to all other fungicides as a seed treatment for the eradication of neck rot.[6]

In Europe, sets are dipped in an aqueous suspension of a dicarboximide fungicide, such as procymidone to control neck rot. There is no fungicide approved for this use in the U.K. although in experiments,[66] considerable reduction in disease in stored bulbs was obtained by dusting sets with iprodione before planting.

b. Preharvest Treatments

Although *B. allii* progressively invades the foliage of onion crops, field applied fungicides, provide little control of the storage phase of the disease. The reason for this is not certain but it is most probable that in the absence of plant symptoms, the timing of the spray application lacked the precision to achieve effective control. Dithiocarbamate sprays or dusts, for example, successfully controlled the symptom producing *Botrytis squamosa* (leaf rot or blast) in U.S. so increasing yields[78] but in some instances weekly sprays (June to August) of the dithiocarbamate fungicides maneb, mancozeb and nabam, resulted in an increase in the incidence of neck rot in New York State.[79] Similarly, field sprays of benomyl were ineffective in the U.K.[80] but in East Germany, seed treatment in conjunction with a single application of benomyl just before the death of the foliage, improved the control of neck rot in stored bulbs.[61]

Accelerated drying of maturing onion foliage with desiccants (neodecanoic acid mixtures) increased the incidence of neck rot on stored bulbs in two of the three cultivars so treated.[81]

c. Post-Harvest Treatments

i. Fungicides

Bulb dips or sprays applied immediately before placing the onions in store in New York State were ineffective in controlling neck rot.[82]

In experiments, fungicide dusts (dicloran and captan) and desiccants (calcium chloride) applied to the "topped" (foliage cut away) necks were only effective if applied to the cut necks up to 1 h after inoculation with *B. allii* spores.[12] There appears therefore to be little advantage in the use of post-harvest chemical measures to control this fungus.

ii. Drying

Effective field and post-harvest drying treatments can be critical in the control of onion neck rot. In 1925, Walker[9] demonstrated the importance of sealing the necks of topped onions at 32 to 34°C to prevent invasion by *B. allii*. The effect was twofold; dried onion neck tissues did not support hyphal growth and the high temperatures were lethal to the pathogen. The growth of the fungus *in vitro* is prevented by temperatures of 30°C[69] and above[7,83] (Figure 3). Mycelial growth of *B. allii* in inoculated onions in New Zealand was prevented in the bulbs maintained at 36°C. Forced ventilation at 36 to 38°C for 7 d is used to cure onions in that country.[30] Onion leaves maintained at 30°C (100% RH) were not readily invaded by the fungus and, although 5% of the leaves became infected, none produced sporing colonies.[84]

The sun's radiant energy is sufficient to field dry onions in warm, dry climates. Supplementary artificial heat is probably necessary only 1 year in 5 in Oregon[11] whereas in the cool climate of the U.K., post-harvest drying is necessary in most years.

Drying temperatures in excess of 40°C have been tested in several countries.[10,11,25,83,85] However, drying temperatures in excess of 38°C severely damaged bulbs in New Zealand, since physiological breakdown can occur at such temperatures.[86] There was no damage to bulbs and neck rot was considerably reduced in onions in stores in the U.S. when heated air at 34 to 35°C was forced (224 m^3 air h^{-1} t onions $^{-1}$) through the bulk of stored bulbs over a 10 to 14 d curing period.[87] The method was modified successfully for pallet box storage in New York State, where 46% of the crop was stored by this method.[88] A similar system was developed for direct-harvested mechanically topped onions in the U.K., in the mid 1970s, where air at 30°C is blown through onion bulb stacks at 425 m^3 h^{-1} t onions^{-1} for 3 to 5 days in the first stage of drying and then the heated air is recirculated for 10 to 14 d before temperatures are allowed to fall to ambient for the remainder of the storage period.[89] Because the incidence of neck rot in U.K. onions was low as a result of seed treatment[18,58] it was difficult to evaluate the practical value of high temperature (30°C) drying on the disease.[90] Co-operative research with the U.K. Agricultural Development and Advisory Service demonstrated that post-harvest drying in 30°C gave the greatest reduction of neck rot in topped and field inoculated onions where the crop was removed to store within 48 h of inoculation.[91]

Treatment within 24 h was necessary to reduce *B. allii* in the necks of bulbs where infrared irradiation was the experimental energy source.[92] Gamma irradiation can be applied on a commercial scale to onions, but it is unlikely, however, that stored produce will withstand it at doses necessary for disease control.[93]

At the present time, forced air drying at temperatures of 30°C and above is the most effective treatment for the control of neck rot where infection sources include contaminated soil, onion dumps, and overlapping onion crops. Where seed treatment reduces seed-borne inoculum to a low incidence,[18] forced air drying at 30°C provides a useful additional level

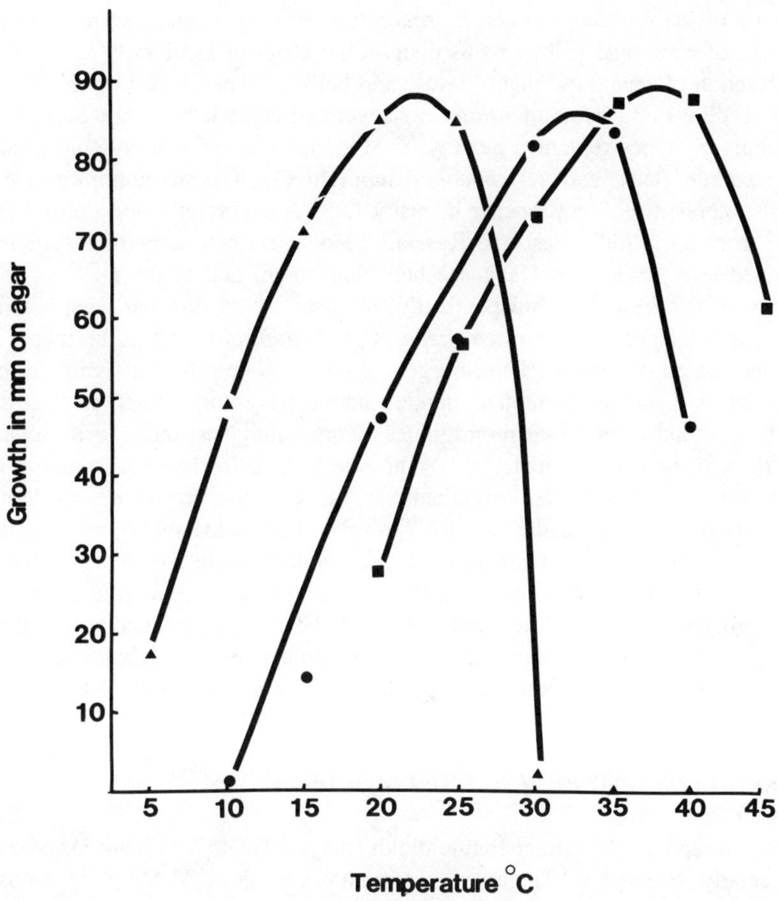

FIGURE 3. Growth response of onion bulb rotting fungi to temperature, *in vitro, Botrytis allii* (▲), *Aspergillus niger* (●), *A. fumigatus* (■). (Maude, R. B., Bambridge, J. M., and Hayden, N. J., unpublished data).

of control aimed at the direct harvested crop where mechanical removal of the foliage provides a wounded neck which requires to be dried and sealed against minimal levels of inoculum.

d. Resistance

Plant resistance is not a strategy presently employed in the control of neck rot. No major gene resistance to *Botrytis* spp. is known[94] but the possibility exists for combining minor genes by breeding to develop polygenic resistance to the pathogen.

Information on the chemical nature of resistance indicates that in the dry skins of coloured onion bulbs this may be related to the presence of the phenol protocatechuic acid[95] and catechol.[96] If, however, fungal invasion of the inner fleshy storage tissue is direct, via a wound in the dry skins, the differential resistance response is not maintained.[97] *B. allii*, in comparison with *Colletotrichum circinans* (smudge) and *Aspergillus niger* (black mold), was least affected by the volatile and nonvolatile chemicals present in the fleshy storage tissues of bulb onions.[97] However, *B. allii* occurred less frequently on pungent than mild onion cultivars and colored pungent cultivars were less diseased than white ones.[98] The presence of the volatile sulfides in the fleshy storage scales and, to a lesser extent, the phenols in the dry outer skins were considered to be responsible for the differences between these types.[99]

Sclerotia of the related pathogen *B. squamosa* (leaf rot or blast) were more numerous on the necks of white and yellow bulbs than on those of other colors.[100]

The reaction of seedlings, plant leaves, and bulbs to inoculations by *B. allii* indicated that most seedlings and bulbs of *Allium cepa* were susceptible[101,102] however, some bulbs of "Rijnsburger" types remained healthy.[101] Moderate resistance was demonstrated in the related *A. porrum* (leek) and *A. schoenoprasum* (chives). The susceptibility in *A. cepa* to *B. allii* also applied to *B. squamosa* in many U.S. commercial onion cultivars[100] and in selections from the Middle East and Russia.[103] However, resistance to *B. squamosa* was demonstrated in *A. fistulosum* (Japanese bunching onion) and in the interspecific hybrids (*A. cepa* × *A. fistulosum*).[100] Independently, resistance to *B. allii* has been obtained in *A. fistulosum*, in interspecific crosses (*A. cepa* × *A. fistulosum*), and in the inbred synthetic *A. cepa* lines based on certain "Rijnsburger" types.[102] Using the leaf-symptom-producing *B. squamosa* as a marker pathogen, nondestructive laboratory infection tests have been developed as an aid to breeding programmes incorporating resistance to *B. sqamosa,* and possibly *B. allii*, into *Allium* material.[103] These tests revealed reactions ranging from susceptibility to resistance in *A. fistulosum* cultivars and a high degree of resistance in *A. cepa* (synthetic lines) × *A. fistulosum* crosses,[103] results which were subsequently confirmed in field trials.[104] Laboratory infection tests of excised bulb scale tissues have been used to identify bulbs of commercial Japanese cultivars with resistance to *B. allii* from which self-pollinated progenies were produced and retested. Laboratory tests failed to reveal consistent significant levels of resistance and hence it was concluded that probably it would be difficult to develop resistant inbreds other than by the introduction of resistant genes from other species.[105]

B. *ASPERGILLUS* AND *PENICILLIUM* MOLDS

1. Disease Occurrence

Aspergillus spp. are high temperature molds (range 30 to 45°C, Figure 3) some of which, and particularly *A. niger* v. Tieghem — optimum growth at 32.5°C, cause post harvest blemishes and, more seriously, rotting of onions grown and stored in hot climates,[2] for sample, in Texas,[34] Egypt,[37] Sudan,[106] and India.[107] They can also cause problems in temperate areas such as the U.K. where onions receive a period of high temperature drying in store.[69,108]

Losses in Texas stores between 1974—1976 of yellow bulb onions stored for 8 months at 27°C and 70% RH amounted to 22 to 76%.[34] Losses of 33% (1979) and 61% (1980) due to *A. niger* were recorded during summer storage in Japan.[109]

Penicillium spp. also occur under high temperature conditions causing a blemish of the ribs of the skins of stored bulbs but less often causing breakdown of the storage tissues.

2. *Aspergillus Niger* (Black Mold)

a. The Causal Pathogen

A. niger infects the necks of onions and occasionally penetrates the sides of bulbs and bases of bulbs where damage to the dry skins has exposed the storage scales. The fungus occurs as a primary pathogen[69,106,110] although rotting may be enhanced by the presence of other organisms.[110,111] Infection of onion scales results in the production of a white mycelium in which are produced the characteristic brown to sooty black conidial heads — hence, the name black mold.

b. Infection Sources

i. Seeds and Sets

A. niger was recovered from the seeds and sets in North America and these were considered to be potential sources of the disease on bulbs.[112] The fungus was also commonly isolated from seed samples (*cv.* "Pusa Red") in India where it caused a severe damping-off of seedlings.[113]

ii. Soil

The fungus present in the soil is considered to be a major source of *A. niger* in Texas, (*cv.* "Grano") with subsequent progress of infection related to rainfall before harvest and high relative humidity in store.[34]

iii. Onion Crops

The fungus has been recovered from the leaves of growing onion crops in Australia[114] and the U.K.[115] and later from the bulbs when placed in store. The source of the fungus on the leaves of the plants was not determined in these cases.

c. Factors Affecting the Growth of the Fungus Relating to Infection of Onion Bulbs

i. Germination and Growth of the Fungus

Optimum temperature and humidity for germination and growth of *A. niger* spores are reported as being about 30°C at 100% RH and approx. 40°C at 93% RH.[116] Figure 3 illustrates the growth response on agar of *Aspergillus* and *Botrytis* spp. at 100% RH and a range of temperatures. It is apparent that the temperature optimum for growth of *A. niger* in culture (at 100% RH) was closely correlated with the storage temperatures which incited disease expression in U.K. stores in recent years.

ii. Infection

A. niger has been shown to be a weak pathogen of onion bulb scales; phenolic substances in brown skinned onion increased their susceptibility to the pathogen.[97] In the U.K., the fungus failed to infect intact bulb storage scales but readily invaded wounded scales maintained at relative humidities in excess of 80% and temperatures of about 30°C.[86]

3. Other *Aspergillus* spp.

a. A. fumigatus Fres. (Blue/Green Mold)

A. fumigatus occurs in the air spora of the U.K. Large concentrations of conidia are produced in warm conditions such as fermenting where self-heating occurs in hay, straw, or compost.[117] The fungus is present in the mycoflora of onion leaves[112] but the post-harvest drying conditions do not usually produce the required temperature and humidity, conducive to rotting by the pathogen. Nevertheless, it has been observed occasionally on bulbs from U.K. onion stores, particularly on those that have been exposed to very high drying temperatures. This response correlates well with its growth in culture (Figure 3).

b. A. alliaceus Thom. and Church (Yellow Mold)

A. alliaceus occasionally occurs on garlic[118] and onions[119,120] imported from countries with hot climates. When present, the progress of infection is essentially similar to that of *A. niger* except that the fungus produces heads of yellow conidia and black sclerotia on affected tissue. High temperatures of 28 and 32°C resulted in the most rapid infection progress but both higher (40°C) and lower temperatures (20°C) caused only slight decay.[118]

4. Disease Control

It has been suggested that the control of storage molds may become difficult in U.K. stores with the increased use of high temperature drying.[108] Provided drying is done according to recommendations[73,89] forcing air at a temperature of 30°C through onions with RH of $<80\%$ at 428 $m^3\ h^{-1}\ tonne^{-1}$, it appears unlikely that *A. niger* will cause bulb rotting (Figure 3). In addition, effects of high temperature on these fungi is limited because onion stack temperatures are reduced to ambient levels (on average 7°C over winter in the U.K.) after the relatively short drying treatment. In hot climates, ambient temperatures remain high during crop production and during storage. In Texas, similar numbers of dried (37.8°C) and undried bulbs were infected with *A. niger*[121] and black mold was reduced only if air humidity was kept low (approx. 36% RH) during high temperature drying.

Chemical treatments have been shown to be beneficial. *Aspergillus* mold of shallots and was reduced when bulbs were field dried for 6 d then received a further 6 d controlled temperature (37.8°C) drying in store and before or after post-harvest drying they were dipped in a fungicide (2,4,5-trichlorophenoxide) for 15 min.[122] Post-harvest bulb rotting was reduced by 16 to 17% if topped onions were dusted with calcium carbonate.[109] Black mold of onions in transit was reduced when bulbs were stored in containers fumigated with nitrogen trichloride at 430 mg m^{-3}.[123] In the past (1919), in-furrow seed treatment with formaldehyde was recommended for the control of black mold in the U.S.[124]

5. *Penicillium* spp. (Blue and Green Mold)

a. The Causal Organisms and Disease

Various *Penicillium* spp. produce blue mold on a range of fruits, bulbs, roots, and seeds throughout the world. Cut surfaces of onion bulbs dipped in copper sulphate were rapidly colonized by species of *Penicillium*.[125] These fungi also rotted stored flower bulbs.[126] *Penicillium* molds sometimes affect the ribs of onion leaf scales blemishing bulb onions which have been subjected to high temperature. Otherwise *Penicillium* spp. cause soft watery rots on the surface, producing conidiophores with broom-like phialides of blue green conidia. Such symptoms were caused by *P. cyclopium* Westling in stored English, Spanish and Dutch onions in the U.K.[127]

P. corymbiferum Westling causes a serious field and storage disease of garlic in California, U.S.[128] In Egypt *P. citrinum* Thom. and *P. funiculosum* Thom. were frequently isolated from diseased stored bulbs of local cultivars.[35]

b. Infection Sources

i. Seeds and Cloves

Penicillium spp. are often found contaminating onion seed but the significance of this is not yet known.[129]

Primary inoculum of *P. corymbiferum* is probably provided by infected garlic cloves in the planting material as the fungus survives poorly in the soil.[128] Contamination of the cloves probably occurs on damaged areas caused when separating cloves for planting.

ii. Growth of Penicillium and Infection

Penicillium spp. isolated from U.K. bulb onions grew, optimally in culture at 25°C. Coincidentally, garlic plants from cloves stored at 25°C were more severely infected than those from cloves stored at 5°C.[128]

As with *A. niger, Pennicillium* blue mold spp. may be controlled in U.K. stores by attention to the post-harvest temperature/humidity drying regimes (Section 4 and Figure 4). Previously,[122] post-harvest mercurial dips followed by in store drying procedures were proposed for the treatment of garlic. The use of mercury is no longer acceptable but bulb treatments using less toxic fungicides with less mammalian toxicity might improve disease control.

C. OTHER FUNGI

1. Disease Occurrence

Fungi which invade the roots and possibly the basal plates of the bulbs may cause problems when onion crops are harvested and stored. These include *Sclerotium cepivorum*

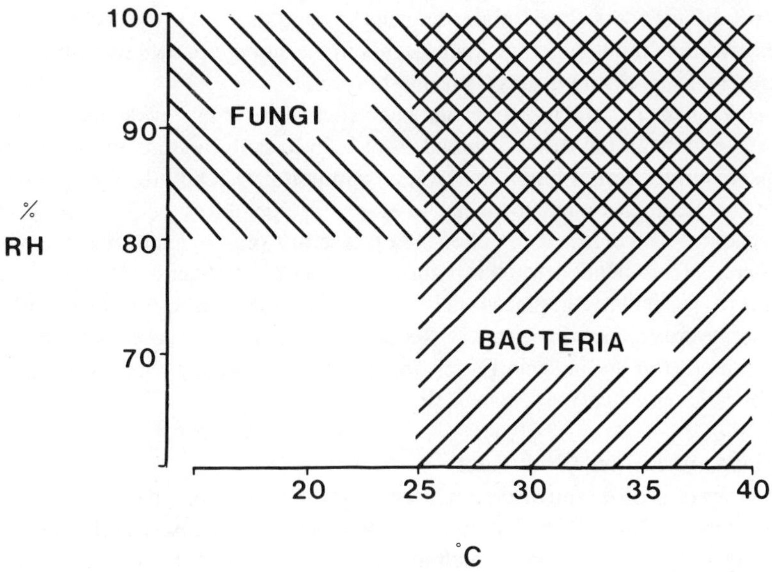

FIGURE 4. Diagram of effect of temperature (°C) and humidity (% RH) on the expression of fungal and bacterial storage diseases of onions. (Taylor, J. D. and Maude, R. B., unpublished data).

Berk. (onion white rot), *Fusarium oxysporum* f. sp. *cepae* (Hanz.) Snyder and Hansen (*Fusarium* basal rot), and *Pyrenochaeta terrestris* (Hansen) Gorenz (pink root). *Colletotrichum circinans* (Berk.) Vogl. (smudge) affects the necks and outer bulb scales of onions. Infected sets and possibly seeds, are additional sources of two of these fungi i.e. *F. oxysporum* f. sp. *cepae*[130] and *C. circinans*.[131] Accounts of the general biology of these fungi have been published.[65,132] The fungi and the storage problems they cause are described briefly here.

2. *Sclerotium cepivorum* (White Rot)

The disease is widespread in green (salad) onions in the U.K. and appears to have increased in bulb onion crops in recent years.[133] The fungus is soil-borne and the disease develops over a temperature range of 10 to 20°C. No post-harvest problems have been reported in the U.K. but the fungus has caused the breakdown and decay of stored bulbs in Australia[134] (see chapter on Root Disease).

White rot, unlike neck rot (*B. allii*), affects the base of onion bulbs where it produces a white mycelium containing numerous small black sclerotia, 0.3 to 0.5 mm in diameter.[135] The application of dicarboximide fungicides to infected organic soils in Canada[136] and to sandy loam soils in the U.K.[137] provided effective preharvest control of the disease on salad onions. Recently, the accelerated degradation of dicarboximides in the soil[138] has resulted in the withdrawal of one fungicide, iprodione, making the future of this method of control uncertain at the present time. Post-harvest dusting of bulbs with the fungicide dicloran has met with some success in Australia.[134]

3. *Fusarium oxysporum* f. sp. *cepae* (*Fusarium* Basal Rot)

Basal rot has been ascribed to various *Fusarium* spp. in the past but the causal pathogen has now been shown to be *F. oxysporum* f. sp. *cepae*. The fungus occurs in the soils of many areas becoming prevalent in those places where onions are grown for a number of years with little or no rotation of crops. It is a disease of warm climates with optimum temperatures for development of 28 to 32°C.[133]

Basal rot affects roots, basal plates and the storage tissues of the lower part of bulbs. It produces a whitish mold which grows upwards in the storage tissues of bulbs. No sclerotia are produced and this distinguishes it from white rot.

The mechanism of field infection has been reviewed.[139] No evidence was found[139] to support the view[140] that the disease spread from bulb to bulb in store and it was concluded that post-harvest infection of bulbs was of minor importance. Only the bulbs which became infected in the field continued to decay in store. Temperatures of 30 to 35°C with high humidity encouraged the development of the disease in store.[141] Therefore, the post-harvest drying systems recommended for the reduction of neck rot (Section II. 5.c.ii) could also reduce basal rot particularly in places such as Georgia (USSR) where *Fusarium* and *Botrytis* post-harvest rots are equally important.[142] Reduction in basal rot in stored bulbs by fumigation has been reported from India.[143] In Egypt, regular daily aeration of onions resulted in a low incidence of *F. oxysporum* f. sp. *cepae*.[141]

4. *Pyrenochaeta terrestris* (Pink Root)

P. terrestris is a soil-borne disease widely distributed affecting many monocotyledonous and dicotyledonous plants.[65,132] Optimum temperatures for disease development are 24 to 28°C.[133] It invades weakened roots which turn pink, shrivel, and die and causes loss of bulb yield. The disease is not common in stored bulb onions; it is not uncommon, however, to find pink root in association with *Fusarium* in affected onion base plates.[144,145] In this relationship, it may increase post-harvest spoilage of the onions. Some resistance in onions to both pathogens has been reported,[65,132,146-148] but complete resistance to pink-root is found in *A. fistulosum*.

5. *Colletotrichum circinans* (Smudge)

Smudge disease of onions was discovered in the U.K. in 1851[149] and at present occurs widely in Europe and North America. Infection occurs through the cuticle of the outer bulb scales and the onion neck. Minute black stromata develop which produce concentrically arranged acervuli. The pathogen causes blemishes on the bulbs of fresh marketed or stored onions and devalues the produce. The disease rarely affects the brown skinned cultivars of "Rijnsburger" type onions produced in the U.K. but white skinned cultivars are particularly vulnerable.[65,132] Post-harvest drying systems in use in the U.K. and U.S. should effectively reduce smudge incidence.

III. BACTERIAL ROTS OF STORED ONIONS

A. OCCURRENCE OF BACTERIAL PROBLEMS

Most bacteria producing post-harvest rots of onions are opportunist organisms; some are primary pathogens and others secondary, following fungal, insect, or nematode attack. In the latter case, the bacteria are masked by the other organisms. Thus when neck rot (*B. allii*) was controlled in U.K. stored bulb onions, the extent of damage caused by bacterial rotting was revealed.[150] Where bacteria are the primary disease agents they may cause considerable crop damage.[151]

High temperatures favor bacterial decay of onions and as a result bacterial problems may be more severe in warmer climates. Although the prevalent species may vary between different climatic environments, bacteria generally are widely distributed and local increases in incidence of specific organisms may relate directly to changes in the crop or store environment.

In the temperate climates of the U.K., an *Erwinia herbicola*-like organism and a *Lactobacillus* sp. were the main bacteria present in decaying bulbs from commercial stores.[152] Bacterial isolates of these species from U.K. onions, when reinoculated, infected healthy bulbs. *Pseudomonas alliicola* Starr and Burkh. (slippery skin disease) has been isolated from bulb onions in the U.K. but it is not common.[152] It has been reported in the U.K. in onions imported from Spain.[153] *P. alliicola* was first recognized as a primary bacterial pathogen of onions in America in 1899.[154] Details of the biology of this bacterium were published in 1942.[155,156] *Pseudomonas cepacia* Burkh. (sour skin disease) is a primary bacterial pathogen of bulb onions occurring at times in American,[157] Canadian,[158] and Italian stores.[159] The organism has not been reported from U.K. bulbs. *Pseudomonas aeruginosa* (Schroeter) Migula has been shown to cause an internal brown rot similar to slippery skin *(P. alliicola)* in Australian bulbs.[152] In warmer climates, soft rot bacteria (*Erwinia* and *Pseudomonas* spp.) are important, causing over 10% losses in stored Texan yellow onions.[34] On average, *P. alliicola, P. cepacia,* and *Erwinia aroidea* (Towns.) were reported in 90% of isolations from Egyptian onions.[35]

B. SYMPTOMS AND ORGANISMS

1. *Pseudomonas alliicola* (Slippery Skin)

Affected bulbs may feel sound but when cut open some of the main storage scales are brown and water-soaked having a cooked appearance.[155] When pressure is applied to the base of the bulb, the center core may be ejected, hence the same slippery skin.[132] Infection progresses downward in the storage scales and bulbs become soft and rotten. Mature bulbs may rot completely within 10 d of infection at room temperature.[155]

P. alliicola is a Gram-negative motile bacterium having one to several polar flagella. Cultural and biochemical identification procedures have been published.[155,156]

2. *Pseudomonas cepacia* (Sour Skin)

P. cepacia attacks only the outer storage scales of onion.[157] The affected scales are yellow and slimy and the bulbs have a vinegary, sour smell, hence the name ''sour skin''. *P. cepacia* is a Gram-negative, nonspore-forming rod with rounded ends. The bacterium is motile having 1 to 3 polar flagella and is 1.9×0.8 μm. Cultural and biochemical identification procedures are published.[157]

3. *Lactobacillus* and *Erwinia* Bacteria

a. Lactobacillus spp.

The bacterium caused severe water soaking of the internal tissues of onion bulbs but only at high temperatures (approx. 35°C).[152] Serological and biochemical tests indicated similarities to *L. plantarum,*[160] a homofermentative bacillus on herbage silage.[161]

b. Erwinia spp.

This organism was the commonest bacterium isolated from rotting U.K. onions.[152] It caused severe water soaking and discoloration of onion bulbs. Biochemically, the bacterium, was similar to *E. herbicola* but strong serological relationships were not obtained.

C. DISEASE SOURCES AND BIOLOGY

The sources of bacteria causing post-harvest rotting of bulb onions are not well defined. The primary pathogens, *P. alliicola* and *P. cepacia* have been isolated from American soil in which onions had been grown;[162] *P. cepacia* has also been recovered from organic muck soil[163] and from irrigation water in U.S.[164]

P. alliicola produced long, sunken, white, tongue-shaped lesions on seed-stalks and leaves of onions in field crops in Hungary.[165] Abundant rainfall in June and July increased disease severity in field and bulb crops in Bulgaria.[166] In India, the disease caused a severe stalk rot, inflorescence decay and failure to set seeds in seed production crops.[167] *P. alliicola* clearly affects the growing onion crop which increases in susceptibility with age. It is considered that the pathogen infected onion bulbs close to harvest.[155]

Pseudomonas cepacia may not be as aggressive as *P. alliicola* in onion crops. Inoculated unwounded plants failed to develop infection.[168] However, stab inoculated into the junction of the leaf blade and sheaths, *P. cepacia* caused infection[168,169] supporting the suggestion[157] that the bacterium may gain entry during field topping of onions. *P. cepacia* declined rapidly in inoculated leaves so that dried diseased leaves are unlikely to be a potent source of the pathogen in the field.[170]

P. aeruginosa like the sour skin bacterium did not infect unwounded plants[151] and it was probable that abrasion of onion leaves under conditions of high rainfall and low temperature was necessary for infection of plants to occur in nature.

Lactobacillus and *Erwinia herbicola* bacteria occur was part of the leaf mycoflora of field onions and it is presumed that they may act as opportunist pathogens when climatic factors and the condition of onion plants are favorable for invasion.

D. DISEASE CONTROL

Post-harvest diseases are caused by bacteria some of which are specific to onions *(P. alliicola, P. cepacia)* and others which are facultative pathogens (*Lactobacillus, Erwinia* spp.) with a wider host range. High temperature drying systems provide conditions suitable for development of *Lactobacillus*-like organisms (Figure 4). Greater understanding of their sources of distribution and of their life-cycles is necessary before sensible strategies for their control can be developed. For example, with *Pseudomonas synringae* pv. *phaseolicola* (halo-blight of bean), studies of seed transmission[171] and field spread of bacterium[172] resulted in the development of successful disease control strategies.[173] However, apart from *P. alliicola,* these onion bacteria may not be that predictable because of the lack of information on their crop biology. Such data is necessary to decide the degree of risk these organisms present to the successful storage of onion bulbs.

IV. CONCLUSIONS

Post-harvest diseases of bulb onions, if uncontrolled, reduce the storage life and therefore the value of the crop. An understanding of the biology of the organisms involved is a prerequisite for disease control. The fungi and bacteria causing storage problems are crop infecting organisms mainly of seed-borne or soil-borne origin. They may be primary or secondary pathogens but generally their incidence in store relates to the amount of source inoculum and the climatic factors during emergence and growth of the crop. In these respects, the bacteria are less understood than the fungi.

Simple remedies can be effective with some of the diseases, for example, seed treatment control of onion neck rot (*B. allii*). In other storage diseases, where inoculum is soil-borne, e.g., white rot, pink root, basal rot and smudge, chemical and physical field treatments can be useful. Rotations, covering or burial of onion trimmings from processing facilities with soil, and avoidance of infection bridging by sensible spatial separation of crops are useful broad strategies for the containment of many diseases. The use of relatively modern high temperature drying systems can be effective in inhibiting the growth of temperate fungal pathogens and rendering the necks of onions dry and unsuitable for fungal growth. Problems arise, however, when drying temperatures are too low and humidities too high facilitating the growth of temperate fungi (Figure 4) or where temperatures and humidities are too high favouring the development of thermolabile molds and bacteria (Figure 4). The fungal problems can be prevented by correct management of relative humidity but the bacteria respond less well and are not controlled (Figure 4). A very similar range of high temperature organisms occurs as part of the mycoflora of tropical onions causing similar problems. However, new advances in disease control strategies will require further information on the crop biology of the post-harvest fungal and bacterial pathogens of bulb onions.

REFERENCES

1. **Anon.,** Onions, in *FAO Prod. Yearbook 1985,* 39, 174, 1986.
2. **Thompson, A. K., Booth, R. H., and Proctor, F.J.,** Onion storage in the tropics, *Trop. Res.,* 14, 19, 1972.
3. **Tackholm, V. and Drar, M.,** Flora of Egypt. *Bull. Fac. Sci. Egypt Univ.,* 3, xii, 64, 1954.
4. **Fenwick, G. R. and Hanley, A. B.,** The genus *Allium, CRC Crit. Rev. Food Sci. Nutr.,* 22, 199, 1985.
5. **Meer, van der,** Onions in modern Egypt, *Onion World,* 2, 105, 9, 1986.
6. **Maude, R. B.,** *Post-Harvest Pathology of Fruits and Vegetables* Dennis, C., Ed., Academic Press, London, 1983.
7. **Munn, M. T.,** Neck-rot disease of onion, *Bull. N.Y. State Agric. Expt. Stn.,* 437, 455, 1917.
8. **Walker, J. C.,** Two undescribed species of *Botrytis* associated with the neck rot diseases of bulb onions. *Phytopathology,* 15, 708, 1925.
9. **Walker, J. C.,** Control of mycelial neck rot of onion by artificial curing, *J. Agric. Res.,* 30, 365, 1925.
10. **Hoyle, B. J.,** Onion curring — a comparison of storage losses from artificial, field, and non-cured onions, *Proc. Am. Soc. Hortic. Sci.,* 52, 407, 1948.
11. **Vaughan, E. K., Cropsey, M. G., and Hoffman, E. N.,** Effects of field-curing practices, artificial drying and other factors in the control of neck rot in stored onion. *Oreg. State Univ. Agric. Expt. Stn., Stn. Bull.,* 77, 21, 1964.
12. **Kaufman, J. and Lorbeer, J. W.,** Control of *Botrytis* neck rot of onions by fungicidal dusts and desiccant chemicals, *Plant Dis. Rep.,* 51, 696, 1967.
13. **Samuel, G.,** Summary of plant disease records in South Australia for the two years ending June 30th, 1930, *J. Agric. South Aust.,* 34, 746, 1931.
14. **Hughes, I. J.,** Onion diseases in Queensland. *Queensl. Agric. J.,* 96, 607, 1970.
15. **Massee, G.,** An onion disease, *Gard. Chron.,* 16, 120, 1894.
16. **Croxhall, H. E.,** Some factors influencing loss of onion bulbs during storage, *Long Ashton Res. Stn. Rep. 1945,* 143, 1946.
17. **Maude, R. B. and Presly, A. H.,** Neck rot *(Botrytis allii)* of bulb onions. I. Seed-borne infection and its relationship to the disease in the onion crop. *Ann. Appl. Biol.,* 86, 163, 1977.
18. **Maude, R. B. and Presly, A. H.,** Neck rot *(Botrytis allii)* of bulb onions II. Seed-borne infection in relationship to the disease in store and the effect of seed treatment, *Ann. Appl. Biol.,* 86, 181, 1977.
19. **McLarty, H. R.,** Report of the dominion field laboratory of plant pathology, Summerland, B. C. Rept. Dominion Botanist for the year 1928, Div. of Botany, Can. Dept. of Agric., 142, 1929.
20. **Jarvis, W. R.,** Onion neck rot. *Minist. Agric. Food Ont.* Fact sheet, 019, 1976.
21. **Rod, J.,** Evaluation of the storability of Czechoslovak cultivars of onion (*Allium cepa* L.), *Sbornik UVTIZ Zahrdnictvi,* 5-6, 1978-1979, 99, 1978.
22. **Hellmers, E.,** *Botrytis* on *Allium* species in Denmark. *Botrytis allii* Munn and *B. globosa* Raabe, *Medd. Vet Hojsk. planteepat. Afd., Khb.,* 25, 51, 1943.
23. **Sorauer, P.,** Das verschimmeln der speiszwiebeln, *Oesterr. Landw. Wehnbl,* 147, 1876.
24. **Brautigam, S.,** *Botrytis allii* Munn an Zwiebelsaatgut, *Nachrichten Wehnbl Bl. dt Pflschutsdienst,* Berlin, 31, 195, 1977.
25. **Doorn, A. M. van, Koert, J. L., and Kreyger, J.,** Investigations on the occurrence and control of neck rot (*Botrytis allii* Munn) in onions, *Versl. Landbouwkd. Onderz. A,* 68.7, 83, 1962.
26. **Netzer, D. and Dishon, I.,** Occurrence of *Botrytis allii* in onions for seed production in Israel, *Plant Dis. Rep.,* 50, 21, 1966.
27. **Hemmi, T. and Niwa, S.,** On gray mould neck rot of stored onions, *Forsch. Plfkr. Kyoto,* 3, 234, 1937.
28. **Shidla, L. A.,** Fungi of the genus *Botrytis* in Lithuania, Part 1. Species and their distribution on agricultural and medicinal plants, *Trudy Akademiia nauk litouskoi SSR Seriia B,* 1, 23, 1971.
29. **Brien, R. M.,** List of plant diseases recorded in New Zealand, *N.Z. Dep. Sci. Ind. Res. Bull.,* No. 67, 39, 1939.
30. **Harrow, K. M. and Harris, S.** Artificial curing of onions for control of neck rot (*Botrytis allii* Munn), *N.Z. J. Agric. Res.,* 12, 529, 1969.
31. **Roed, H.,** *Botrytis* (grey mould) on *Allium cepa* and *Allium ascalonicum* in Norway, *Acta Agric. Scand.,* 1, 20, 1951.
32. **Siemaszko, W.,** Grey mould of onion - *Botrytis allii* Munn, *Rocz. Nauk Roln.,* 21, 12, 1929.
33. **Perisic, M., Babovic, M., and Markovic, S.,** Contribution to the study of *Botrytis allii* the parasite of onion, *Archiv. Za Poljoprivredna Nauke,* 30, 143, 1977.
34. **Miller, M. E. and Dillon, R. C., Jr.,** Survey of bulb diseases on onions in South Texas, *J. Rio Grande Valley. Hortic. Soc.,* 33, 25, 1979.
35. **Hussein, F. N., Abd-Elrazik, A., Darweish, F. A., and Rushdi, M. H.,** Survey of storage diseases of onions and their incitants in upper Egypt, *Egypt J. Phytopathol.,* 9, 15, 1977.

36. **Jones, H. A. and Mann, L. K.,** *Onions and their Allies,* Interscience Publishers Inc., New York, 1963, 268.
37. **Walker, J. C.,** *Botrytis* neck rots of onions, *J. Agric. Res.,* 33, 893, 1926.
38. **Owen, J. H., Walker, J. C., and Stahmann, M. A.,** Variability in onion neck rot fungi, *Phytopathology,* 40, 749, 1926.
39. **Morgan, D. J.,** Numeric taxonomic studies of the genus *Botrytis* II. Other *Botrytis* taxa, *Trans. Br. Mycol. Soc.,* 56, 327, 1971.
40. **Hennebert, G. L.,** Les *Botrytis* des *Allium, 15th Int. Symp. Phytopharmacy and Phytiatry,* 851, 1963.
41. **Hennebert, G. L.,** *Botrytis* and *Botrytis*-like genera, *Persoonia,* 7, 183, 1973.
42. **Fresenius, G.,** *Botrytis aclada* Fresen., *Beitrage zur Mykologie,* 16, Henrich Ludwig Bronner, Frankfurt, 1850.
43. **Poeteren, N. van,** Onderzoek over het koprot in de uien van de oogst 1938, *Verslagen en mededelingen plantenziektenkundige dienst Wageningen,* 90, 1939.
44. **Moore, W. C.,** Diseases of crop plants (1943—1946), *Bull. Minist. Agric. Fish. Food, London,* 139, 90, 1948.
45. **Presly, A. H.,** Studies on *Botrytis* spp. occuring on onions *(Allium cepa)* and leeks *(Allium porrum), Plant Pathol.,* 34, 422, 1985.
46. **Hickman, C. J. and Ashworth, D.,** The occurrence of *Botrytis* spp. on onion leaves with special reference to *B. squamosa, Trans Br. Mycol. Soc.,* 26, 153, 1943.
47. **Ellis, M. B.,** Dematiaceous hyphomycetes, Commonw. Mycol. Inst. Kew England, 1971, 608.
48. **Maude, R. B.,** Onion neck rot, *Minist. Agric. Fish. Food, London,* Leafl. no. 779, 1981.
49. **Maude, R. B.,** *Botrytis allii* Munn, synonym *B. aclada* Fres., *European Handbook of Plant Diseases,* Smith, I. M., Archer, S. A., Junez, J., Lelliot, R., and Phillips, D. H., Eds., Blackwell Scientific Publication Ltd., Oxford, U.K., 1987.
50. **Presly, A. H.,** Methods of inducing sporulation of some *Botrytis* species occurring on onions and leeks, *Trans. Br. Mycol. Soc.,* 85, 621, 1985.
51. **Wood, R. K. S.,** The biology and control of diseases caused by *Botrytis* spp., *Proc. 1st Br. Insectic. Fungi. Conf.,* 309, 1961.
52. **Yarwood, C. E.,** *Botrytis* infection of onion leaves and seed stalks, *Plant Dis. Rep.,* 22, 428, 1938.
53. **Blodgett, E. C.,** Observations on blasting of onion seed heads in Idaho, *Plant Dis. Rep.,* 30, 77, 1946.
54. **Tichelaar, G. M.,** *Botrytis allii, Jaarverslag Pziektenkundig Onderzoek,* p. 30, 1970.
55. **Tichelaar, G. M.,** Studies on the biology of *Botrytis allii* on *Allium cepa, Neth. J. Plant Pathol.,* 73, 157, 1967.
56. **Bochow, H. and El Mosallamy, H. M.,** Investigations on the infection behaviour of *Botrytis allii* Munn in *Allium cepa* L., *Arch. Phytopathol. Pflanzenschutz,* 15, 103, 1979.
57. **Stewart, A. and Mansfield, J. W.,** Fungal development and plant response in detached onion, onion bulb scales and leaves inoculated with *Botrytis allii, B. cinerea, B. faba, and B. squamosa, Plant Pathol.,* 33, 401, 1984.
58. **Maude, R. B.,** The correlation between seed-borne infection by *Botrytis allii* and neck rot development in store, *Seed Sci. Technol.,* 11, 829, 1983.
59. **Maude, R. B., Bambridge, Judith M., Presly, A. H., and Phelps, Kathleen,** Effects of field environment and cultural practices on incidence of *Botrytis allii* (neck rot) in stored bulb onions, *Q. Spec. Izzaz. Viticolt. Enolog.,* 2, 241, 1986.
60. **Janyska, A. and Rod, J.,** The causes of spreading of *Botrytis allii* Munn in onions under storage conditions, *Bull. Vyzkumny Slechtitelsky Ustav Zelinarsky Olomouc,* 21/22, 70, 1979.
61. **Bochow, H. and Bottcher, H.,** Zur Bekampfung von *Botrytis allii* Munn durch Einsitz von Fungizicken, *Nachrichtenbl. Dtsch. Pflanzenschutzdienst (Berlin),* 32, 135, 1978.
62. **Ellerbrock, L. A. and Lorbeer, J. W.,** Etiology and control of onion flower blight, *Phytopathology,* 67, 155, 1977.
63. **Ellerbrock, L. A. and Lorbeer, J. W.,** Sources of primary inoculum of *Botrytis squamosa, Phytopathology,* 67, 363, 1977.
64. **Lorbeer, J. W.,** personal communication, 1985.
65. **Walker, J. C.,** *Diseases of Vegetable Crops,* McGraw-Hill, New York, 1952, 529.
66. **Anon.,** Control of neck rot in onions grown from sets, *Technical Conference of Advisory Plant Pathologists - MAFF,* 1981.
67. **Smith, F. G. and Jones, A. G.,** Production of onion sets, *Agriculture, (London),* 66, 298, 1957.
68. **Lachman, W. H. and Michelson, L. F.,** Effects of warm storage on the bolting of onions grown from sets, *Am. Soc. Hortic. Sci.,* 75, 495, 1960.
69. **Maude, R. B.,** Storage rots of onions, *Rep. Natl. Veg. Res. Stn. 1983,* 64, 1984.
70. **Lorbeer, J. W. and Tichelaar, G. M.,** A selective medium for the assay of *Botrytis allii* in organic and mineral soils, *Phytopathology,* 60, 1301 (Abstr.), 1970.

71. **Kritzman, G. and Netzer, D.,** A selective medium for isolation and identification of *Botrytis* spp. from soil and onion seed, *Phytoparasitica,* 6, 3, 1978.
72. **Maude, R. B., Bambridge, Judith M., and Presly, A. H.,** The persistence of *Botrytis allii* in field soil, *Plant Pathol.,* 31, 247, 1982.
73. **Anon.,** Dry bulb onions, *Hort. Enterpr.,* Bklt. I M.A.F.F., 73, 77, 1978.
74. **Maude, R. B.,** Neck rot *(Botrytis allii)* in the autumn sown bulb onion crop, *Rep. Natl. Veg. Res. Stn. 1976,* 94, 1976.
75. **Derbyshire, D. M. and Shipway, M. R.,** Control of post-harvest deterioration in vegetables in the UK, *Outl. Agric.,* 9, 246, 1978.
76. **Tichelaar, G. M.,** *Botrytis allii, Jaarverslag Plziektenkundig onderzoek* for 1970, p. 32, 1971.
77. **Janyska, A. and Rod, J.,** The effect of the treatment of onion (*Allium cepa* L.) seed on the yields and on the occurrence of neck rot, *Zahradnictvi,* 7, 197, 1980.
78. **Newhall, A. G. and Rawlins, W. A.,** Control of onion blast and mildew with carbamates, *Phytopathology,* 42, 212, 1952.
79. **Kaufman, J., Lorbeer, J. W., and Friedman, B. A.,** Relationship of fungicides and field spacings to *Botrytis* neck rot of onions grown in New York, *U.S. Dep. Agric., Agric. Res. Serv. Publ.,* 51-1, 6, 1964.
80. **Maude, R. B. and Presly, A. H.,** Neck rot of onions, *Rep. Natl. Veg. Res. Stn. 1973,* p. 95, 1974.
81. **Pendergrass, A., Isenberg, F. M., St. John, L. E., Jr., and Lisk, D. J.,** A foliage drying harvest aid for onions, *HortScience,* 4, 294, 1969.
82. **Lorbeer, J. W. and Kawamoto, S. O.,** Annual progress report on research supported by the Orange County Vegetable Improvement Co-operative Association Incorporated, Cornell University, Ithaca, New York, 1963.
83. **Doorn, A. M. van,** Control of onion neck rot by artificial curing, *Mededelingen Directeur van te tuinbouw,* 18, 250, 1955.
84. **Presly, A. H.,** *Botrytis* Species on Overwintered Salad Onions, Ph.D. thesis, University of Birmingham, 1978.
85. **Guzman, V. L. and Hayslip, N. C.,** Effect of time of seeding and varieties on onion production and quality when grown in two soil types, *Fla. State Hortic. Soc. Proc.,* 75, 156, 1962.
86. **Maude, R. B., Taylor, J. D., Munasinghe, H. L., Bambridge, Judith M., and Spencer, Angela,** Storage rots of onions, *Rep. Natl. Veg. Res. Stn. 1984,* 64, 1985.
87. **Gunkel, W. W., Lorbeer, J. W., Kawamoto, S. O., Kaufman, J., and Smith, Jr., H. A.,** Recent development on artificial heating a method of control of *Botrytis* neck rot in bulk stored onions, *Ann. Prog. Rep., N.Y. Farm Electr. Counc.,* 57, 1973.
88. **Gunkel, W. W., Lorbeer, J. W., Bensin, R. F., and Smith, Jr., H. A.,** Application of the artificial heating method to control *Botrytis* neck rot in pallet box stored onions, *Ann. Prog. Rep. N.Y. Farm Electr. Counc.,* 88, 1976.
89. **Anon.,** Dry bulb onions. Effects of various handling and drying practices on storage v7/40, *Rep. Kirton Exp. Hortic. Stn.,* 14, 53, 1977.
90. **Shipway, M. R.,** Developments in onion drying and storage, *Rep. Kirton Exp. Hortic. Stn.,* 33, 1980.
91. **Maude, R. B., Shipway, M. R., Presly, A. H., and O'Connor, D.,** The effects of direct harvesting and drying systems on the incidence and control of neck rot (*Botrytis allii*) in onions, *Plant Pathol.,* 33, 263, 1984.
92. **Rosberg, D. W. and Johnson, H. B.,** Post-harvest infra-red irradiation for the control of gray mold neck rot disease of onions, *Phytopathology,* 48, 345, 1958.
93. **Josephson, E. S. and Peterson, M. S.,** Preservation of Food by Ionizing Radiation, Vol. 3, CRC Press, Boca Raton, FL, 1983.
94. **Jarvis, W. R.,** *Botryotinia* and *Botrytis* species. Monograph No. 15, Research Branch, *Can. Dept. Agric.,* 195, 1977.
95. **Angell, H. R., Walker, J. C., and Link, K. P.,** The relation of protocateuchuic acid to disease resistance in the onion, *Phytopathology,* 20, 431, 1930.
96. **Link, K. P. and Walker, J. C.,** The isolation of catechol from pigmented onion scales and its significance in relation to disease resistance in onions, *J. Biol. Chem.,* 100, 379, 1933.
97. **Hatfield, W. C., Walker, J. C., and Owen, J. H.,** Antibiotic substances in onion in relation to disease resistance, *J. Agric. Res.,* 77, 115, 1948.
98. **Owen, J. H., Walker, J. C., and Stahmann, M. A.,** Pungency, color and moisture supply in relation to disease resistance in the onion, *Phytopathology,* 40, 292, 1950.
99. **Walker, J. C., Owen, J. H., and Stahmann, M. A.,** Relative importance of phenols and volatile sulphides in disease resistance in the onion, *Phytopathology,* 40, 30, 1950.
100. **Bergquist, R. R. and Lorbeer, J. W.,** Reaction of *Allium* spp. and *Allium cepa* to *Botryotinia (Botrytis) squamosa, Plant Dis. Rep.,* 55, 394, 1971.
101. **Meer, W. P. van der, Bennekom, J. L. van, Giessen, A. C. van der,** Testing onions (*Allium cepa* L.) and other *Allium* species for resistance to *Botrytis allii* Munn, *Euphytica,* 19, 152, 1970.

102. **Maude, R. B., Presly, A. H., Miller, Judith M., and Large, Angela,** Neck rot of bulb onions, *Rep. Natl. Veg. Res. Stn. 1979,* 60, 1980.
103. **Currah, Lesley, and Maude, R. B.,** Laboratory tests for leaf resistance to *Botrytis squamosa* in onions, *Ann. Appl. Biol.,* 105, 277, 1984.
104. **Maude, R. B., Presly, A. H., and Currah, L.,** Field tests of resistance to *B. squamosa* in the genus *Allium. Rep. Natl. Veg. Res. Stn. 1985,* 62, 1986.
105. **Miyaura, K., Shinada, Y., and Gabelman, W. H.,** Selection for resistance of onions to *Botrytis allii* by seed inoculation method, *HortScience,* 20, 769, 1985.
106. **Musa, Sulaka K., Habish, H. A., Abdala, A. A., and Adlan, A. B.,** Problems of onion storage in the Sudan, *Trop. Sci.,* 15, 319, 1973.
107. **Venkatarayan, S. V. and Delvi, M. H.,** Black mould of onions in storage caused by *Aspergillus niger, Curr. Sci.,* 20, 243, 1951.
108. **Stow, J. R.,** Effects of humidity on losses of bulb onions *(Allium cepa)* stored at high temperature, *Exp. Agric.,* 11, 81, 1975.
109. **Tanaka, K. and Nonaka, F.,** Studies on the rot of onion bulbs caused by *Aspergillus niger* and its control by lime applications, *Bull. Fac. Agric., Saga Univ.,* 51, 47, 1981.
110. **Machacek, J. E.,** The black mold of onion caused by *Aspergillus niger, Phytopathology,* 19, 733, 1929.
111. **Maude, R. B., Bambridge, J. M., and Spencer, A.,** Storage rots of onions, *Rep. Natl. Veg. Res. Stn. 1983,* 64, 1984.
112. **van Pelt, W.,** A new fungus disease causing serious damage in storage houses, *Ohio, Agric. Exp. Stn. Res. Bull.,* 1, 152, 1917.
113. **Gupta, R. P., Srivastava, P. K., Srivastava, V. K., and Paudey, U. B.,** Note on fungi associated with onion seeds, their pathogenicity and control, *Seed Res.,* 12, 98, 1984.
114. **Anon.,** Plant Diseases. Notes contributed by the Biological Branch, *Agric. Gaz. N.S.W.,* 1, 199, 1939.
115. **Maude, R. B., Taylor, J. D., Bambridge, J. M., and Reader, S. L.,** Storage rots of onions, *Rep. Natl. Veg. Res. Stn. 1985,* 63, 1986.
116. **Bonner, J. T.,** A study of the temperature and humidity requirements of *Aspergillus niger, Mycologia,* 40, 723, 1948.
117. **Mullins, J., Harvey, R., and Seaton, A.,** Sources and incidence of airborne *Aspergillus fumigatus* (Fres.), *Clin. Allergy,* 6, 209, 1976.
118. **Walker, J. C. and Murphy, A.,** Onion-bulb decay caused by *Aspergillus alliaceus, Phytopathology,* 24, 289, 1934.
119. **Ekstrand, H.,** A fungal disease new to Sweden on imported eating onions, *Vaxtskyddsnotiser,* 4, 63, 1948.
120. **Anon.,** *Aspergillus alliaceus - Q. Rep. Pl. Prot. Comm. for S.E. Asia Pacific Region,* Jan.-Mar., April-June, 1961.
121. **Miller, M. E. and Dillon, R. C., Jr.,** Effects of artificial drying and controlled atmosphere storage on control of black mold *(Aspergillus niger)* of onions, *Phytopathology,* 69, 5330, 1979.
122. **Aycock, R. and Jenkins, J. M., Jr.,** Methods of controlling certain diseases of shallots, *Plant. Dis. Rep.,* 44, 934, 1960.
123. **Pryor, D. H.,** Reduction of post-harvest spoilage in fresh fruits and vegetables destined for long distance shipment, *Food Technol.,* 4, 56, 1950.
124. **van Pelt, W.,** Onion diseases as found in Ohio, *Ohio Agric. Exp. Stn. Res. Bull.,* 1, 70, 1919.
125. **Dillon Weston, W. A. R. and Taylor, R. E.,** Development of *Penicillium* on the cut surfaces of certain vegetables, *Nature, (London),* 151, 54, 1943.
126. **Moore, W. C.,** Diseases of bulbs, *Minist. Agric. Fish. Food London* Leafl., 117, 176, 1949.
127. **Wijeratnam, R. S. W. and Lowings, P. H.,** Watery rot of stored onions caused by *Penicillium cyclopium* Westling, *Plant Pathol.,* 27, 100.
128. **Smalley, E. B. and Hansen, H. N.,** *Penicillium* decay of garlic, *Phytopathology,* 52, 666, 1962.
129. **Kudrina, V. N.,** Changes in the microflora and germination of vegetable seeds during storage, *Izverstiia Timiriazevskaia sel'skokhoziaistue-nnaia Akademiia,* 2, 175, 1967.
130. **Szatala, O.,** *Fusarium* rot of onions in Hungary, *Ann. Inst. Prot. Plant Hung.,* 9, 301, 1964.
131. **Behr, L.,** On a total loss of germinating onions caused by *C. dematium* f. *circinans, Zbl. Bakt.,* 116, 552, 1963.
132. **Sherf, A. F. and Macnab, A. A.,** *Vegetable Diseases and Their Control,* John Wiley & Sons, New York, 1986.
133. **Anon.,** Pest control in tropical onions, *Trop. Develop. Res. Inst.,* London, 1986, 109.
134. **Hughes, I. K.,** Onion storage rot control, *Queensl. J. Agric. Anim. Sci.,* 27, 391, 1970.
135. **Anon.,** Onion neck rot, *Minist. Agric. Fish. Food* London Leafl., 779, 1981.
136. **Utkhede, R. S. and Rahe, J. E.,** Evaluation of chemical fungicides for control of onion white rot, *Pestic. Sci.,* 10, 414, 1979.
137. **Entwistle, A. R., Munasinghe, H. L., Jonas, Lyn B., and Haynes, Alison J.,** White rot disease of onions - autumn sown bulb onions, *Rep. Natl. Veg. Res. Stn. 1979,* 67, 1980.

138. **Walker, A., Brown, D. A., and Entwistle, A. R.,** Enhanced degradation of iprodione and vinclozolin in soil, *Pestic. Sci.*, 17, 183, 1986.
139. **Lorbeer, J. W. and Stone, K. W.,** Reaction of onion to *Fusarium* basal rot, *Plant Dis. Rep.*, 49, 522, 1965.
140. **Walker, J. C. and Larson, R. H.,** Onion diseases and their control, *U.S. Dept. Agric., Agric, Res. Serv. Publ.*, No. 208, 27, 1959.
141. **Ashour, W. A., Ali, M. D. H., Morsy, A. A., and Dirab, M. M. M.,** Effect of temperature, moisture and aeration on the development of basal rot of onion during storage, *Agric. Res. Rev.*, 51, 163, 1973.
142. **Sardzhveladze, S., Dolidze, M., and Kirimelashvili, N.,** The development of diseases of onion during storage under different conditions, *Jr. Nil Zemledeliya Gruz. SSR*, 23, 141, 1976.
143. **Dang, J. K. and Thakur, D. P.,** Controlling storage rot on onion bulbs caused by *Fusarium solani* by fumigation, *Indian Phytopathol.*, 25, 359, 1973.
144. **Davis, G. N. and Henderson, W. J.,** The interrelationship of the pathogenicity of a *Phoma* and a *Fusarium* on onions, *Phytopathology*, 27, 763, 1937.
145. **Hess, W. M.,** Pink root of onion caused by *Pyrenochaeta terrestris, Diss. Abstr.*, 23, 1478, 1962.
146. **Marlatt, R. B. and McKittrick, R. T.,** Pink root resistant onions in Arizona, *Plant Dis. Rep.*, 42, 1310, 1958.
147. **Mehr, A. E., O'Brien, M. J., and Davis, E. W.,** Pathogenicity of *Fusarium oxysporum* f. sp. *cepae* and its interaction with *Pyrenochaeta terrestris* on onion, *Euphytica*, 11, 197, 1962.
148. **Wooliams, G. E.,** Resistance of onion varieties to *Fusarium* basal rot and to pink root, *Can. Pl. Dis. Surv.*, 46, 101, 1966.
149. **Berkeley, M. J.,** *Gard. Chron.*, 11, 595, 1851.
150. **Taylor, J. D., Dudley, C. L., and Littlejohn, I. H.,** Bacterial rots of stored onions, *Rep. Natl. Veg. Res. Stn. 1979*, 76, 1980.
151. **Cother, E. R., Darbyshire, B., and Brewer, J.,** *Pseudomonas aeruginosa:* Cause of internal brown rot of onion, *Phytopathology*, 66, 828, 1976.
152. **Taylor, J. D.,** Bacterial rots of stored onion, *Rep. Natl. Veg. Res. Stn. 1974*, 116, 1975.
153. **Roberts, P.,** A soft rot of imported onions caused by *Pseudomonas alliicola* (Burkh.) Starr and Burkh. *Plant Pathol.*, 22, 98, 1973.
154. **Stewart, F. C.,** A bacterial rot of onion, *Bull. N.Y. State Agric. Expt. Stn.*, 164, 209, 1899.
155. **Burkholder, W. H.,** Three bacterial plant pathogens: *Phytomonas cayophylli* sp. n., *Phytomonas alliicola* sp. n., and *Phytomonas manihotus* (Arthand - Berthet et Bondar) Vilgas, *Phytopathology*, 32, 141, 1942.
156. **Starr, M. P. and Burkholder, W. H.,** Lipolytic activity of phytopathogenic bacteria determined by means of spirit blue agar and its taxonomic significance, *Phytopathology*, 32, 598, 1942.
157. **Burkholder, W. H.,** Sour skin, a bacterial rot of onion bulbs, *Phytopathology*, 40, 115, 1949.
158. **Creelman, D. W.,** Diseases of vegetable crops, *Can. Pl. Dis. Surv.*, 47, 44, 1967.
159. **Bazzi, G.,** Identification of *Pseudomonas cepacia* on onion bulbs in Italy, *Phytopathol. Z.*, 95, 254, 1979.
160. **Taylor, J. D. and Holden, C. M.,** Bacterial rots of stored onions, *Rep. Natl. Veg. Res. Stn. 1976*, 104, 1977.
161. **Keddie, R. M.,** The properties and classification of Lactobacilli isolated from gas and silage, *J. Appl. Bacteriol.*, 22, 403, 1959.
162. **Kawamoto, S. O. and Lorbeer, J. W.,** Soft rot bacteria associated with onion decay, *Phytopathology*, 57, 341, 1967.
163. **Ballard, R. W., Palleroni, N. J., Doudoroff, M., Stanier, R. Y., and Mandel, M.,** Taxonomy of aerobic Pseudomonads: *Pseudomonas cepacia, P. marginata, P. alliicola* and *P. caryophylli, J. Gen. Microbiol.*, 60, 199, 1970.
164. **Erwin, R. D. and Vaughan, E. K.,** Bacterial soft rot of onions — a role of irrigation water and identification of the causal organisms, *Am. Phytopathol. Soc. Proc.*, 1, 47, (Abstr.), 1974.
165. **Hevesi, M. and Viranyi, F.,** An unknown symptom on onion plants caused by *Pseudomonas alliicola* (Burkholder) Starr et Burkholder, *Acta Phytopathol. Acad. Sci. Hung.*, 10, 281, 1975.
166. **Vicanov, M.,** Effect of harvest dates and storage of onion slippery skin injection (*Pseudomonas alliicola* Burk.) on onion bulbs, *Gradinar. Lozar. Nauka*, 13, 63, 1976.
167. **Swarup, J., Magarkoti, M. S., and Saksena, H. K.,** A new bacterial rot of onion caused by *Pseudomonas alliicola* in India, *Indian J. Mycol. Plant Pathol.*, 3, 187, 1974.
168. **Kawamoto, S. O. and Lorbeer, J. W.,** Infection of onion leaves by *Pseudomonas cepacia, Phytopathology*, 64, 1440, 1974.
169. **Kawamoto, S. O. and Lorbeer, J. W.,** Histology of onion leaves infected with *Pseudomonas cepacia, Phytopathology*, 62, 1266, 1972.
170. **Kawamoto, S. O. and Lorbeer, J. W.,** Multiplication of *Pseudomonas cepacia* in onion leaves, *Phytopathology*, 62, 1263, 1972.
171. **Taylor, J. D., Dudley, C. L., and Presly, (nee Gray), L.,** Studies of halo-blight seed infection and disease transmission in dwarf beans, *Ann. Appl. Biol.*, 93, 267, 1979.

172. **Taylor, J. D.,** Field studies on halo-blight of beans *(Pseudomonas phaseolicola)* and its control by foliar sprays, *Ann. Appl. Biol.*, 70, 191, 1972.
173. **Taylor, J. D., Phelps, K., and Dudley, C. L.,** Epidemiology and strategy for control of halo-blight of beans, *Ann. Appl. Biol.*, 93, 167, 1979.

INDEX

A

B

C

M

Q

R

S

Y

Z

AUTHOR INDEX

A

B

C

D

E

F

G

H

I

J

K

L

M

T

U

V

W

Y

Z